Advances in Experimental Medicine and Biology

Volume 1420

Advances in Experimental Medicine and Biology provides a platform for scientific contributions in the main disciplines of the biomedicine and the life sciences. This series publishes thematic volumes on contemporary research in the areas of microbiology, immunology, neurosciences, biochemistry, biomedical engineering, genetics, physiology, and cancer research. Covering emerging topics and techniques in basic and clinical science, it brings together clinicians and researchers from various fields.

Advances in Experimental Medicine and Biology has been publishing exceptional works in the field for over 40 years, and is indexed in SCOPUS, Medline (PubMed), EMBASE, BIOSIS, Reaxys, EMBiology, the Chemical Abstracts Service (CAS), and Pathway Studio.

2021 Impact Factor: 3.650 (no longer indexed in SCIE as of 2022)

Jorge S. Burns
Editor

Potency Assays for Advanced Stem Cell Therapy Medicinal Products

Springer

Editor
Jorge S. Burns
University of Ferrara
Ferrara, Italy

ISSN 0065-2598 ISSN 2214-8019 (electronic)
Advances in Experimental Medicine and Biology
ISBN 978-3-031-30039-4 ISBN 978-3-031-30040-0 (eBook)
https://doi.org/10.1007/978-3-031-30040-0

This Springer imprint is published by the registered company Springer Nature Switzerland AG
The registered company address is: Gewerbestrasse 11, 6330 Cham, Switzerland

In memoriam
Paolo Bianco (1955–2015)
Luc Sensebé (1953–2020)

Preface

The potency assay serves as a hub where academic biological insights intertwine with industrial processing knowledge, and the chapters of this book pivot on this busy juncture. Experienced contributors from academia, clinical research centres and industry provide detailed reflections on the topic and are thanked enormously for having done so despite the extraordinary challenges presented by the Covid-19 pandemic.

The editor's first chapter draws from the history of the early phases of vaccine discovery and current implementation, to highlight the sometimes-similar convoluted aspects found in the art of stem cell therapy. Risk-based therapy progressed to ethically sound effective intervention with the discovery of the mechanisms of action. Potency assays underscore this principle and form a vital part of the medical revolution being presented by new advanced therapy medicinal products.

The industrial sector has played a large part in driving pragmatic progress to overcome cell-based therapy obstacles. Whilst head of research and development at Innovacell AG, Marco Thurner and his team, including Raffaela Torggler, Eva Margreiter and Rainer Marksteiner, co-contributors of Chap. 2, investigated potency assay development for clinical use. Marco led development of a potency assay for human skeletal muscle-derived cells during clinical phase drug development and as a prerequisite to market approval application (MAA). Measurement of the activity of acetylcholinesterase (AChE), expressed throughout skeletal and nervous tissue, was successfully elaborated as a potential potency assay for human skeletal muscle-derived cells (aSMDC) that are used to treat patients with fecal incontinence.

The research team of Joaquim Vives at the largest research teaching hospital in Catalonia explores application of human multipotent stromal cells (hMSC) and the optimisation of potency assays to assess the immunomodulative potential of clinical-grade hMSC. Co-authored with Sílvia Torrents and Marta Grau-Vorster, Chap. 3 provides an overview to the many diverse and challenging aspects facing potency assay development for advanced therapy medicinal products (ATMP). The team has also studied the stability of multipotent stromal cell-based products and excipients that could play a key role extending the shelf-life of the critical quality attributes (CQA) of the final product. Broad clinical experience has included cell-banking strategies for the production of clinical grade mesenchymal stromal cells from different tissues. Notable examples of potency assays from approved therapies are overviewed in Chap. 9.

Moustapha Kassem a scientist, physician and endocrinologist based at the University Hospital of Odense, Denmark, has spearheaded the development of fully differentiating immortalised cell strains of human bone marrow-derived multipotent stromal cells (hBM-MSC), ideal for obtaining data concerning microarray gene expression analysis, microRNA regulation and proteomic phenotypes for identifying molecular signaling pathways directly associated with osteogenic differentiation. Chapter 4 describes extensive characterisation of the osteogenic biomarkers of human bone marrow-derived multipotent stromal cells (hBM-MSC) and their relevance for potency assays.

The unmet clinical need of cartilage regeneration for joint damage and osteoarthritis has also been an area of intense research. Consistent with a growing appreciation that cell-secreted factors can be of therapeutic benefit, Lucienne A. Vonk's research team at the University Medical Center Utrecht in the Netherlands demonstrated that extracellular vesicles derived from hBM-MSC can promote cartilage regeneration in vitro. Chapter 5 provides insights into establishing potency assays that not only function with whole cells, but also acellular products derived from the cells.

Raghavan Chinnadurai at the Mercer University School of Medicine, USA, has explored multiparametric analysis of hBM-MSC for the purposes of characterising their potency to modulate the immune system. Secretome analysis has been compared with quantitative RNA-based gene array analysis targeting immunomodulatory and homing properties of MSC. In Chap. 6, he describes advanced technologies for potency assay measurement, discussing how diverse complementary approaches can enhance prospects for establishing specific potency biomarkers.

Developing Lab-on-chip devices for biomedical diagnostics, Despina Moschou at the University of Bath applies microfabrication and microelectronics with a view to meet the ASSURED criteria; Affordable, Sensitive, Specific, User friendly, Rapid analysis, Equipment-free and Delivered at point of care. As described with Sotirios Papamatthaiou in Chap. 7, adoption of a printed circuit board (PCB) platform would achieve these aims and be readily scalable for existing industrial platforms. Adapting the Lab-on-PCB approach to take advantage of alternative technologies would be particularly advantageous for high performance, efficient, cost-effective potency assays.

In Chap. 8, Claire Roddie, Associate Professor in Hematology at University College London (UCL) and consultant Hematologists at UCL Hospital, together with Juliana Dias and Amaia Cadiñanos-Garai, describes potency assays for one of the most significant novel approaches to cell-mediated therapy, the use of chimeric antigen receptor T cells (CAR-T) that are genetically engineered to produce a tailored T-cell receptor for use in immunotherapy. Their pursuit of adoptive cell therapies involves pre-clinical development of novel CAR-T projects where the development of potency assays can be particularly challenging given the complexity of the therapeutic modality.

Juli Mansnérus, Postdoctoral Researcher at the Faculty of Law, University of Helsinki, has published extensively on ATMP and ethical challenges of personalized medicine. Both Juli and co-author Waltter Roslin have partaken in the DECIDER project, exploring diagnostic tools and treatments for ovarian cancer using AI methods. This project has received funding from the

European Union's Horizon 2020 research and innovation programme under grant agreement No 965193 for DECIDER. Chapter 10 provides an important legal perspective on the integrity of potency assays as a basis safe clinical intervention, with appreciation of the manner, whereby ATMP regulations were set up as a *lex specialis* introducing particular provisions to the existing pharmaceutical legislation with respect to authorisation, supervision and pharmacovigilance of ATMP to ensure they are safe and effective.

It is significant that ATMPs are usually developed by academia or within hospitals and involve small medium enterprise (SME) companies rather than big pharmaceutical companies that predominantly develop conventional medicines. However, the whole cell therapy sector, like potency assays, is in continuous evolution, and in the final Chap. 11, an Editorial forward-looking perspective is presented, regarding many innovative technological developments, institutional roles and guidance contributing to potency assay development in the future. A renewed focus on potency assays will help establish capabilities and standards for scientifically sound reportable data to correlate product-specific biological activity with therapeutic activity and streamline the strategic development of advanced medicines with more cost-effective success.

Ferrara, Italy Jorge S. Burns

Contents

Contributors

Jorge S. Burns Department of Environmental and Prevention Sciences, University of Ferrara, Ferrara, Italy

Amaia Cadiñanos-Garai USC/CHLA Cell Therapy Program, Keck School of Medicine of USC, University of Southern California (USC), Los Angeles, CA, USA

Raghavan Chinnadurai Department of Biomedical Sciences, Mercer University School of Medicine, Savannah, GA, USA

Juliana Dias UCL Cancer Institute, University College London, London, UK

Royal Free Hospital London, NHS Foundation Trust, London, UK

Marta Grau-Vorster Banc de Sang i Teixits, Edifici Dr. Frederic Duran i Jordà, Barcelona, Spain

Transfusion Medicine Group, Vall d'Hebron Research Institute, Universitat Autònoma de Barcelona, Barcelona, Spain

Moustapha Kassem University Hospital of Odense, University of Southern Denmark, Odense, Denmark

Danish Stem Cell Center, University of Copenhagen, Copenhagen, Denmark

College of Medicine, King Saud University, Riyadh, Saudi Arabia

Juli Mansnérus University of Helsinki, Helsinki, Finland

Eva Margreiter Innovacell AG, Innsbruck, Austria

Rainer Marksteiner Innovacell AG, Innsbruck, Austria

Despina Moschou Centre for Biosensors, Bioelectronics and Biodevices (ToC3Bio) and Department of Electronic & Electrical Engineering, University of Bath, Bath, UK

Sotirios Papamatthaiou Centre for Biosensors, Bioelectronics and Biodevices (ToC3Bio) and Department of Electronic & Electrical Engineering, University of Bath, Bath, UK

Claire Roddie UCL Cancer Institute, University College London, London, UK

Department of Haematology, UCL Hospital, London, UK

Waltter Roslin University of Helsinki, Helsinki, Finland

Marco Thurner Innovacell AG, Innsbruck, Austria

Finnegan, Henderson, Farabow, Garrett & Dunner LLP, Munich, Germany

Raffaela Torggler Innovacell AG, Innsbruck, Austria

Sílvia Torrents Banc de Sang i Teixits, Edifici Dr. Frederic Duran i Jordà, Barcelona, Spain

Transfusion Medicine Group, Vall d'Hebron Research Institute, Universitat Autònoma de Barcelona, Barcelona, Spain

Joaquim Vives Banc de Sang i Teixits, Edifici Dr. Frederic Duran i Jordà, Barcelona, Spain

Musculoskeletal Tissue Engineering Group, Vall d'Hebron Research Institute (VHIR), Universitat Autònoma de Barcelona, Barcelona, Spain

Departament de Medicina, Universitat Autònoma de Barcelona, Barcelona, Spain

Lucienne A. Vonk Department of Orthopaedics, University Medical Center Utrecht, Utrecht, The Netherlands

About the Editor

Jorge S. Burns received his degree in Cellular Pathology at the University of Bristol and PhD at the University of Wales College of Medicine in 1992, authoring the first UK manuscript to describe the use of retroviral vectors in epithelial cells. At the Ludwig Institute for Cancer Research, University College London Branch, a postdoctoral fellowship in Breast Cancer research provided technology insights leading to an appointment at renowned proteome research facilities at the University of Southern Denmark in Odense. Subsequent industrial experience was obtained with management of a cell culture facility supporting mass spectrometry for immune therapy target discovery at MDS Proteomics A/S. Joining the academic research team led by Prof. Moustapha Kassem at Odense University Hospital brought focus on human bone marrow-derived stromal cells and their osteogenic differentiation potential, plus participation in the Innovation Consortium 3-D scaffolds. Preclinical stem cell characterisation studies progressed to translational regenerative medicine at the University of Modena in Italy, upon joining the EU Framework 7 Reborne consortium, whereby bone therapy clinical trials required exploration of novel osteogenic potency assays. To pursue development of potency assay biosensors, Prof. Burns obtained an award for project number 154/25.11.2016, P_37_221/2015, "A novel GRaphene Biosensor Testing Osteogenic Potency; capturing best stem cell performance for regenerative medicine" (GRABTOP), from the Ministry of Research and Innovation in Romania, Operational Program Competitiveness Axis 1 Section E, co-financed from European Regional Development Fund "Investments for your future". The multidisciplinary laboratory inaugurated at the Faculty of Medical Engineering, University Politehnica of Bucharest, tested the applicability of the nanomaterial graphene and its diverse functionalised forms. Current research interests revolve around improved methods of in vivo mimicry at the Department of Environmental and Prevention Sciences at the University of Ferrara.

The Art of Stem Cell-Based Therapy

1

Jorge S. Burns

Yet an experiment, were you to try it, could free you from your cavil–and the source of your arts' course springs from experiment.

Dante Alighieri, The Divine Comedy, Paradiso, Canto II, lines 94-96, c. 1304-1321. English translation by Allen Mandelbaum.

1.1 The Dawn of a New Era

Centuries before Robert Hooke's description of cells as observation XVIII in *Micrographia*, 1655, the importance of experiment had already been highlighted in one of the most important poems of the Middle Ages. Just as Dante's use of a Florentine vernacular and around 90 neologisms marked a radical shift from writing poetry in Latin, so too cellular therapy has represented a paradigm shift in medicine whereby advanced therapy medicinal products (ATMP) can revolutionise the medical treatment of numerous traumatic pathologies of unmet medical need, literally making *La Vita Nuova*, the new life.

However, potency assays achieve far more than freedom from one's cavil or petty objections, they represent crucial experiments at the hub of the comprehensive complexity surrounding cell therapy [23]. Moreover, numerous factors beyond biological and scientific considerations underly the increasing significance and importance that potency assays currently accrue. Many of the issues surrounding Potency assays today have been encountered in historical situations where medicine progressed in the face of risk, when therapeutic approaches were adopted without there necessarily being a full understanding of the mechanisms responsible for a beneficial effect. A brief account of the ancient therapeutic procedure of variolation can highlight the many facets involved in the development and establishment of new therapeutic approaches and the emerging critical role of potency assays.

1.2 Lessons from a Past Disease

The word variola was introduced as a term for Smallpox by Bishop Marious of Avenches in AD 570, derived from the Latin word *varius* meaning 'stained' or from *varus*, denoting 'mark on the skin'. *Small pockes* was terminology used in England at the end of the fifteenth century (*pocke* meaning sack) and would distinguish the devastating disease from syphilis, then referred to as the great pockes. As early as 430 BCE it had been appreciated that Smallpox survivors were protected from a recurrence of the disease and could nurse the afflicted. The origins of the most successful approach to combat Smallpox, termed inoculation (from Latin inoculare, 'to graft')

J. S. Burns (✉)
Department of Environmental and Prevention Sciences, University of Ferrara, Ferrara, Italy
e-mail: js.burns@unife.it

J. S. Burns (ed.), *Potency Assays for Advanced Stem Cell Therapy Medicinal Products*, Advances in Experimental Medicine and Biology 1420, https://doi.org/10.1007/978-3-031-30040-0_1

Fig. 1.1 Variolation for smallpox originated in the Far East and later spread to Europe and Africa. This Japanese memorial relief by Fumio Saita in front of the Asakura Ishikai Hospital depicts a patient receiving a smallpox variolation in 1790 from the physician Ogata Shunsaku (1748–1810). His inspiration came from reading the 60th volume of the 'Imperially Commissioned Golden Mirror of the Orthodox Lineage of Medicine' a compilation of medical writings of the Han dynasty (202 BCE–220 CE) published in China in 1742. Ogata went on to publish the booklet *Shutō hitsujun ben* (種痘必順弁), *pox essentials*, describing how variolation ensured gentle smallpox in 1793. Original photograph by Wolfgang Michel-Zaitsu 'Between East and West-Variolation in Early Modern Japan': Proceedings of the International Symposium on the History of Indigenous Knowledge (ISHIK 2022), Kaifeng, China

were ancient, practiced in Africa, India, China and Japan long before its introduction to Europe in the eighteenth century (Fig. 1.1). Perhaps as early as the tenth century CE, the Chinese knew that by opening pustules of a Smallpox patient and drying the matter with a little cotton subsequently transferred to a recipient's nostrils, it was possible to transmit a relatively mild form of the disease that was prophylactically protective, helping avoid death in an epidemic. By the fifteenth century, the documented methods, in effect, a form of cell-based therapy, indicated careful ritualisation; 'nasal insufflation' involved use of silver blowpipes, right nostril for boys, left for girls. Those with relatively mild Smallpox symptoms, (possibly because they were infected with *variola minor* as opposed to *variola major*), were favoured donors of the scabs that were then left to dry for some time and ground to a powder or exposed to hot steam and various herbs or a grain of musk. Although unknown at the time, it is now appreciated that such procedures would damage virus particles, helping attenuate the infectious dose that was riskily being administered via the same route as natural Smallpox infections. In contrast to the Asian and African inhaled variolation procedures, Europeans and Americans inoculated via a puncture to the skin, a route avoiding a potential swift spread of the disease in the lungs, promoting a slower viral progress that would favour a more effective immune system defense. The geographical East-West discrepancy in inoculation routes largely reflects the pivotal role played by Lady Mary Wortley Montague, born to an aristocratic family in 1689, in bringing the procedure of variolation to the West from Turkey [18]. Her motivation was high; in 1713, her only and younger brother died of Smallpox aged just 20. Two years later she contracted the disease herself and against expectation survived, although she was left badly scarred. When her husband was appointed British

Ambassador to Turkey, she unconventionally accompanied him there, possibly primed by her attending physicians, Fellows of the Royal Society who likely knew of the folk practice of inoculation in distant Turkey, where Smallpox had already devastated the country during the Ottoman era. In Constantinople, Lady Montagu promptly approached professional inoculators, at that time predominantly women. Just 2 weeks after her arrival she already wrote to her father, a close friend and at least one of her former attending physicians, recounting the essential point that Smallpox inoculation could confer a mild protective form of the disease and avoid fatalities. She wrote 'I am going to tell you a thing that would wish yourself here' … indicative of an enthusiasm that would prove highly influential. Her husband's premature recall back home hastened an opportune Smallpox inoculation of her 5-year-old son before leaving Constantinople. Overseeing the elderly Greek woman who performed the inoculation, was Charles Maitland, a Scottish surgeon appointed to the Embassy. This would prove prudent, Smallpox outbreaks were becoming frequent in England and just 3 years later, a 1721 epidemic in Boston and London spurred Lady Montagu to also have her 4-year-old daughter receive inoculation. This was performed by an apprehensive, but persuaded and experienced Charles Maitland, who invited three members of the Royal College of Physicians as witnesses. Such was Lady Montagu's aristocratic influence that Princess Caroline wished to have her children inoculated. Concerns for unqualified practices were high, so 6 convicted prisoners and 11 orphans were first variolated in an experiment to deem the procedure safe before inoculating royal children. Maitland subsequently published a 40-page book 'Account of inoculating the Smallpox' and the procedure became fashionable. Thomas Nettleton, a physician inoculator, reported in 1722 that there were about one in five deaths among Smallpox patients within parts of Yorkshire and the surrounding area, yet there were no deaths in 61 people he had inoculated. From 897 inoculations performed in 1729 there were only 17 deaths. These results indicated variolation was helpful at stemming mortality when contracting natural Smallpox, but the procedure was not completely innocuous, indeed variolated patients risked spreading Smallpox to others as well as death. There was criticism and concern for the procedure in Parliament, yet over the following years variolation grew in popularity until an improved, safer alternative was made available. Edward Jenner, tutored by the renowned surgeon and experimental scientist John Hunter and fellow of the Royal Society, chose to practice medicine in Berkeley. There he became familiar with countryside lore that dairy-maids were protected from Smallpox after having suffered from cowpox, a far less-aggressive disease. Numerous observations and experiments as a country physician, eventually led to the invention of cowpox vaccination against Smallpox. Jenner's observations were not immediately accepted by the Royal Society, but he persisted with private publications providing details of how to distinguish Cow Pox lesions from other similar pustular lesions of other (unknown) cause [49]. Inoculation was initially accompanied by a highly individualised preparation of diet and therapy, yet by the late 1760s all inoculated patients followed a similar preparative regimen. Vaccines, although predominantly sourced via Jenner and his colleagues, would diversify, becoming mixed with other isolates since there was no ability to characterise viruses as the molecular genetic level. Eventually, the vaccine used in the US as part of a worldwide Smallpox eradication programme, declared successful in 1980, was called Vaccinia and was manufactured from infected calf skin in the US by the company Wyeth [59]. Of note, subsequent analysis of the Wyeth vaccine has suggested it may have been predominantly derived from vaccinia virus strains distinct from the Cow Pox strains discovered by Jenner [71]. What transpired over the dramatic course of Smallpox history was not only the evolution of a highly individualised therapy to a more successful standardised technique, but also the emergence of new institutions that replaced an old physician-orientated individualised monopoly, in favour of a new generation of medical practitioners.

1.3 Potency Assays for Modern Disease Intervention

Notable aspects in Smallpox vaccine history included the urgent need to treat a devastating health condition, influential key individuals championing the introduction of new therapeutic approaches, careful insights from repeated experimentation and widespread treatment during the latter phases of therapeutic development facilitated by a better understanding of the underlying molecular biology. Medical intervention unfolded over centuries to improve safety and efficacy, up to the point of eradicating the disease from the global population by 1980 [82]. The importance of Chemistry, Manufacturing and Control (CMC), plus need for accelerated quantitative testing of the Critical Quality Attributes (CQA) of a vaccine product were dramatically highlighted during the COVID-19 pandemic [2]. With unprecedented speed, hundreds of laboratories worldwide generated SARS-CoV-2 virus-specific vaccines. Beyond established platforms of live attenuated virus (LAV), inactivated virus, recombinant proteins and protein-based virus-like particles (VLPs), a new technological era of market-approved mRNA vaccines was introduced [19]. With rapid innovation, some latest mRNA vaccine designs already differ from vanguard versions by introducing a self-amplifying capacity that allows for smaller doses, with a freeze-drying preparation process that avoids the need for refrigeration pertaining to liquid mRNA vaccines [14]. Following Emergency Use Authorisation (EUA), the new mRNA-based vaccine platforms have proved to be very successful. Product specific tests have included in vitro bioanalytical batch release and characterisation assays for the antigen and lipid nanoparticle component encapsulating the mRNA protecting it from RNAses. There remains a rare risk that individuals may experience hypersensitivity and even anaphylactic reactions to components of the vaccine drug product, including adjuvants and stabilisers or cryo-protectants. It has been proposed that specifically designed potency assays may draw on experience gained in nanomaterial research to ensure such components do not interfere with potency [74], certainly it is demanding to develop tests for potency of vaccines produced by new technologies [64]. Much is being learned whilst innovative therapy is underway, information that can feed into knowledge-based potency assays regarding the mechanism of action of the vaccine to correlate with a clinically relevant immune response. Of necessity, the evolution of SARS-CoV-2 potency assays has to be appreciated as a work in progress with technology that accelerates non-clinical immunoassays providing a complementary route to establish functionally meaningful potency assays. World Health Organisation (WHO) endorsement of international antibody reference standards for use in immunological assays can enable better comparison of data generated across vaccine trials [35].

1.4 From Viral Vaccination to Safe Therapy with Cells

Clearly, a long complex history surrounds the apparently 'straightforward' situation of a defined virus target and an established vaccination-based therapeutic strategy. An enormous evolution in vaccine development has yielded remarkable accomplishments in the response to the COVID-19 pandemic [63]. Fundamentally, early dubious practices of variolation were promptly replaced by the safer treatment by vaccination and now advanced technologies of analysis and measurement have enabled far more rigorous and informative science, providing therapy for novel viral targets. Certainly, in the case of viral vaccination, an understanding of key molecular mechanisms, has been vital for defining key targets and viral vulnerabilities with notable success [42, 54]. Assays are continuously being developed to maintain integral Chemistry Manufacturing and Control (CMC) with quantitative testing of the critical quality attributes (CQA) of a vaccine product [65].

Many of the above principles remain relevant for the alternative field of stem cell-based therapy [85]. However, much more complexity governs the factors to be considered when contemplating a potency assay for advanced therapy medicinal products (ATMP) encompassing gene therapies, somatic cell therapies and tissue engineered products. Drawing similarity to the history of vaccination, early studies have offered numerous reports of beneficial effects of applied 'stem' cells, but there remain major hurdles with regard to an ability to control and measure the therapeutic procedures involved, to help confirm the results. The potency assay serves as a challenging yet vital means of improving stem cell-based therapy. When rigorously upheld, it serves as one of the most effective means of avoiding indiscriminate unproven treatments, ensuring ATMPs, in particular those involving the use of stem cells, can be introduced safely and efficiently. Regulatory authorities in both Europe and the USA [28] prioritise potency assays as integral components of manufacturing processes with good reason, ultimately they serve to accelerate patient access to more trustworthy innovative therapies [22]. An increasing degree of oversight and enforcement has become necessary to protect people from misinformation and unscrupulous profiteering stem cell clinics [1, 7, 39, 69, 75]. The European Medical Agency (EMA), the USA Food and Drug Administration (FDA), the Japanese Pharmaceutical and Medical Devices Agency (PMDA) and Korean Ministry of Food and Drug Safety (MFDS) among others, provide guidelines on conditions for which stem cell-based therapies are approved [32], concurring that well-designed clinical trials are necessary to ensure an acceptable quality of therapy [41]. An important role is also being played by major science-based research organisations, particularly the International Society for Stem Cell Research (ISSCR) and International Society for Cell and Gene Therapy (ISCT). They provide forums for international research and authoritatively educate all stakeholders. Published guidelines and committee statements have highlighted topics such as clarification of cell nomenclature [78]; minimal criteria for characterisation [20]; broad ethical implications [6]; inclusivity and diversity [21]; derivation, banking and distribution of cell lines [40]; hospital practices supporting externally manufactured ATMP [9]; ethical considerations in application of the European Union hospital exemption rule [15]; investor perspectives [48]; and potency assays [11, 80]. The latest updated 2021 ISSCR guidelines for stem cell research and clinical translation provided scientifically rigorous and ethically justifiable oversight policies [5, 44] and served as a very helpful comparator for regulations governing related research activities in Japan [83]. Stimulating highest level debate, they were not without criticism [8, 30].

It can be apt to borrow the terminology of watchmakers, whereby functions in addition to telling the time, e.g. display of date or a chronograph dial, are termed complications. The 'complications' involved when defining a Potency assay can be simplified when there is minimal and streamlined handling of stem cells prior to application, however a recent survey of such procedures between laboratories suggested they lacked congruency [61]. Beyond microbial contamination, cultured cells risk functional degradation and impairment of subsequent function [10, 38, 68]. Reducing complication also assists with a desired standardisation and sustainability of procedures. Conversely, extending complication can be of great benefit, since there exists groundbreaking proof of principle that ex vivo genetic modification of epidermal stem cells can potentiate them to treat the potentially lethal genetic disease of Junctional Epidermolysis Bullosa [16, 27]. Also, since autologous stem cell expansion under cGMP culture conditions is usually a prerequisite to reach a clinical dose, there is an opportunity to precondition the naïve cells to enhance their inherent function [51].

Bringing harmony to a narrative of vaccine development and cell therapy, in circumstances when a vaccine shield was unavailable or failed to prevent severe SARS-CoV-2 induced symptoms, stem/stromal cell-based therapy may reduce the risk of mortality in patients with critical COVID-19 [13, 31, 33, 34, 62], principally caused by a hyperactive pro-inflammatory immune component [45]. This reflects that beyond early notions therapeutic stem cells were principally regenerative 'building blocks', differentiating to the required cell types needed to reconstitute damaged tissue, strong evidence has subsequently indicated the stem cells secrete factors that have a feedback immunomodulatory role via cell-cell interactions. This extends the therapeutic scope of stem cell-based treatments to immune-mediated conditions [46] including facilitation of allogeneic transplantation without immunosuppression [79]. Preconditioning strategies [72, 76] and methods for characterising immuno-modulatory potency both in vivo [12] and ex vivo [17, 43, 70] are underway, complemented by development of high-throughput on-chip technologies [67].

1.5 From Stem Cell Safety and Efficacy to Potency

Regulatory authority recommendations place information about presumed mechanisms of action (MoA) and pathophysiological disease processes as secondary to the fundamental concerns of safety and efficacy, the latter prioritising the simple question of whether a defined outcome is improved by the treatment. ATMP product potency testing seeks to establish a correlation between a measured property of the product and a desired clinical effect, demonstrable with either in vivo or in vitro assays. There is potential for significant flexibility and staging so that specific potency assays may be adequately evaluated on a case-by-case basis during product development. Clinical trials progress through four major phases. Phase I places emphasis on safe drug interaction with the human body; Phase II, accurate dose and initial efficacy data with observation of any side effects; Phase III, evaluation of safety and efficacy; Phase IV after formal approval there is determination of public safety of the new product [50]. Manufacturers are expected to have defined the potency assay acceptance criteria before the initiation of pivotal Phase III clinical trials. The sponsor/investigator must provide the regulatory authorities with comprehensive documentation on the number of participants in the clinical studies and the required number may vary according to the intervention, rare diseases usually requiring fewer participants than common diseases. Before attending, volunteers for each clinical trial should be informed of the enrolment criteria, possible side effects and the advantages of the study. There is constructive debate on the elements needed for adequate informed consent to implement regulated clinical trials of cell products ethically and responsibly [47, 73, 81]. Regulatory authorities are responsive to the ongoing concerns for necessary information as novel cell-based therapies move from bench to bedside [37]. Guidelines and regulations are issued on what information needs to be made to the public, scientific and clinical community at the onset of a clinical trial to protect human subjects, provide financial disclosure by clinical investigators, invoke institutional review boards and provide data protection of electronic records and electronic signatures. It is not unusual that one does not know the full mechanism of action of a cell-based therapeutic at early-stage clinical trials, at that phase potency assays may have only a limited potential to comparably assess different cell lots and lines [3]. However, if for phase III trials patients were educated about the potency assay for their particular intervention, this could enhance knowledge for informed consent and help avoid unproven therapies that risk damaging the public perception of stem cell research and regenerative medicine [4].

1.6 Complementing Stem Cell-Based Therapy Art with Standard Operating Procedures

This book seeks to provide a timely overview of what is to be understood by potency assays and its associated terminology, focusing on stem cell-based ATMP.

Well-characterised epithelial stem cells became the first approved stem cell-based therapies in Europe [53] providing limbal stem cell derived long-term corneal regeneration [57]. In contrast, the most extensively studied cell type under consideration for therapeutic applications, commonly yet controversially termed Mesenchymal Stem Cells (MSC), has proved arduous to comprehend. Although apparently accessible from a number of tissue sources [25, 60], MSC remain phenotypically enigmatic [36, 58] and of debated nomenclature [77]. Consequently, this book places emphasis on MSC since they represent an excellent extensively studied cell type for exposing the many critical aspects of the potency assay and its important role in establishing a genuine understanding of the role played by stem cells in therapeutic intervention.

In addition, novel cell-based therapies using chimeric antigen receptor (CAR) redirected T cells, efficacious in the treatment of leukemia/lymphoma, represent a therapeutic approach gaining prominence, with new potency assay challenges for extending this promising therapy to solid tumours. Notably, engineered MSC may have a complementary role in enhancing efficacy when extending CAR T-cell therapy to solid tumours [84].

Novel stem cell-based therapies present many challenges [66] and potency assays have always been influenced by many factors [55]. Foremost is the source of stem cells, that may originate from diverse tissue sources and undergo a range of preparation procedures. Ideally, for consistent quality under current good manufacturing practice (cGMP) each element of the process needs to be carefully documented and sourced so there can be traceability, accountability and reproducibility [24]. Important considerations include manufacturing process details related to cell processing, expansion and formulation and whether these can adopt a suitable closed and automated workflow system [29].

The patients receiving variolation were often initially subjected to a preparative phase aimed at maximising the chances that they would respond well to the procedure. In the case of stem cell-based therapy, this may not always be possible. The therapeutic cells obtained for autologous cell therapy are subject to the health condition of the patient at the time and there may be a limited opportunity to control this, especially in circumstances involving urgent treatment of an unexpected traumatic injury. This underscores the importance of a prompt potency assay in determining whether the sourced cells, subject to heterogeneity and specific contexts, are indeed fit for purpose.

The standard operating procedure involved in manufacturing the ATMP therapeutic and choice of administration route are also subject to numerous options and choices and these will retrospectively impinge on the potency assay, as it needs to remain relevant to the type of therapy envisaged. The same type of cells, sourced and prepared in the same way, may have common aspects of quality control and characterisation, yet require distinct potency assays according treatment modality to account for different attributes responsible for specific mechanisms of action. Quests for suitable potency assay biomarkers that may predictively indicate appropriate stem cell function remain susceptible to the dynamic cellular responses to different microenvironments. Phenotypic expression within the parameters of an in vitro assay may not necessarily prove consistent with that of the same cells in their in vivo microenvironment. Nonetheless, targeted research has revealed helpful insights into MSC heterogeneity [26] and candidate biomarkers for surrogate potency assays [52].

Over the course of decades, academic scientific researchers, physicians and hospital staff have gained important insights that can help with optimisation of procedures to minimise time and costs, improve assay development and establish how best to deliver a therapeutic impact from biological properties and functions. Research towards stem cell-based therapy has benefited from keen detailed observation and creativity, rendering it an art among scientists and physicians. This is increasingly being complemented by an objective and pragmatic approach within Industry, aiming to accelerate translation of scientific observations to widespread ATMP application. Changes in scale and automation with rigorously maintained protocols have sought to address the many current challenges and provide a sustainable workflow [56]. Early and constant engagement with regulatory agencies throughout the manufacturing process can help ensure timely provision of the necessary documentation during clinical trials. Since potency assays constitute a critical step for the release of drug products, the assays are stringently validated. Figure 1.2 illustrates the terminology for the key parameters sought in analytical procedures according to the

Fig. 1.2 ICH guidance for key parameters validating the analytical procedures of potency assays. (https://database.ich.org/sites/default/files/Q2_R1__Guideline.pdf)

International Council of Harmonization (ICH) guidelines.

Clinical studies are conducted to collect safety and effectiveness information to support marketing applications for a new drug product. A clinical study sponsor obtains authorisation through filing for an investigational new drug (IND) in the USA or an equivalent investigational medicinal product (IMP) in Europe, necessary to commence human clinical trials in the jurisdiction. Successful completion of Phase I, II and III clinical trials can allow manufacturing and marketing of the ATMP. Phase IV trials represent a post-marketing surveillance following approval of the product. Additional regulations govern the final post-marketing and commercial approval of the Biological License Application (BLA), defined by the FDA as a request for permission to introduce a biological product into interstate commerce, regulated under Code of Federal Regulations Title 21 (21 CFR 600–680) [29]. Such procedures ensure rigorous clinical testing and ultimately accelerate the provision of novel safe medicinal cell products.

References

1. Aboalola D, Badraiq H, Alsiary R, Zakri S, Aboulola N, Haneef L, Malibari D, Baadhaim M, Alsayegh K (2022) An infodemic of misinformation on stem cell therapy among the population of Saudi Arabia: a cross-sectional study. Front Med 9:789695
2. Acter T, Uddin N, Das J, Akhter A, Choudhury TR, Kim S (2020) Evolution of severe acute respiratory syndrome coronavirus 2 (SARS-CoV-2) as coronavirus disease 2019 (COVID-19) pandemic: a global health emergency. Sci Total Environ 730:138996
3. Anderson AJ, Cummings BJ (2016) Achieving informed consent for cellular therapies: a preclinical translational research perspective on regulations versus a dose of reality. J Law Med Ethics 44:394–401
4. Ansari Z, Kuriyan A, Albini T (2017) Unproven stem cell therapy for macular degeneration. Oncotarget 8:90636
5. Anthony E, Lovell-Badge R, Morrison SJ (2021) New guidelines for stem cell and embryo research from the ISSCR. Cell Stem Cell 28:991–992
6. Assen LS, Jongsma KR, Isasi R, Tryfonidou MA, Bredenoord AL (2021) Recognizing the ethical implications of stem cell research: a call for broadening the scope. Stem Cell Reports 16:1656–1661
7. Bauer G, Elsallab M, Abou-El-Enein M (2018) Concise review: a comprehensive analysis of reported adverse events in patients receiving unproven stem cell-based interventions. Stem Cells Transl Med 7:676–685
8. Baylis F (2021) ISSCR guidelines fudge heritable human-genome editing. Nature 594:333
9. Bersenev A, Gustafson MP, Hanley PJ (2022) ISCT survey on hospital practices to support externally manufactured investigational cell-gene therapy products. Cytotherapy 24:27–31
10. Boland LK, Burand AJ, Boyt DT, Dobroski H, Di L, Liszewski JN, Schrodt MV, Frazer MK, Santillan DA, Ankrum JA (2019) Nature vs. nurture: defining the effects of mesenchymal stromal cell isolation and culture conditions on resiliency to palmitate challenge. Front Immunol 10:1080
11. Bravery CA, Carmen J, Fong T, Oprea W, Hoogendoorn KH, Woda J, Burger SR, Rowley JA, Bonyhadi ML, Van't Hof W (2013) Potency assay development for cellular therapy products: an ISCT review of the requirements and experiences in the industry. Cytotherapy 15:9–19.e9
12. Chinnadurai R, Bates PD, Kunugi KA, Nickel KP, DeWerd LA, Capitini CM, Galipeau J, Kimple RJ (2021) Dichotomic potency of IFNγ licensed allogeneic mesenchymal stromal cells in animal models of acute radiation syndrome and graft versus host disease. Front Immunol 12:708950
13. Chouw A, Milanda T, Sartika CR, Kirana MN, Halim D, Faried A (2022) Potency of mesenchymal stem cell and its secretome in treating COVID-19. Regen Eng Transl Med 8:43–54
14. Cohen J (2022) First self-copying mRNA vaccine proves itself in pandemic trial. Science 376:446
15. Cuende N, Ciccocioppo R, Forte M, Galipeau J, Ikonomou L, Levine BL, Srivastava A, Zettler PJ (2022) Patient access to and ethical considerations of the application of the European Union hospital exemption rule for advanced therapy medicinal products. Cytotherapy 24:686–690
16. De Rosa L, Enzo E, Zardi G, Bodemer C, Magnoni C, Schneider H, De Luca M (2021) Hologene 5: a phase II/III clinical trial of combined cell and gene therapy of junctional epidermolysis bullosa. Front Genet 12:705019
17. DeVeaux SA, Ogle ME, Vyshnya S, Chiappa NF, Leitmann B, Rudy R, Day A, Mortensen LJ, Kurtzberg J, Roy K, Botchwey EA (2022) Characterizing human mesenchymal stromal cells' immune-modulatory potency using targeted lipidomic profiling of sphingolipids. Cytotherapy 24:608–618
18. Dinc G, Ulman YI (2007) The introduction of variolation 'A La Turca' to the West by Lady Mary Montagu and Turkey's contribution to this. Vaccine 25:4261–4265
19. Dolgin E (2021) The tangled history of mRNA vaccines. Nature 597:318–324
20. Dominici M, Le Blanc K, Mueller I, Slaper-Cortenbach I, Marini F, Krause D, Deans R, Keating A, Prockop D, Horwitz E (2006) Minimal criteria for defining multipotent mesenchymal stromal cells. The International Society for Cellular Therapy position statement. Cytotherapy 8:315–317
21. Fears R, Akutsu H, Alentajan-Aleta LT, Caicedo A, Campos de Carvalho AC, Čolić M, Cornish J, Cossu G, Debré P, Dierckxsens G, El-Badri N, Griffin G,

Chingo-Ho Hsieh P, Inamdar MS, Kumar P, Abraham CM, Maciulaitis R, Al Mahtab M, O'Brien FJ, Pepper MS, Meulen VT (2021) Inclusivity and diversity: integrating international perspectives on stem cell challenges and potential. Stem Cell Reports 16:1847–1852
22. Fernández-Santos M, García-Arranz M, Andreu E, García-Hernández A, López-Parra M, Villarón E, Sepúlveda P, Fernández-Avilés F, García-Olmo D, Prosper F, Sánchez-Guijo F, Moraleda JM, Zapata A (2022) Optimization of mesenchymal stromal cell (MSC) manufacturing processes for a better therapeutic outcome. Front Immunol 13:918565
23. Galipeau J, Krampera M, Leblanc K, Nolta JA, Phinney DG, Shi Y, Tarte K, Viswanathan S, Martin I (2021) Mesenchymal stromal cell variables influencing clinical potency: the impact of viability, fitness, route of administration and host predisposition. Cytotherapy 23:368–372
24. García-Muñoz E, Vives J (2021) Towards the standardization of methods of tissue processing for the isolation of mesenchymal stromal cells for clinical use. Cytotechnology 73:1–10
25. Gonzalez-Vilchis RA, Piedra-Ramirez A, Patiño-Morales CC, Sanchez-Gomez C, Beltran-Vargas NE (2022) Sources, characteristics, and therapeutic applications of mesenchymal cells in tissue engineering. Tissue Eng Regen Med 19:325–361
26. Han ZC, Du WJ, Han ZB, Liang L (2017) New insights into the heterogeneity and functional diversity of human mesenchymal stem cells. Biomed Mater Eng 28:S29–S45
27. Hirsch T, Rothoeft T, Teig N, Bauer JW, Pellegrini G, De Rosa L, Scaglione D, Reichelt J, Klausegger A, Kneisz D, Romano O, Secone Seconetti A, Contin R, Enzo E, Jurman I, Carulli S, Jacobsen F, Luecke T, Lehnhardt M, Fischer M, Kueckelhaus M, Quaglino D, Morgante M, Bicciato S, Bondanza S, De Luca M (2017) Regeneration of the entire human epidermis using transgenic stem cells. Nature 551:327–332
28. Iglesias-Lopez C, Obach M, Vallano A, Agustí A (2021) Comparison of regulatory pathways for the approval of advanced therapies in the European Union and the United States. Cytotherapy 23:261–274
29. Jayaraman P, Lim R, Ng J, Vemuri MC (2021) Acceleration of translational mesenchymal stromal cell therapy through consistent quality GMP manufacturing. Front Cell Dev Biol 9:648472
30. Johnston J, Baylis F, Greely HT (2021) ISSCR: grave omission of age limit for embryo research. Nature 594:495
31. Kaffash Farkhad N, Sedaghat A, Reihani H, Adhami Moghadam A, Bagheri Moghadam A, Khadem Ghaebi N, Khodadoust MA, Ganjali R, Tafreshian AR, Tavakol-Afshari J (2022) Mesenchymal stromal cell therapy for COVID-19-induced ARDS patients: a successful phase 1, control-placebo group, clinical trial. Stem Cell Res Ther 13:283
32. Kidpun P, Ruanglertboon W, Chalongsuk R (2022) State-of-the-art knowledge on the regulation of advanced therapy medicinal products. Per Med 19:251
33. Kirkham AM, Bailey AJM, Monaghan M, Shorr R, Lalu MM, Fergusson DA, Allan DS (2022) Updated living systematic review and meta-analysis of controlled trials of mesenchymal stromal cells to treat COVID-19: a framework for accelerated synthesis of trial evidence for rapid approval-FASTER approval. Stem Cells Transl Med 11:675–687
34. Kirkham AM, Monaghan M, Bailey AJM, Shorr R, Lalu MM, Fergusson DA, Allan DS (2022) Mesenchymal stem/stromal cell-based therapies for COVID-19: first iteration of a living systematic review and meta-analysis: MSCs and COVID-19. Cytotherapy 24:639–649
35. Knezevic I, Mattiuzzo G, Page M, Minor P, Griffiths E, Nuebling M, Moorthy V (2021) WHO International Standard for evaluation of the antibody response to COVID-19 vaccines: call for urgent action by the scientific community. Lancet Microbe 3:e235
36. Langrzyk A, Nowak WN, Stępniewski J, Jaźwa A, Florczyk-Soluch U, Józkowicz A, Dulak J (2018) Critical view on mesenchymal stromal cells in regenerative medicine. Antioxid Redox Signal 29:169–190
37. Laurencin CT, McClinton A (2020) Regenerative cell-based therapies: cutting edge, bleeding edge, and off the edge. Regen Eng Transl Med 6:78–89
38. Lavrentieva A, Hoffmann A, Lee-Thedieck C (2020) Limited potential or unfavorable manipulations? Strategies toward efficient mesenchymal stem/stromal cell applications. Front Cell Dev Biol 8:316
39. Leask F (2019) Spotting 'unproven' stem cell therapies in the wild. BioTechniques 67:253
40. Levine BL, Munsie M, Levine AD, Ikonomou L, International SFCGTCOTEOCAGT (2022) The peril of the promise of speculative cell banking: statement from the ISCT Committee on the Ethics of Cell and Gene Therapy. Cytotherapy 24:977–978
41. Lim F (2021) Proven and unproven cell therapies – what we have learned so far. ISBT Sci Ser 16:213–218
42. Lin DY, Gu Y, Wheeler B, Young H, Holloway S, Sunny SK, Moore Z, Zeng D (2022) Effectiveness of Covid-19 vaccines over a 9-month period in North Carolina. N Engl J Med 386:933–941
43. Lipat AJ, Cottle C, Pirlot BM, Mitchell J, Pando B, Helmly B, Kosko J, Rajan D, Hematti P, Chinnadurai R (2022) Chemokine assay matrix defines the potency of human bone marrow mesenchymal stromal cells. Stem Cells Transl Med 11(9):971–986
44. Lovell-Badge R (2021) Stem-cell guidelines: why it was time for an update. Nature 593:479–479
45. Manoharan R, Kore RA, Mehta JL (2022) Mesenchymal stem cell treatment for hyperactive immune response in patients with COVID-19. Immunotherapy 14:1055
46. Markov A, Thangavelu L, Aravindhan S, Zekiy AO, Jarahian M, Chartrand MS, Pathak Y, Marofi F, Shamlou S, Hassanzadeh A (2021) Mesenchymal stem/stromal cells as a valuable source for the treatment of immune-mediated disorders. Stem Cell Res Ther 12:192

47. Master Z, Smith C, Tilburt JC (2020) Informed consent for stem cell–based interventions. JAMA 323:893–893
48. Moody J, Milligan WD, St Onge M, Goonewardene A, Rivers P (2021) Cell and gene therapy: a snapshot of investor perspectives. Cytotherapy 23:256–260
49. Morabia A (2018) Edward Jenner's 1798 report of challenge experiments demonstrating the protective effects of cowpox against smallpox. J R Soc Med 111:255
50. Mousaei Ghasroldasht M, Seok J, Park HS, Liakath Ali FB, Al-Hendy A (2022) Stem cell therapy: from idea to clinical practice. Int J Mol Sci 23:2850
51. Ocansey DKW, Pei B, Yan Y, Qian H, Zhang X, Xu W, Mao F (2020) Improved therapeutics of modified mesenchymal stem cells: an update. J Transl Med 18:42
52. Ofiteru AM, Becheru DF, Gharbia S, Balta C, Herman H, Mladin B, Ionita M, Hermenean A, Burns JS (2020) Qualifying osteogenic potency assay metrics for human multipotent stromal cells: TGF-β2 a telling eligible biomarker. Cells 9:E2559
53. Pellegrini G, Lambiase A, Macaluso C, Pocobelli A, Deng S, Cavallini G, Esteki R, Rama P (2016) From discovery to approval of an advanced therapy medicinal product-containing stem cells, in the EU. Regen Med 11:407
54. Pilishvili T, Gierke R, Fleming-Dutra KE et al (2021) Effectiveness of mRNA Covid-19 vaccine among U.S. health care personnel. N Engl J Med 385:e90
55. Pimpaneau V, Gianelli F, Trouvin JH, Poiseau AD (2015) The challenges of potency assay development for cell-based medicinal products in Europe. Regul Rapp 12:5–10
56. Priesner C, Hildebrandt M (2022) Advanced therapy medicinal products and the changing role of academia. Transfus Med Hemother 49:158
57. Rama P, Matuska S, Paganoni G, Spinelli A, De Luca M, Pellegrini G (2010) Limbal stem-cell therapy and long-term corneal regeneration. N Engl J Med 363:147–155
58. Renesme L, Pierro M, Cobey KD, Mital R, Nangle K, Shorr R, Lalu MM, Thébaud B (2022) Definition and characteristics of mesenchymal stromal cells in preclinical and clinical studies: a scoping review. Stem Cells Transl Med 11:44–54
59. Riedel S (2005) Edward Jenner and the history of smallpox and vaccination. Bayl Univ Med Cent Proc 18:21–25
60. Rojewski MT, Weber BM, Schrezenmeier H (2008) Phenotypic characterization of mesenchymal stem cells from various tissues. Transfus Med Hemother 35:168–184
61. Ruoss S, Walker JT, Nasamran CA, Fisch KM, Paez CJ, Parekh JN, Ball ST, Chen JL, Ahmed SS, Ward SR (2021) Strategies to identify mesenchymal stromal cells in minimally manipulated human bone marrow aspirate concentrate lack consensus. Am J Sports Med 49:1313–1322
62. Sadeghi B, Roshandel E, Pirsalehi A, Kazemi S, Sankanian G, Majidi M, Salimi M, Aghdami N, Sadrosadat H, Samadi Kochaksaraei S, Alaeddini F, Ringden O, Hajifathali A (2021) Conquering the cytokine storm in COVID-19-induced ARDS using placenta-derived decidua stromal cells. J Cell Mol Med 25:10554–10564
63. Saleh A, Qamar S, Tekin A, Singh R, Kashyap R (2021) Vaccine development throughout history. Cureus 13:e16635
64. Sanyal G (2022) Development of functionally relevant potency assays for monovalent and multivalent vaccines delivered by evolving technologies. npj Vaccines 7:1–10
65. Sanyal G, Särnefält A, Kumar A (2021) Considerations for bioanalytical characterization and batch release of COVID-19 vaccines. npj Vaccines 6:53
66. Schneider CK, Celis P, (CAT) TCFAT (2010) Challenges with advanced therapy medicinal products and how to meet them. Nat Rev Drug Discov 9:195–201
67. Schneider RS, Vela AC, Williams EK, Martin KE, Lam WA, García AJ (2022) High-throughput on-chip human mesenchymal stromal cell potency prediction. Adv Healthc Mater 11:e2101995
68. Shin EY, Yoon YJ, Lee JE, Shim SH, Park GH, Lee DR (2021) Identification of putative markers that predict the in vitro senescence of mesenchymal progenitor cells. Cells 10:1301
69. Sipp D, Caulfield T, Kaye J, Barfoot J, Blackburn C, Chan S, Luca MD, Kent A, McCabe C, Munsie M, Sleeboom-Faulkner M, Sugarman J, Zimmeren EV, Zarzeczny A, Rasko J (2017) Marketing of unproven stem cell–based interventions: a call to action. Sci Transl Med 9:eaag0426
70. Skibber MA, Olson SD, Prabhakara KS, Gill BS, Cox CS (2022) Enhancing mesenchymal stromal cell potency: inflammatory licensing via mechanotransduction. Front Immunol 13:874698
71. Smith KA (2011) Edward jenner and the small pox vaccine. Front Immunol 2:21
72. Srinivasan A, Sathiyanathan P, Yin L, Liu TM, Lam A, Ravikumar M, Smith RAA, Loh HP, Zhang Y, Ling L, Ng SK, Yang YS, Lezhava A, Hui J, Oh S, Cool SM (2022) Strategies to enhance immunomodulatory properties and reduce heterogeneity in mesenchymal stromal cells during ex vivo expansion. Cytotherapy 24:456–472
73. Sugarman J, Barker R, Charo R (2019) A professional standard for informed consent for stem cell therapies. JAMA 322:1651
74. Szebeni J, Storm G, Ljubimova JY, Castells M, Phillips EJ, Turjeman K, Barenholz Y, Crommelin DJA, Dobrovolskaia MA (2022) Applying lessons learned from nanomedicines to understand rare hypersensitivity reactions to mRNA-based SARS-CoV-2 vaccines. Nat Nanotechnol 17:337–346
75. Turner L (2020) Preying on public fears and anxieties in a pandemic: businesses selling unproven and

unlicensed "stem cell treatments" for COVID-19. Cell Stem Cell 26:806
76. Uberti B, Plaza A, Henríquez C (2022) Preconditioning strategies for mesenchymal stromal/stem cells in inflammatory conditions of livestock species. Front Vet Sci 9:806069
77. Viswanathan S, Shi Y, Galipeau J, Krampera M, Leblanc K, Martin I, Nolta J, Phinney DG, Sensebe L (2019) Mesenchymal stem versus stromal cells: International Society for Cell & Gene Therapy (ISCT®) Mesenchymal Stromal Cell committee position statement on nomenclature. Cytotherapy 21:1019–1024
78. Viswanathan S, Ciccocioppo R, Galipeau J (2021) Consensus International Council for Commonality in Blood Banking Automation–International Society for Cell & Gene Therapy statement on standard nomenclature abbreviations for the tissue of origin of mesenchymal stromal cells. Cytotherapy 7:393–395
79. Wang X, Wang K, Yu M, Velluto D, Hong X, Wang B, Chiu A, Melero-Martin JM, Tomei AA, Ma M (2022) Engineered immunomodulatory accessory cells improve experimental allogeneic islet transplantation without immunosuppression. Sci Adv 8:eabn0071
80. Watt SM (2022) The long and winding road: homeostatic and disordered haematopoietic microenvironmental niches: a narrative review. Biomater Transl 3:31
81. Wiese DM, Wood CA, Braid LR (2022) From vial to vein: crucial gaps in mesenchymal stromal cell clinical trial reporting. Front Cell Dev Biol 10:867426
82. World Health Organization (1980) The global eradication of smallpox: final report of the Global Commission for the Certification of Smallpox Eradication, Geneva, December 1979. World Health Organization
83. Yui H, Muto K, Yashiro Y, Watanabe S, Kiya Y, Kamisato A, Inoue Y, Yamagata Z (2022) Comparison of the 2021 International Society for Stem Cell Research (ISSCR) guidelines for "laboratory-based human stem cell research, embryo research, and related research activities" and the corresponding Japanese regulations. Regen Ther 21:46–51
84. Chan LY, Dass SA, Tye GJ, Imran SAM, Wan Kamarul Zaman WS, Nordin F (2022) CAR-T Cells/-NK Cells in Cancer Immunotherapy and the Potential of MSC to Enhance Its Efficacy: A Review. Biomedicines 10:804. https://doi.org/10.3390/biomedicines10040804
85. Hoang DM, Pham PT, Bach TQ, Ngo ATL, Nguyen QT, Phan TTK, Nguyen GH, Le PTT, Hoang VT, Forsyth NR, Heke M, Nguyen LT (2022) Stem cell-based therapy for human diseases. Signal Transduct Target Ther 7:272. https://doi.org/10.1038/s41392-022-01134-4

Potency Assay Development: A Keystone for Clinical Use

2

Raffaela Torggler, Eva Margreiter,
Rainer Marksteiner, and Marco Thurner

2.1 Potency Assays as Part of Cell-Based ATMP Quality Control Testing

2.1.1 What Is a Potency Assay?

It is indisputable that medicines for human use must undergo strict quality control testing to ensure a safe, stable and efficacious product. This task becomes more challenging when developing and manufacturing highly complex medicines such as cell-based Advanced Therapy Medicinal Products (ATMPs) [1, 2]. These products often contain cells as a drug substance (i.e. active ingredient) that have been isolated, expanded and/or differentiated or even grown to form tissue-like structures in vitro before use in a patient. Cell-based ATMPs can be divided into somatic cell therapy medicinal products and tissue engineered products, depending on their functional principle in the body [1]. Somatic cell therapy medicinal products exert a pharmacological, immunological or metabolic action to treat, prevent or diagnose a disease, whereas tissue engineered products are intended to regenerate, repair or replace a human tissue [1]. As part of the quality control strategy of any cell-based ATMP, it must be demonstrated that the drug substance is biologically active in the manner needed for a clinical outcome and thus its potency has to be evaluated [3–6]. Basically, potency can be described as a measure of the product's biological activity that is necessary for the desired therapeutic effect. More precisely, regulatory authorities define potency as 'the quantitative measure of biological activity based on the attribute of the product which is linked to the relevant biological properties' [7] or 'the specific ability or capacity of the product, as indicated by appropriate laboratory tests or by adequately controlled clinical data obtained through the administration of the product in the manner intended, to effect a given result' [8]. A potency assay measures the biological activity representing the desired mechanism of action (MoA) of an ATMP in a quantitative manner. Usually, a product-specific attribute that is directly or indirectly linked to the biological activity is detected. An example for such an attribute is an enzyme whose activity is associated with the relevant biological function of the ATMP such as the ability to interact with existing cells for tissue repair. The challenge when work-

R. Torggler · E. Margreiter · R. Marksteiner
Innovacell AG, Innsbruck, Austria
e-mail: Raffaela.Torggler@innovacell.com; Eva.Margreiter@innovacell.com; Rainer.Marksteiner@innovacell.com

M. Thurner (✉)
Innovacell AG, Innsbruck, Austria

Finnegan, Henderson, Farabow, Garrett & Dunner LLP, Munich, Germany

J. S. Burns (ed.), *Potency Assays for Advanced Stem Cell Therapy Medicinal Products*, Advances in Experimental Medicine and Biology 1420, https://doi.org/10.1007/978-3-031-30040-0_2

ing with such complex systems is to identify an attribute that is quantifiable and represents the relevant MoA of the ATMP.

2.1.2 Regulatory Requirements for a Potency Assay

One requirement for marketing authorisation of ATMPs is the demonstration of all relevant information on the characterisation of the product. Potency testing is part of the product characterisation and control strategy, which explains the importance of being able to measure potency for marketing authorisation and thus of establishing a potency assay during product development [9–12]. To release a product on the market, specifications for potency and other product quality characteristics must be defined and clearly stated [7]. Potency should be measured in a quantitative manner, making it easier to set precise acceptance criteria. Each batch must meet these specifications for its release as part of product quality control testing, so potency testing plays a central role for drug release [3–6]. Product characteristics beside potency that must be tested involve identity, purity, sterility and viability [3, 10, 13]. Regulatory guidelines also describe the importance of potency assays for validating the manufacturing process, demonstrating batch-to-batch consistency as well as determining the stability and shelf-life of a product [3–5].

As ATMPs vary in drug substance and desired MoA, no uniform potency assay is available. This implies that a potency assay needs to be developed individually for every ATMP and the application it is used for. Thus, the evaluation of a potency assay by the regulatory agencies must also occur on a case-by-case basis. A potency assay used for marketing authorisation must comply with the appropriate regulations and guidelines [3–5]. This includes fulfilling the following key points for a potency assay; the resulting data

(i) are quantitative, enabling the precise specification of acceptance criteria for product release
(ii) indicate the relevant biological activity and reflect the expected MoA of the ATMP
(iii) give an indication of the overall product quality as potency depends on other quality parameters such as purity, identity, and viability
(iv) correlate with the drug dose (e.g. number of potent cells)
(v) are controlled by appropriate standards, reference materials and/or other controls
(vi) allow establishing stability specifications
(vii) can (ideally) be linked to clinical efficacy meaning that higher biological activity indicated by the potency assay result leads to a better clinical outcome, and
(viii) must be validated to guarantee specificity, accuracy, precision, linearity, range and robustness of the assay according to general rules on method validation (Fig. 2.1) [5, 14].

2.2 Approach to Develop a Potency Assay for Cell-Based ATMPs

2.2.1 General Considerations

The development of a potency assay for cell-based ATMPs is a challenging but at the same time essential task. Being able to measure the product's potency in a quantitative manner is not only a prerequisite for market approval application and a crucial part of quality control but also valuable for clinical use, including its role in estimating clinical efficacy and defining the effective dose of an ATMP.

Some general considerations must be kept in mind when establishing a potency assay. Although it may not be possible to immediately meet all key requirements of a potency assay, such as linking potency to clinical efficacy, it should be initiated during early stages of product development [3, 5, 15]. This allows the collection of valuable data from pre-clinical or early clinical to late clinical stages as well as evaluation of the impact of manufacturing changes on the product's quality already during

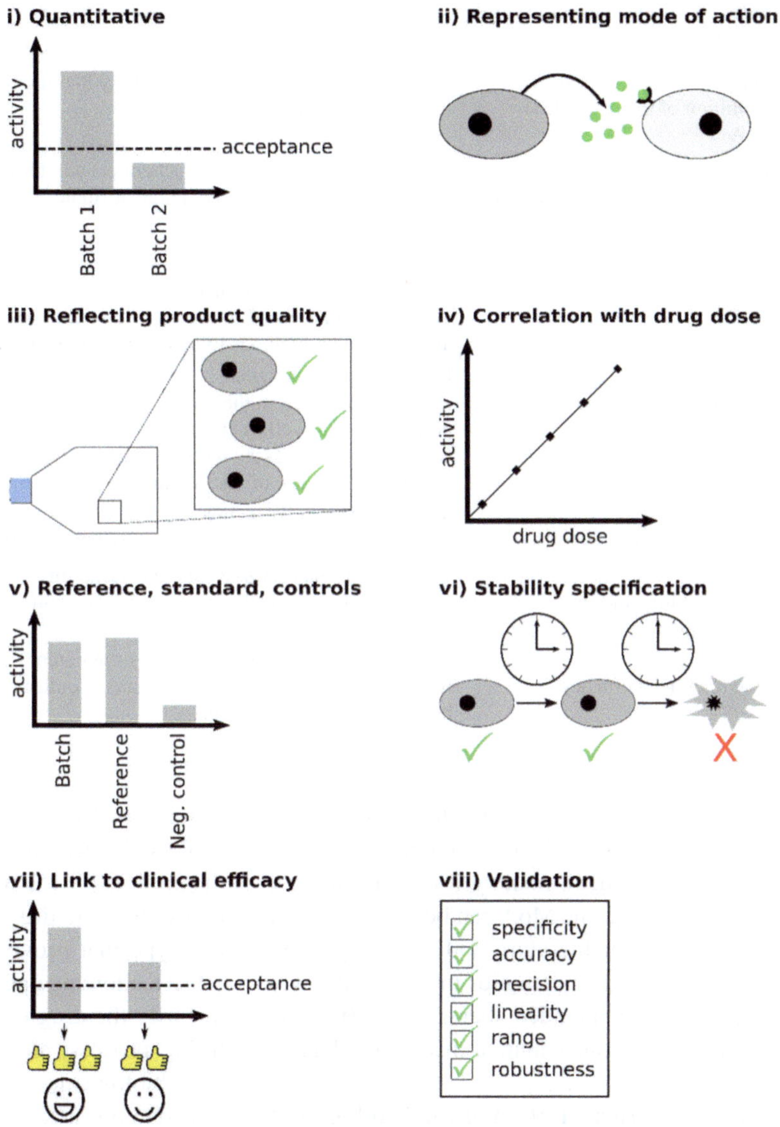

Fig. 2.1 The key requirements for potency assays. A potency assay must provide quantitative data that correlate with the drug dose and reflect the desired MoA as well as the overall product quality. Proper controls must be included when performing the experiment. A potency assay can be used to determine product's stability specification and ideally to estimate clinical efficacy. The successful establishment of a potency assay according to these criteria and the validation of the assay are required for marketing authorisation of an ATMP

its development. The potency assay might need to be adapted or re-designed during product development, as increased understanding of the ATMP can introduce new opportunities to improve the potency assay. It is therefore advisable to identify and characterise more than one attribute indicating the biological activity and appropriate detection method suitable for a potency assay [5]. Constant refinement of the potency strategy throughout the clinical development results in a well-established assay that fulfils the key requirements and reflects the clinical experience.

In addition to these general considerations, several steps are suggested when developing a potency assay, as outlined in this section and in Table 2.1. It is important to mention that the steps do not have to be performed in this exact order as some steps can be interchanged or done in parallel.

Step 1 – Definition of the (Expected) MoA

First, the MoA or at least the expected MoA of the ATMP must be defined. MoA describes the mechanism how an ATMP causes the desired therapeutic effect in the patient. However, this

Table 2.1 Steps to develop a potency assay

Steps	Short description
Step 1: Definition of the (expected) MoA	Define the intended biological activity that the product has to exhibit for its (expected) therapeutic effect
Step 2: Characterisation of the product and definition of possible attributes for potency testing	Determine the physical, (bio-)chemical and biological characteristics of the product's drug substance to define attributes that directly or indirectly represent the intended MoA
Step 3: Design of the potency assay	Decide whether a biological assay, non-biological analytical assay or multiple assays are most suited for measuring potency of the respective ATMP. Develop and implement methods to quantitatively measure defined attributes
Step 4: Evaluation of the potency assay	Define appropriate controls and standards. Test the relationship between the results obtained from the potency assay and the drug dose (e.g. number of potent cells) as well as critical quality parameters as already set in place. Optimise potency assay methodology for routine use. Draft standard operating procedures and use of instruments
Step 5: Validation of the potency assay	Validate specificity, accuracy, precision, linearity, range and robustness of the proposed potency assay according to general guidelines for method validation
Step 6: Test variability and set specification	Test batch-to-batch variability and variability throughout the manufacturing process. Set specifications including acceptance criteria based on the collected data
Step 7: Link to clinical efficacy	Test whether the result provided by the potency assay can be linked to the intended therapeutic effect in patients. Confirm, if possible, by formal hypothesis-driven procedure

MoA can be complex, multifactorial and is often not fully understood or characterised during ATMP development, making it difficult to specify the MoA [5]. The MoA can be specified using knowledge from literature and by studying the product's biological activities in vitro and in vivo concerning the intended therapeutic effect, for example, in surrogate animal disease models.

Step 2 – Characterisation of the Product and of Possible Attributes for Potency Testing

A potency assay needs to represent the MoA and indicate the relevant biological activity of the product. To develop such an assay, the product as well as the materials and methods used during the manufacturing process must be characterised thoroughly [5]. Variabilities in the manufacturing process (e.g. cultivation times) and materials the drug substance is brought in contact with (e.g. growth factors) might alter the product-specific attributes that are linked to the product's biological activity. These possible effects should be considered when defining attributes suspected to represent the MoA. Moreover, it is necessary to acquire a deep understanding of biological as well as physical and (bio-)chemical properties of the product [4, 15, 16]. These data are required to first define the drug substance (e.g. the relevant cell type that provokes the intended therapeutic effect), allowing one to subsequently identify and quantify the drug substance during the manufacturing process; and second, to specify the relevant biological activity by means of attributes linked to this activity. The identification of physical and (bio-)chemical properties that are directly involved in, or indirectly linked to, the intended biological activity of the product is valuable for developing an analytical potency assay [3]. An example for an indirect association could be an enzyme that is only expressed and active during fulfilment of the desired biological activity in vitro [14]. The most crucial step here is to clearly demonstrate that these attributes are associated with the desired biological activity. While physical and (bio-)chemical characteristics show that the requirements to provoke the intended biological effect are given, biological characterisation refers to the analysis of the product's relevant function in a living biological system, for example measurement of cell migration in vivo in animals. Thus, direct analysis of the biological activity or indirect analysis by physical or (bio-)chemical properties can be considered. Taken together, the characterisation analysis should result in the identification of parameters to define the drug substance, and of attributes that are linked to the biological activity and potentially relevant to the proposed MoA [15].

Step 3 – Design of the Potency Assay

After a comprehensive characterisation of the product, defining the MoA and attributes that possibly represent the MoA, the next step is to design an assay allowing measurement of these attributes in a quantitative manner. It is important to consider that the potency assay should comply with the respective regulations and the current good manufacturing practice [4, 5, 17]. Potency tests can be grouped into biological, non-biological analytical, or multiple assays [5].

Biological assays measure the product's ability to elicit the intended effect in vivo using animal studies or in vitro using organ, tissue or cell culture systems. These assays demonstrate that the product is biologically active in the context of a biological system, serving as a good indication that the product will provoke the desired effect also in the patient. However, providing quantitative data is more complicated and biological assays are often themselves complex and time-consuming [4]. This can be very problematic if the product stability does not allow waiting for the result of the potency assay for days or weeks before lot release. Owing to these difficulties, it is not always feasible to develop a biological assay measuring the product's potency. In this case, potency needs to be evaluated in non-biological analytical assays. Here, surrogate markers, i.e. physical or (bio-)chemical parameters that are linked to the intended biological activity, are measured in a quantitative manner. This requires extensive characterisation of the product and appropriate tests to clearly demonstrate the correlation between the measured parameter and the biological activity [5]. If potency cannot be measured adequately by a single biological or non-biological analytical assay, multiple assays have to be utilised. This might be the case if the biological activity can only be sufficiently indicated by the measurement of more than one parameter that cannot be tested in a single potency assay. These multiple assays, also called an assay matrix, can consist of biological and/or analytical assays that provide quantitative data and optionally additional qualitative data.

Overall, the aim of this step is to find analysis methods suitable for quantitative measurement of the product-specific attributes that indicate the biological activity and reflect the desired MoA. Although the quantitative nature of potency assays is described as a key requirement, quantification of biological activities is not always feasible. In these cases, semi-quantitative assays might be accepted, although defining the acceptance criteria and validating the assay might be more difficult. As potency testing is ultimately part of product quality control necessary for batch release, the potency assay needs to be validated, and several tests to evaluate the assay must be conducted. Therefore, it is advisable to define a range of appropriate attributes and more than one analysis method to possess at least one potency assay that fulfils all requirements, and ideally indicates clinical efficacy [5]. Potency assays that are not suitable for batch release, for example potency assays that are not fully validated or in vivo assays that are too time-consuming, can still provide useful information for product characterisation and further development.

Step 4 – Evaluation of the Potency Assay

A potency assay that quantitatively measures product-specific attributes indicating the relevant biological activity fulfils the main requirements that are necessary to describe it as a potency assay. However, additional considerations and tests are required to show the fulfilment of the other key requirements to generally improve the potency assay and to establish it as a practicable method for routine use.

One of the core requirements for potency testing is the inclusion of reference material, standards and/or other controls when performing the assay to ensure that the assay has performed as expected [4, 5]. Without these controls, it is impossible to distinguish whether the result obtained is caused by a biological event or a technical artifact. A negative result can indicate either that the product does not exhibit the desired biological activity or that the assay failed due to technical problems. Only if detected as such, technical errors can be corrected, indicating a need to repeat the experiment. As every ATMP and potency assay is different, the necessary controls must be considered individually for each potency assay during the assay development process. Reference materials, for example, can be either well characterised materials, such as cell lines that are similar to the product, or a product batch with demonstrated activity and performance. The provision of the latter is often not feasible as it requires large amounts of the product that are stable and storable over a longer period without losing their quality properties [15]. The crucial role of these controls in a standardised process shows the need for validation of the controls themselves including the evaluation of their stability. Especially when a new batch of the reference material is required, its performance must be compared to the original batch.

In addition, it is necessary to show that the result obtained correlates with the dose of the product (e.g. number of potent cells). Generally, the results provided by the potency assay should be used alongside other quality parameters such as viability, purity and identity. It is advisable to perform tests showing these dependencies. An example for a quality parameter is the number of cell doublings indicating the proliferative age of a cell culture. Depending on the ATMP, the desired biological activity might decrease with progressive cell divisions. If this is the case, it should affect the result measured by the potency assay. Moreover, as potency is a crucial indicator for stability studies, it is important to show that the potency assay can be used to determine the shelf life of a product and to validate the product's stability.

To generally optimise a potency assay, sources of variability should be omitted by establishing and subsequently following a detailed standard operating procedure [5]. Developers should consider whether to use externally supplied kits or established in-house methods and procedure. The long-term supply and quality of crucial reagents used in the potency assay should be ensured. As the potency assay is used as a routine assay for drug release, it should be as cheap, fast and simple to use as possible.

Step 5 – Validation of the Potency Assay

A potency assay must be validated to be suitable in routine use for drug release procedures. During the validation process, the specificity, accuracy, precision, linearity, range and robustness of a potency assay are tested [5, 15, 18]. To analyse these validation characteristics, statistical methods must be applied and methods must be fully described. Detailed definitions and descriptions of how to gain and present the validation results are outlined in ICH Topic Q2 (R1) [18] and are summarised in the following paragraphs. A re-validation is necessary if the procedure of the potency assay or the manufacturing process, including the composition of the product is changed [18]. Modifications to the workflow of a potency assay require a comparability study between the original and the modified assay [5].

Specificity means that only the drug substance (e.g. relevant cell type) can elicit a positive response in the potency assay. No positive signal should be detectable when measuring impurities, product matrices alone (i.e. material or cells that stabilise or support the drug substance) or progenitor cells of the relevant cell type. Specificity can be demonstrated, for example, when the drug substance alone provokes the same response as the drug substance spiked with an appropriate amount of impurities such as irrelevant cell types. This also shows that the result of a potency assay depends on the purity of the product, another important quality parameter.

Accuracy is also termed 'trueness'. An assay is considered accurate if the measured value is sufficiently close to a conventional true value or reference value. One way to demonstrate accuracy in a validation process is to measure a reference standard of known behaviour with the newly developed potency assay and to show that this elicits the expected response.

Precision means that the measurement of the same sample multiple times under prescribed conditions leads to similar results. Precision can be divided into categories of repeatability, intermediate precision and reproducibility. Repeatability can be tested by measuring one sample under the same operating conditions within a short time frame. Intermediate precision reveals within-laboratory variation, for example, by performing the same measurement procedure on different days or by different operators. The highest level of precision is obtained when one expresses the variation of measurements executed by different laboratories.

To demonstrate linearity, there needs to be a direct correlative relationship between the obtained value and the amount of active substance that is detected and measured in the assay. This substance can be the drug substance (e.g. relevant cell type) and/or an appropriate standard solution. For example, if the potency assay detects the end-product of an enzymatic reaction, a serial dilution from a stock solution of this end-product can be used to demonstrate linearity of the assay.

Range can be described as the interval between the lower and upper amounts of measurable substance in which the assay is sufficiently accurate, precise and linear. This can be specified during the linearity study.

It is also recommended to test the robustness in the validation process. An assay is considered robust when it remains unaffected by small modifications.

The FDA also lists system suitability as a relevant parameter to be validated [5]. This can be described as a test to ensure that the system performs as expected. In addition, the modifications for validating qualitative or semi-quantitative assays are described by the FDA [5].

Step 6 – Test Variability and Set Specifications

During the assay development, it is important to test if and how the biological activity indicated by the potency assay varies between different product batches. Optimally, batch-to-batch variability analysis is performed during the manufacture of clinically used product batches. Based on all relevant data collected from tests to establish, evaluate and validate the potency assay as well as from pre-clinical and clinical studies, specifications should be set. During early clinical development it is not necessary to already define strict specifications, as the establishment of the potency assay might not be completed [5, 15]. In addition, data acquired during this stage of clinical testing is valuable for optimising or adapting the potency assay. The more the potency assay stands the test of time throughout clinical development, the more the specifications can be tightened to finally set well-defined acceptance criteria for drug release.

Step 7 – Link to Clinical Efficacy

Ideally, the activity measured by the potency assay should provide a link to clinical efficacy [4, 5]. This means that a batch with promising potency results should lead to a better treatment outcome than a batch with only moderately good results. This association is important to define the effective dose and acceptance criteria for batch release. Pharmacological information that is derived during preclinical and clinical studies can be screened for any connection to putative potency attributes analysed during manufacture of batches used in those studies. Post-hoc subgroup definition for retrospective analysis between efficacy/safety outcomes and potency analysis can be performed in early clinical phases where sample sizes are low. Information derived in this way might be utilised for hypothesis-driven testing of potency attributes and clinical efficacy in large sample clinical trials, ultimately validating the potency assay for clinical use.

2.2.2 Progressive Implementation of a Potency Assay

Before a product can be released onto the market, establishment of the potency assay must be completed and fully validated with defined specifications set. However, development of the potency assay already starts much earlier and should occur ideally in parallel with the product development. The further the product development progresses, the more sophisticated the potency assay becomes. Thus, implementation of the potency assay in product development should occur progressively [5]. Although maybe not completely established, a potency assay should be already implemented during pre-clinical and early clinical studies. Potency testing in these phases can provide fundamental data valuable for further product development. When implementing the potency assay in late phase studies, it is important that the potency assay is already sufficiently mature to provide reliable and accurate results, to gather meaningful clinical data and to link potency to clinical efficacy. During these late development phases, defined acceptance criteria must be set to ensure that the batches used in the study are biologically active and can be consistently manufactured. In addition, potency testing is applied to determine the stability of the product. A validated potency assay with defined acceptance criteria that fulfils all the requirements can be finally used to support product market approval.

2.3 Clinical Value of Potency Assays

2.3.1 Relationship Between Potency and Clinical Efficacy

Clinical efficacy is a measure of how successful a treatment is in achieving a desired therapeutic effect. The only way to determine clinical efficacy of a newly developed product is to perform clinical studies. No in vivo animal or in vitro study can predict whether the proposed effect, such as significant improvement of symptoms, is achieved when treating human patients with the product. One requirement for gathering reliable and meaningful data during the clinical studies is to ensure that the product is consistently manufactured from batch to batch and fulfils predefined quality parameters and stability expectations. Assessing the product's potency plays a central role, in turn highlighting the important clinical value of potency assays [4, 5]. Demonstrating consistency in the manufacturing process is also an essential control to show that the process is stable and performs as expected for every batch. In addition, clinical studies should confirm the clinical efficacy of a product that is manufactured according to a specific procedure and that exhibits measurable parameters with pre-defined specifications. Assuming successful manufacturing procedures and consistent potency test performance, clinically destined sample batches are expected to result in similar clinical outcomes.

A potency assay measures the product's biological activity usually by analysing a product-specific attribute that is directly involved in or indirectly linked to the biological activity required for the product's mode of action. Although biological activity is a requirement for clinical efficacy, it does not guarantee the success of the treatment. This means that measuring the product's potency does not replace evaluation of the product's clinical efficacy in blinded, randomised and placebo controlled clinical trials. Still, a link between potency and clinical efficacy can be established by showing that a batch with promising, good potency results is more efficacious treatment than a batch with only moderate potency results. Thus, potency testing can be expected to demonstrate that the product has the potential to provoke the desired effect also in the patient.

2.3.2 Potency and Defining an Effective Dose for ATMPs

Finding the dose that is required to induce the desired effect is a central task during clinical development of ATMPs. Generally, clinical trials comparing different product doses are carried out to evaluate optimal safety and efficacy balance to find the effective dose; described as the amount of drug substance in a product required to achieve the desired effect [10]. For market authorisation it is important to determine the minimal effective dose, i.e. the lowest dose capable of eliciting the desired effect [3, 10]. Identifying the minimal effective dose is also important to keep the cost of the final product as low as possible. In cases of cell-based ATMPs, a higher dose per product might go along with increased cell cultivation time and effort, which indeed results in higher manufacturing costs.

For cell-based ATMPs, the dose is often indicated as the number of cells or as the cell content (e.g. protein content) that meet pre-defined acceptance criteria in terms of viability, identity, purity and potency. To design meaningful dose-finding clinical studies, it is of utmost importance to ensure potency consistency between batches. Special care must be taken when alterations in the manufacturing process are required to produce different doses (e.g. low and high cell count), since this might have an impact on the potency of the ATMP. This underlines the necessity of understanding the extent to which manufacturing processes may influence ATMP potency and of having an appropriate potency assay in place to control for batch-to-batch sample consistency. Besides determining the dose-response correlation with consistent potency per dose in clinical studies, considering varying potencies per dose might help optimise the safety and efficacy outcome. Ideally, the result obtained from a potency assay is linked to clinical efficacy, providing valuable data for dose definition. In addition, potency test results should correlate with the dose per formulation, thereby helping to define the effective dose [3]. This highlights the importance of a potency assay for dose definition.

Developers of ATMPs might consider specifying potency as potency per unit of content (e.g. protein content or number of cells) or potency per batch. If a batch exhibits low potency, increasing the cell number is only possible within the developer's pre-defined range for product release or, following market approval, only within the label. Especially for cell-based ATMPs, increasing product amount by further proliferation of cells might lead to their alteration, for example, via senescence or differentiation, and this could ultimately result in a failure in product release. In case a specific potency per cell might be required to provoke the desired effect, lower potency per cell may not necessarily be compensated by simply increasing the cell number. Thus, defining a range of effective potency per product unit could help in estimating the range of product amount possibly required. Product or process optimisation studies are useful to find ways to increase potency per unit of content (e.g. cell number). This might imply that a lower dose (e.g. fewer cells) is required to elicit the desired effect. However, also here the dose can only be reduced within its pre-defined range. Further reduction of the dose might lead to alterations in clinical efficacy as well as safety and can be expected to require additional regulatory approval.

Taken together, potency assays are valuable tools for clinical use as they demonstrate manufacturing consistency, provide a link to clinical efficacy and play an important role in defining the effective dose.

2.4 Potency Assay in Product and Process Development

Potency testing is conducted on the final product and a result within a pre-defined acceptance range is required for release of the product for clinical use. However, in-process potency testing is useful to control the quality of the product already during upstream manufacturing steps and to identify critical methods and materials. Even if the quality flaws are so severe that the production must be stopped, early detection of such problems is important as money and time can be saved. Data derived from in-process potency testing can be used to foster process optimisation and further process development, leading to a process assuring consistent derivation of high-quality products.

In addition to its role in characterising the product and controlling the product's quality, a potency assay is valuable for a range of other studies necessary to optimise existing and develop new products and processes. This involves product stability, comparability and compatibility studies. The role of potency testing during these studies is summarised in the following sections.

2.4.1 Stability Studies

Stability can be defined as the period of time during which the quality of the product remains within pre-defined specifications [16]. It is crucial to know the in-use as well as the storage stability of the final product for its clinical implementation. In-use stability describes how long a product is stable once it starts to be in use, for example, after thawing of cryopreserved cells. The ability to store the product is often necessary due to feasibility issues (e.g. to comply with the time limits of supply chains or transport chains). In this case, it is important to determine the product's storage stability to clarify whether it is possible to store the product at all; and if it is, for how long, and which storage procedure can be applied [19]. The latter are the key questions to find out whether the product can be preserved (e.g. frozen), and whether the quality of the product remains intact after preservation (e.g. freezing and thawing) [20]. In addition, the stability of the product must be ensured during the shipping procedure [21]. Based on these data, the product's shelf life under storage conditions and in use can be defined [15, 19]. It is advisable to investigate the stability not only of the final product but also of intermediates. This provides important information on the period of production and storage opportunities of intermediates.

To determine its stability, the quality of the product is measured over time or before and after the proposed storage conditions. As potency is a central indicator for the product's quality, a well-established potency assay is indispensable for stability studies [15, 19]. Although viability tests provide an easy and fast indication of cell-based ATMP stability and might be required for biological activity, they do not demonstrate the biological activity required for the product to be efficacious [16]. Therefore, viability alone is not a good measure of stability. The importance of a potency assay for stability studies must be already considered during assay development. The FDA states that a potency assay should provide data to establish the period beyond which the product no longer exhibits its desired result [5]. Stability studies provide meaningful information to help establish the procedure of the manufacturing process. For this reason, it is advisable to study stability by measuring potency already during early product development.

2.4.2 Comparability Studies

Comparability is achieved when changes in the manufacturing process do not lead to relevant alterations of the product's quality properties, safety or efficacy; that is to say, when the product produced in the modified way is comparable to the original product [22]. These changes may involve intentional modifications to optimise the manufacturing process, but also other alterations such as a different supplier of critical material [16].

Before any changes can be implemented in the manufacturing process, their impact on the product's quality as well as on the validity of so far established non-clinical or clinical data needs to be evaluated in comparability studies [10, 22]. These studies involve the determination of several quality parameters of which potency is a very crucial one [15]. Potency testing does not only allow evaluation of whether product manufacture occurred as expected but it also indicates the biological activity, and ideally provides a link to clinical efficacy. To ensure that a product remains safe and efficacious after process changes, potency testing should be supported by additional biological assays measuring biological functions (e.g. cell migration, differentiation, etc.) [15, 16]. This is especially important when potency is measured with a non-biological analytical assay. If pre-clinical (in vitro or in vivo) studies do not sufficiently demonstrate the comparability of the products' quality especially its biological activity, clinical testing may be required to show comparable efficacy and safety. The latter clinical testing would be much more costly and time-consuming. As changes in the manufacturing process during early product development can be evaluated in the next clinical study, comparability testing during development is usually less expensive in terms of time and cost than any modifications required post-market authorisation [15]. A well-established potency assay may reduce the effort required for comparability studies, and can advantageously reduce the risk of incurring costly post-market clinical studies for comparability evaluation due to process changes.

2.4.3 Compatibility Studies

In many cases, the final ATMP does not only consist of the drug substance (e.g. relevant cell type) but also matrix material important for support, stabilisation or delivery. Compatibility studies need to be performed to demonstrate that the drug substance is compatible with these materials, and other substances it comes in contact with such as delivery device or product container [16]. This involves evaluating the quality of the product in its final composition and surrounding material. Potency testing plays also here an important role as it indicates whether the product retains its biological activity.

2.5 State of the Art of Potency Assays for Cell-Based ATMPs

2.5.1 Examples of Potency Assays for Cell-Based ATMPs

Potency assays must be developed individually for each ATMP and for each product-specific MoA, implying that no uniform potency assay is available. Still, potency assays developed for different ATMPs might have some aspects in common. Thus, when planning to develop a potency assay, it is advisable to study potency assays that have been already approved and are already applied on a routine basis.

Although the type and nature of the biological activity varies between different ATMPs, it often involves the stimulation/inhibition of other cells or the replacement of cells that are no longer existing or capable of fulfilling their function [17]. When a cell-based ATMP aims at activating or inhibiting specific cells, the biological activity of the cells within that ATMP could be the expression and/or secretion of relevant biomolecules such as cytokines following infusion or implantation. In this case, potency could be tested by measuring the expression or secretion efficiency in vitro, if it can be clearly demonstrated that the presence of these biomolecules lead to the desired effect on target cells [23]. Another example of biological activity is the ability of ATMP cells to migrate and develop cell networks or to form cell structures either with pre-existing host cells or among themselves [14, 24]. This may play an important role in tissue regeneration. Here, a potency assay should be designed to quantitatively measure cell structure generation, for example by detecting a surrogate marker that is demonstrably only expressed or active when the desired cell structure has formed [14]. Biological activities of stem cell-based ATMPs also involve their self-renewal and differentiation capacity, that should be measured qualitatively in a potency assay [17].

Although the manufacturing process is standardised, potency needs to be determined for every batch separately, as the starting material varies in quality and genetic composition. Biological potency assays directly measure the product's biological activity in the context of a living system while analytical potency assays determine physical or chemical parameters that are involved in or indirectly linked to the intended biological activity. Due to the complexity of biological assays, the product's potency is often determined by measuring these analytical surrogate markers. Widely-used surrogate markers include secretion factors, cell surface markers or activation markers that are commonly detected by methods such as enzyme-linked immunosorbent assay (ELISA), flow cytometry or enzymatic assays [25]. These surrogate markers enable a faster and simpler readout, suitable for batch release testing. However, it is crucial to clearly demonstrate that the result obtained by the analytical potency assay correlates with the intended biological activity. For example, if the secretion efficacy of a relevant factor is measured by an analytical potency assay, it must be demonstrated that this factor is linked to the intended biological activity such as the activation of another cell type. Table 2.2 shows examples of potency assays

Table 2.2 Examples of potency assays for ATMPs mainly based on surrogate markers

Product name (Company) Drug substance MoA	Marker	Detection method	Test to link to biological activity	References
ICEF15 (Innovacell) Human autologous skeletal muscle-derived cells (aSMDC) Formation of skeletal myofibers by cell-to-cell fusion following intramuscular injection	AChE activity	Colorimetric enzymatic assay, absorbance detected by plate reader	Myotube formation efficiency correlates with AChE activity in vitro. High AChE activity was linked to high treatment outcome in faecal incontinence treatment by ICEF15	[14]
MultiStem®(Athersys) Allogeneic bone marrow derived multipotent progenitor cells Treatment of acute myocardial infarction by paracrine activity to promote angiogenesis	Secretion of the angiogenic factors VEGF, IL8 and CXCL5	ELISA	Immunodepletion studies demonstrate that VEGF, IL8 and CXCL5 are necessary for MultiStem-induced angiogenesis determined by tube formation ability in the human umbilical vein endothelial cell angiogenesis assay	[23]
Apligraf® (Organogenesis/ Novartis) Skin-like construct consisting of allogeneic keratinocytes and fibroblasts Treatment of VLU and DFU by creation of physical barrier and paracrine activity	Histological parameters	Histological analysis	In vitro and in vivo tests demonstrate an association between histological parameters and functional outcome	[28]
Dermagraft® (Shire Regenerative Medicine/ Organogenesis) Dermal tissue engineered from allogeneic dermal fibroblasts Treatment of DFU by cell colonisation and paracrine activity	Secretion of regenerative factors such as VEGF	ELISA	Metabolic activity correlated with the VEGF secretion efficiency. High VEGF secretion was detected when metabolic activity of the product was within the therapeutic range	[28–30]
Neo-Urinary™ Conduit (Tengion) Tissue engineered neo-organ based on autologous smooth muscle cells Tissue regeneration by cell migration and paracrine activity involved in cell recruitment	(i) Secretion of VEGF and MCP1 (factors involved in cell recruitment) (ii) Cell migration	(i) ELISA (ii) In vitro cell migration assay	(i) MCP1 secretion increased monocytes recruitment and VEGF secretion was associated with monocyte/macrophage infiltration as shown by a study on vascular transformation (ii) Efficient cellular migration represented a principal MoA	[28, 31, 32]
Provenge® (Dendreon Pharmaceuticals) Autologous matured APCs loaded with tumor-specific antigen Treatment of prostate cancer through the cell killing ability of $CD8^+$ T cells activated by the loaded and matured APCs	Expression of the cell surface marker CD54 (marker for immune cell activation)	FACS	Expression of CD54 on APCs increased during cultivation with an APC activating factor Upregulation of CD54 correlated with improved survival of the patients	[33–35]

(continued)

Table 2.2 (continued)

Product name (Company) Drug substance MoA	Marker	Detection method	Test to link to biological activity	References
ChondroCelect™ (TiGenix) Autologous chondrocytes Treatment of cartilage defects by chondrogenesis and cartilage formation	Expression of marker	PCR-based marker assay	The expression of these markers correlated with in vitro models demonstrating chondrocyte functionality and in vivo testing of cartilage formation such as the ectopic cartilage formation assay. This assay itself could be used as an in vivo potency assay as it was correlated with animal efficacy model	[16]
Prochymal™ (Osiris/ Mesoblast) Allogeneic MSCs derived from bone marrow Treatment of graft-versus-host disease by paracrine activity to downregulate inflammatory responses	Expression of TNFR1 (marker of anti-inflammatory activity)	ELISA	Anti-inflammatory activity of TNFR1	[36, 37]

Abbreviations: *AChE* Acetylcholinesterase, *APC* Antigen-presenting cell, *CXCL5* C-X-C motif chemokine ligand 5, *DFU* diabetic foot ulcer, *ELISA* enzyme-linked immunosorbent assay, *IL8* Interleukin 8, *MCP1* monocyte chemotactic protein 1, *MSC* mesenchymal stem cell, *TNFR1* tumour necrosis factor receptor 1, *VEGF* vascular endothelial growth factor, *VLU* venous leg ulcer

for ATMPs mainly based on surrogate markers, including their detection method and the appropriate test to demonstrate the link to the desired biological activity. The given examples focus on potency assays but do not provide a full description of the product's MoA and therapeutic potential. An overview of potency assays for T cells used in immunotherapy [26] and engineered chimeric antigen receptor (CAR)-T [27] cells are provided by recent reviews.

2.5.2 Challenges to Potency Assay Development for ATMPs

Due to the complexity of ATMPs, significant challenges are associated with the development of potency assays as summarised by the FDA [5] and described in the following section.

First, the MoA can be complex and therefore difficult to define. Still, specifying the MoA is a very crucial steps not only for the development of a potency assay but also for the general understanding of the product, design of clinical studies and correct interpretation of obtained data.

Second, as ATMPs are derived from cells or tissue of human beings, the variability of the material is high to start with. Although the manufacturing process is standardised and sources of variability in the production procedure are avoided wherever possible, the potency of the final product still depends on the quality and genetic constitution of the starting material. This makes it more difficult to set defined specifications and to distinguish whether the variability comes from the manufacturing process or the starting material.

Third, the amount of final product might be limited due to the biological source and nature of the product (e.g. limited proliferation potential of cell), implying that only limited material for testing is available. Increasing product yield is not always possible as this either requires more starting material or a longer manufacturing process and more cell duplications. The latter does not only imply higher manufacturing costs but might also lead to a decrease of the product's quality and biological activity. Thus, potency testing must be feasible with limited testing material.

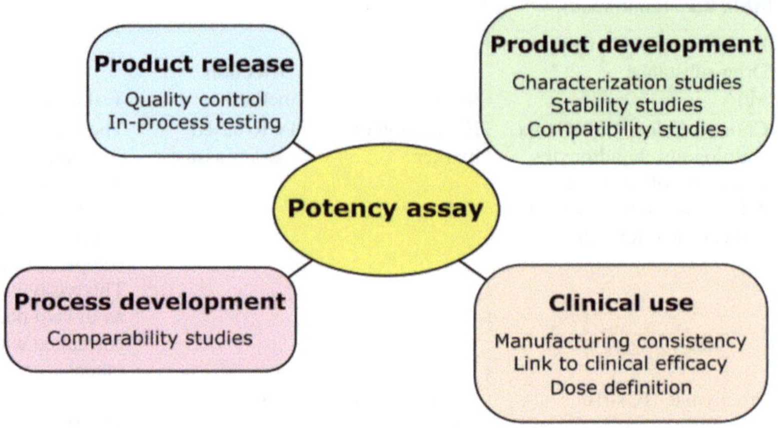

Fig. 2.2 The central role of potency testing in ATMP development and manufacturing. A potency assay is a valuable tool for product quality control, characterisation, development and optimisation strategies as well as for clinical use

Fourth, the stability of ATMPs might be restricted. Potency testing is required for batch release, implying that evaluating the final product's potency must occur within the period over which stability is ensured. Fast potency testing is especially important when cell freezing is not possible.

Fifth, appropriate reference material or standards to demonstrate that the potency assay works as expected might not be available. For example, the use of a potent batch as reference material for further potency assays might be challenging if large amounts are required and its stability must be ensured.

Sixth, the product might be a combined product, meaning that more than one cell type exhibiting different biological activities is ultimately required to elicit the desired therapeutic effect. In this case the development of multiple assays might be required to measure the product's potency.

2.6 Conclusion

Several challenges have to be faced when developing a potency assay for ATMPs but unique therapeutic potential, wide range of applications and the importance of a well-established potency assay make it worth the investment. Being able to measure the product's potency is valuable for batch release (quality control), product development (characterisation, compatibility and stability studies), process development (comparability studies) and clinical use (manufacturing consistency, link to clinical efficacy and dose definition) (Fig. 2.2). Potency testing using different approaches is meaningful throughout early to late phases of product development, implying that it is beneficial to start early and broadly with potency assay development. Use of different approaches to measure the product's potency helps addressing each challenge, such as quality control for batch release or comparability studies, with the most suitable potency assay. Regulatory bodies provide guidelines describing the requirements of a product-specific potency assay, and check whether these requirements are fulfilled, and thus whether the newly developed potency assay can be approved. Ultimately, the product developer and regulatory bodies provide complementary knowledge whilst working together to establish a reliable potency assay to provide a safe and effective product.

References

1. EMA European Medicines Agency (2015) Reflection paper on classification of advanced therapy medicinal products. https://www.ema.europa.eu/en/human-regulatory/marketing-authorisation/advanced-therapies/advanced-therapy-classification. Accessed 4 May 2020
2. FDA U.S. Food and Drug Administration (1998) Guidance for industry: guidance for human somatic cell therapy and gene therapy. https://www.fda.gov/regulatory-information/search-fda-guidance-documents/guidance-human-somatic-cell-therapy-and-gene-therapy. Accessed 14 April 2020

3. EMA European Medicines Agency (2008) Guideline on human cell-based medicinal products. https://www.ema.europa.eu/en/human-cell-based-medicinal-products. Accessed 14 April 2020
4. EMA European Medicines Agency (2016) Guideline on potency testing of cell based immunotherapy medicinal products for the treatment cancer. https://www.ema.europa.eu/en/potency-testing-cell-based-immunotherapy-medicinal-products-treatment-cancer-0. Accessed 14 April 2020
5. FDA U.S. Food and Drug Administration (2011) Guidance for industry: potency tests for cellular and gene therapy products. https://www.fda.gov/regulatory-information/search-fda-guidance-documents/potency-tests-cellular-and-gene-therapy-products. Accessed 25 March 2020
6. CFR - Code of Federal Regulations Title 21 Sec. 610.10. https://www.accessdata.fda.gov/scripts/cdrh/cfdocs/cfCFR/CFRSearch.cfm?fr=610.10. Accessed 14 April 2020
7. International Council for Harmonisation of Technical Requirements for Pharmaceuticals for Human Use (1999) Specifications: test procedures and acceptance criteria for biotechnological/biological products (ICH Q6B). https://www.ich.org/page/quality-guidelines. Accessed 14 April 2020
8. CFR - Code of Federal Regulations Title 21 Sec. 600.3 (s). https://www.accessdata.fda.gov/scripts/cdrh/cfdocs/cfcfr/CFRSearch.cfm?fr=600.3. Accessed 14 April 2020
9. European Commission (2003) Commission Directive 2003/63/EC of 25 June 2003 amending Directive 2001/83/EC of the European Parliament and of the Council on the Community code relating to medicinal products for human use (Text with EEA relevance). https://eur-lex.europa.eu/eli/dir/2003/63/oj. Accessed 14 April 2020
10. European Commission (2009) Commission Directive 2009/120/EC of 14 September 2009 amending Directive 2001/83/EC of the European Parliament and of the Council on the Community code relating to medicinal products for human use as regards advanced therapy medicinal products (Text with EEA relevance). https://eur-lex.europa.eu/eli/dir/2009/120/oj. Accessed 14 April 2020
11. CFR - Code of Federal Regulations Title 21 Sec. 601.2 (d). https://www.accessdata.fda.gov/scripts/cdrh/cfdocs/cfcfr/CFRSearch.cfm?fr=601.2. Accessed 14 April 2020
12. FDA U.S. Food and Drug Administration (2008) Guidance for FDA reviewers and sponsors: content and review of Chemistry, Manufacturing, and Control (CMC) information for human somatic cell therapy Investigational New Drug Applications (INDs). https://www.fda.gov/regulatory-information/search-fda-guidance-documents/content-and-review-chemistry-manufacturing-and-control-cmc-information-human-somatic-cell-therapy. Accessed 14 April 2020
13. CFR - Code of Federal Regulations Title 21 Part 610 Subpart B. https://www.accessdata.fda.gov/scripts/cdrh/cfdocs/cfcfr/CFRSearch.cfm?CFRPart=610&showFR=1&subpartNode=21:7.0.1.1.5.2. Accessed 24 April 2020
14. Thurner M, Asim F, Garczarczyk-Asim D, Janke K, Deutsch M, Margreiter E, Troppmair J, Marksteiner R (2018) Development of an in vitro potency assay for human skeletal muscle derived cells. PLoS One 13(3):e0194561. https://doi.org/10.1371/journal.pone.0194561
15. EMA European Medicines Agency (2019) Draft guideline on quality, non-clinical and clinical requirements for investigational advanced therapy medicinal products in clinical trials. https://www.ema.europa.eu/en/guideline-quality-non-clinical-clinical-requirements-investigational-advanced-therapy-medicinal. Accessed 14 April 2020
16. Bravery CA, Carmen J, Fong T, Oprea W, Hoogendoorn KH, Woda J, Burger SR, Rowley JA, Bonyhadi ML, Van't Hof W (2013) Potency assay development for cellular therapy products: an ISCT* review of the requirements and experiences in the industry. Cytotherapy 15(1):9–19.e9. https://doi.org/10.1016/j.jcyt.2012.10.008
17. EMA European Medicines Agency (2011) Reflection paper on stem cell-based medicinal products. https://www.ema.europa.eu/en/stem-cell-based-medicinal-products. Accessed 14 April 2020
18. International Council for Harmonisation of Technical Requirements for Pharmaceuticals for Human Use (2005) Validation of analytical procedures: text and methodology (ICH Q2 R1). https://www.ich.org/page/quality-guidelines. Accessed 14 April 2020
19. International Council for Harmonisation of Technical Requirements for Pharmaceuticals for Human Use (1995) Quality of biotechnological products: stability testing of biotechnological/biological products (ICH Q5C). https://www.ich.org/page/quality-guidelines. Accessed 14 April 2020
20. Antebi B, Asher AM, Rodriguez LA, Moore RK, Mohammadipoor A, Cancio LC (2019) Cryopreserved mesenchymal stem cells regain functional potency following a 24-h acclimation period. J Transl Med 17:297. https://doi.org/10.1186/s12967-019-2038-5
21. Veronesi E, Burns JS, Murgia A, Candini O, Rasini V, Mastrolia I, Catani F, Paolucci P, Dominici M (2015) cGMP-compliant transportation conditions for a prompt therapeutic use of marrow mesenchymal stromal/stem cells. In: Turksen K (ed) Stem cells and good manufacturing practices: methods, protocols, and regulations. Springer, New York, pp 109–122
22. International Council for Harmonisation of Technical Requirements for Pharmaceuticals for Human Use (2004) Comparability of biotechnological/biological products subject to changes in their manufacturing process (ICH Q5E). https://www.ich.org/page/quality-guidelines. Accessed 14 April 2020

23. Lehman N, Cutrone R, Raber A, Perry R, Van't Hof W, Deans R, Ting AE, Woda J (2012) Development of a surrogate angiogenic potency assay for clinical-grade stem cell production. Cytotherapy 14(8):994–1004. https://doi.org/10.3109/14653249.2012.688945
24. Iqbal F, Szaraz P, Librach M, Gauthier-Fisher A, Librach CL (2017) Angiogenic potency evaluation of cell therapy candidates by a novel application of the in vitro aortic ring assay. Stem Cell Res Ther 8:184. https://doi.org/10.1186/s13287-017-0631-1
25. Stroncek DF, Jin P, Wang E, Jett B (2007) Potency analysis of cellular therapies: the emerging role of molecular assays. J Transl Med 5:24. https://doi.org/10.1186/1479-5876-5-24
26. de Wolf C, van de Bovenkamp M, van de Hoefnagel M (2018) Regulatory perspective on in vitro potency assays for human T cells used in anti-tumor immunotherapy. Cytotherapy 20(5):601–622. https://doi.org/10.1016/j.jcyt.2018.01.011
27. Kiesgen S, Messinger JC, Chintala NK, Tano Z, Adusumilli PS (2021) Comparative analysis of assays to measure CAR T cell–mediated cytotoxicity. Nat Protoc 16:1331–1342. https://doi.org/10.1038/s41596-020-00467-0
28. Guthrie K, Bruce A, Sangha N, Rivera E, Basu J (2013) Potency evaluation of tissue engineered and regenerative medicine products. Trends Biotechnol 31(9):505–514. https://doi.org/10.1016/j.tibtech.2013.05.007
29. Mansbridge J, Liu K, Patch R, Symons K, Pinney E (1998) Three-dimensional fibroblast culture implant for the treatment of diabetic foot ulcers: metabolic activity and therapeutic range. Tissue Eng 4(4):403–414. https://doi.org/10.1089/ten.1998.4.403
30. Mansbridge J (2006) Commercial considerations in tissue engineering. J Anat 209(4):527–532. https://doi.org/10.1111/j.1469-7580.2006.00631.x
31. Roh JD, Sawh-Martinez R, Brennan MP, Jay SM, Devine L, Rao DA, Yi T, Mirensky TL, Nalbandian A, Udelsman B, Hibino N, Shinoka T, Saltzman WM, Snyder E, Kyriakides TR, Pober JS, Breuer CK (2010) Tissue-engineered vascular grafts transform into mature blood vessels via an inflammation-mediated process of vascular remodeling. Proc Natl Acad Sci U S A 107(10):4669–4674. https://doi.org/10.1073/pnas.0911465107
32. Guthrie KI, Sangha N, Genheimer CW, Basu J, Ludlow JW (2013) Migration assay to evaluate cellular interactions with biomaterials for tissue engineering/regenerative medicine applications. In: Basu J, Ludlow JW (eds) Organ regeneration: methods and protocols. Humana Press, Totowa, pp 189–196
33. Higano CS, Small EJ, Schellhammer P, Yasothan U, Gubernick S, Kirkpatrick P, Kantoff PW (2010) Sipuleucel-T. Nat Rev Drug Discov 9(7):513–514. https://doi.org/10.1038/nrd3220
34. Sheikh NA, Jones LA (2008) CD54 is a surrogate marker of antigen presenting cell activation. Cancer Immunol Immunother 57(9):1381–1390. https://doi.org/10.1007/s00262-008-0474-9
35. Patel PH, Kockler DR (2008) Sipuleucel-T: a vaccine for metastatic, asymptomatic, androgen-independent prostate cancer. Ann Pharmacother 42(1):91–98. https://doi.org/10.1345/aph.1K429
36. Kurtzberg J, Prockop S, Teira P, Bittencourt H, Lewis V, Chan KW, Horn B, Yu L, Talano J-A, Nemecek E, Mills CR, Chaudhury S (2014) Allogeneic human mesenchymal stem cell therapy (remestemcel-L, Prochymal) as a rescue agent for severe refractory acute graft-versus-host disease in pediatric patients. Biol Blood Marrow Transplant 20(2):229–235. https://doi.org/10.1016/j.bbmt.2013.11.001
37. Kebriaei P, Isola L, Bahceci E, Holland K, Rowley S, McGuirk J, Devetten M, Jansen J, Herzig R, Schuster M, Monroy R, Uberti J (2009) Adult human mesenchymal stem cells added to corticosteroid therapy for the treatment of acute graft-versus-host disease. Biol Blood Marrow Transplant 15(7):804–811. https://doi.org/10.1016/j.bbmt.2008.03.012

Potency Assays: The 'Bugaboo' of Stem Cell Therapy

3

Sílvia Torrents, Marta Grau-Vorster, and Joaquim Vives

Abbreviations

2D	2-dimensional
3D	3-dimensional
ALP	Alkaline Phosphatase
ATMP	Advanced Therapy Medicinal Products
CAR-T	Chimeric Antigen Receptor-T cells
cGMP	current Good Manufacturing Practice
CLI	Critical Limb Ischaemia
CQA	Critical Quality Attribute
DP	Drug Product
DS	Drug Substance
ECFA	Ectopic Cartilage Formation Assay
ELISA	Enzyme-Linked Immunosorbent Assay
EMA	European Medicines Agency
EU	European Union
FDA	Food and Drug Administration
FIH	First in Human
GTP	Good Tissue Practice
ICH	International Conference on Harmonization
IPC	In-Process Control
ISCT	International Society for Cell and Gene Therapy
MHLW	Ministry of Health, Labour and Welfare
MoA	Mechanisms of Action
MSC	multipotent Mesenchymal Stromal Cells
PMDA	Pharmaceuticals and Medical Devices Agency
QbD	Quality by Design
QC	Quality Control
QMS	Quality Management System
QTPP	Quality Target Product Profile
TPP	Target Product Profile
US	United States of America

S. Torrents · M. Grau-Vorster
Banc de Sang i Teixits, Edifici Dr. Frederic Duran i Jordà, Barcelona, Spain

Transfusion Medicine group, Vall d'Hebron Research Institute, Universitat Autònoma de Barcelona, Barcelona, Spain

J. Vives (✉)
Banc de Sang i Teixits, Edifici Dr. Frederic Duran i Jordà, Barcelona, Spain

Musculoskeletal Tissue Engineering Group, Vall d'Hebron Research Institute (VHIR), Universitat Autònoma de Barcelona, Barcelona, Spain

Departament de Medicina, Universitat Autònoma de Barcelona, Barcelona, Spain
e-mail: jvives@bst.cat

3.1 Potency – What a Cell Can Do

The expected effect of medicines is determined upon production by defining a critical quality attribute (CQA) known as potency [1]. According to the International Conference on Harmonization (ICH) of Technical Requirements for Registration

J. S. Burns (ed.), *Potency Assays for Advanced Stem Cell Therapy Medicinal Products*, Advances in Experimental Medicine and Biology 1420, https://doi.org/10.1007/978-3-031-30040-0_3

of Pharmaceuticals for Human Use, potency is defined as the quantitative measure of biological activity based on the attribute of the product, that is linked to the relevant biological properties [2]. Such activity can be assayed based on the intended biological effect, which should ideally be related to the actual clinical response. Advanced Therapy Medicinal Products (ATMP) pose an unprecedented challenge for drug developers due to the complex nature of this type of medicine [3]. Hence the importance of clearly defining suitable and reliable potency assays to guarantee batch-to-batch consistency for safe cellular products, capable of exercising the intended therapeutic effect. Generally, the measurement of biological activity will become the potency test for the drug substance (DS) and drug product (DP). Adequate potency assays are needed to predict the therapeutic efficacy of cell- and gene-based medicinal products throughout product development programmes and not only after marketing approval [4]. However, it is unlikely that one single assay will capture all biological effects of complex innovative medicines. Therefore, various in vitro or in vivo biological assays may be needed to convincingly measure potency, which in turn increases cost, time and production logistics. This is particularly relevant for small batches of allogeneic products or for single dose of autologous products. Moreover, the lack of relevant animal models is a major drawback for in vivo assays. On the other hand, in vitro assays are limited to the measurement of biochemical or physiological responses at the cellular level. Nonetheless, there has been rapid progress in the development of advanced cellular systems, including organoids and organs-on-a-chip that can recapitulate to some extent the clinical situation in miniaturised formats. Paradoxically, rather than resembling the clinical situation faithfully, there is a risk of generating artefactual environments with these approaches [5]. Nonetheless, it is worth mentioning some current tools for traditional drug testing, mainly for toxicology assessment and the study of mechanism of action (MoA) that may be implemented for use with cell- and gene-based therapies, such as two dimensional (2D) micropatterned co-cultures [6], three dimensional (3D) trans-well co-cultures [7, 8], 3D spheroids and organoids [9, 10], and 3D bioprinted tissues [11].

3.2 Relevance of Potency Assays

Identity of ATMP is commonly taken as a surrogate marker of their potency, particularly in early phase developments. However, this is a simplification in the recognition of the products' attributes that can be misleading provided that typical surface marker expression panels used for identifying cellular populations composing the active ingredient are (a) incomplete, (b) unspecific (in general) and (c) not necessarily linked to what cells can actually do in vivo. Although the expression of specific cell surface markers is extremely useful for a rapid identification of the drug substance (DS) and/or the drug product (DP), its biological response upon specific stimuli or the behaviour within the pathogenic *milieu* cannot be precluded by only considering the expression of such markers. Therefore, suitable potency testing is key for characterising this type of complex product, batch-to-batch consistency and comparability among different manufacturers. A combination of multiple methods may be needed to adequately define potency in functional assays during product development programmes. Certain assays may be needed to control quality amid procedural changes, whereas others are more suitable for product characterisation, comparability and release for clinical use. Preferably, the relevant, validated potency assay should reflect the clinical MoA and form part of the specifications for the DS and/or DP. When an appropriate potency assay is used for the DP, an alternative method (physicochemical and/or biological) may suffice for quantitative assessment at the DS stage. Potency assays are expected to be validated prior to pivotal late-stage clinical trials (Fig. 3.1). In some cases, the measurement of biological activity within a specific range may provide useful information, particularly if the MoA is poorly understood. In any case, it is important to objectively correlate the potency data with the actual pharmaceutical activity with

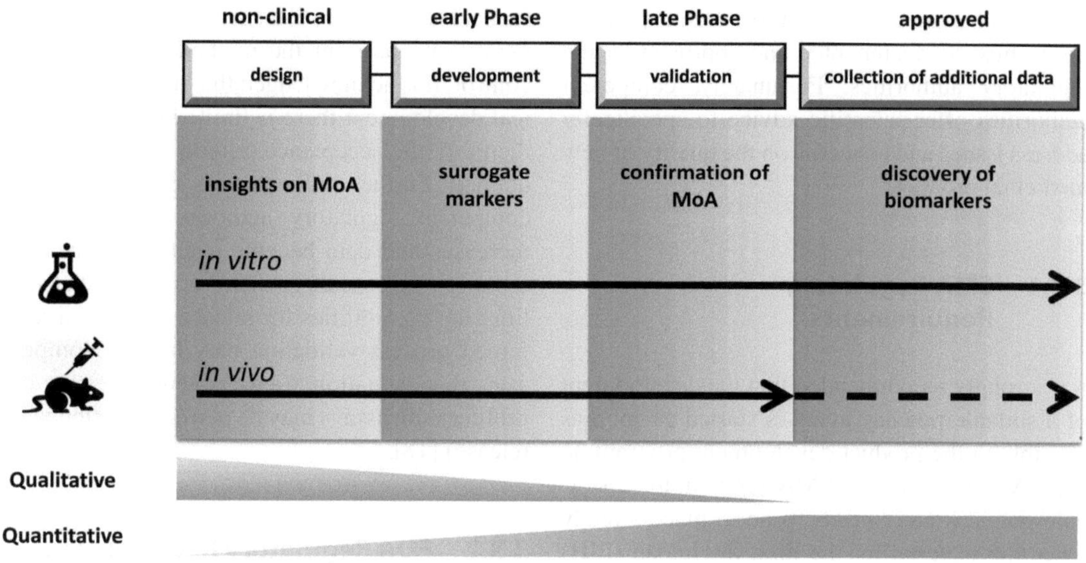

Fig. 3.1 From discovery to validation of potency assays alongside clinical development of cell- and gene-based medicinal products

confirmation in a clinically relevant setting [12]. For instance, the first advanced therapy medicinal product (ATMP) approved by the European Medicine Agency (EMA), ChondroCelect, used an Ectopic Cartilage Formation Assay (ECFA) in experimental mice but failed to demonstrate a valid correlation between the ECFA histology score and actual cartilage repair in an orthotopic goat model [13]. Therefore, the European cell therapy company TiGenix NV (subsequently acquired by Takeda Pharmaceutical Company) opted to define a panel of surrogate molecular markers that this time was demonstrated to correlate well with other relevant in vitro models of chondrocyte functionality, such as the well-described 3D chondrogenic pellet assay [14, 15]. This example illustrates how major efforts are needed to make potency assays relevant and predictive of the clinical effect. Due to the often-incomplete knowledge about the product in early stages of development, one should focus on improving the understanding of a comprehensive set of key aspects such as: (a) the final product conditioning and stability, (b) dosing and dosage of cell-based therapies, (c) intended route of administration, (d) interaction with concomitant treatments, (e) co-morbidities in the patient (particularly for chronic disease conditions) and (f) MoA. An additional concern in this field relates to the limited information available from commercial entities, in which some aspects of the potency assays used are deemed proprietary and not disclosed, making it more difficult to find relevant literature to support decisions.

3.3 Regulatory Requirements

As happens to be the case for all medicines, ATMP need the approval of the competent regulatory authority before commercial use in humans [3]. Regulatory agencies, such as the EMA in Europe, the Food and Drug Administration (FDA) in the US, or the Pharmaceutical and Food Safety Bureau (PFSB), the Japanese equivalent to the FDA, demand potency assays for product release. Although guidance for the development of appropriate assays is provided by regulatory authorities and scientific societies, e.g. International Society for Cell and Gene therapy (ISCT), it can still represent a major hurdle for most developers, especially when challenged to provide a potency assay associated to a MoA that is not necessarily completely understood [15,

16]. In all cases, a justification of the potency assay must be documented and approved by the regulatory authorities. Fortunately, competent authorities offer scientific advice to specifically address issues and concerns on the quality of new medicinal product.

3.3.1 EMA Regulatory Requirements

It is strongly recommended that the development of a suitable potency assay is started as soon as possible in the product development programme [17]. According to the EMA regulations, a suitable potency assay needs to be in place already when the material for the First in Human (FIH) clinical trial is produced and it should be validated prior to phase III clinical trials, unless otherwise justified [18]. At early stages of development, the absence of quantitative limits for biological activity may be acceptable and surrogate potency markers can be considered for release tests but such circumstances need to be appropriately justified. In line with current Good Manufacturing Practice (cGMP) requirements applicable to ATMP, a certain level of flexibility is recognised, so that the manufacturer can implement the measures that are most appropriate to the unique characteristics of the manufacturing process and of the product [19]. Indeed, knowledge about the product's attributes, particularly its potency, evolves together with the procedures used in the product development programme and this often involves adjustment of the manufacturing process as improved data becomes available.

In the case of gene therapy products, the intended gene edition should be demonstrated and the potency assay should cover: (a) the efficiency of gene edition, (b) the level and stability of expression of the therapeutic sequence or its direct activity or deletion and (c) a measure of the resulting functional activity, where possible.

Specifications must be relevant for the performance of the medicinal product and the acceptance criteria for each of the CQA must be based on sound scientific knowledge supported by available information specific to the candidate medicine, the batches used in non-clinical and/or clinical studies and data from stability studies, taking into account the methods used for their control. It is acknowledged that during early clinical development there is limited experience and therefore the acceptance criteria may be broadly defined. Further refinement is expected by the competent regulatory authority as knowledge increases and data become available. When the scarcity of materials or their very short shelf-life limit the possibilities for release controls, a reinforced process validation may help to compensate this situation (e.g. potency testing or proliferation assays may be performed after batch release) [18].

3.3.2 FDA Regulatory Requirements

According to the American FDA, all licensed product shall meet the standards applicable in all tests for potency, sterility, purity and identity [20]. Although specific potency tests are addressed to each specific product, all of them must comply with applicable biologics and cGMP regulations as listed below:

- Indicate potency (biological activity/activities) specific to the product
- Provide test results for product release
- Provide quantitative data
- Meet pre-defined acceptance and/or rejection criteria
- Include appropriate reference materials, standards, and/or controls
- Establish and document the accuracy, sensitivity, specificity and reproducibility of the test methods employed through validation
- Measure identity and strength (activity) of all active ingredients
- Provide data to establish dating periods
- Meet labelling requirements

3.3.3 Japanese FDA Regulatory Requirements

In Japan, the Ministry of Health, Labour and Welfare (MHLW) is in charge of pharmaceutical regulatory affairs. The Pharmaceuticals and

Medical Devices Agency (PMDA) reviews applications for drugs, medical devices, and regenerative medicines, and prepares review reports, whereas MHLW grants marketing authorisation [21]. All new drug/regenerative medical product applications are submitted to the PMDA. For the development of new drugs, sufficient data must be gathered on quality, efficacy and safety of new drugs, in both non-clinical and clinical studies. The Japanese authorities have always put the emphasis on safety and quality issues, rather than efficacy. Remarkably, the pharmaceutical industry in general has been presented as breeding distrust in Japan, following a number of scandals in the past [22]. This changed dramatically with the discovery of induced Pluripotent Stem Cells (iPSC) by Prof. Shinya Yamanaka [23], leading to deregulation and conditional approval of innovative drugs for regenerative medicine [24, 25]. Several guidelines are available in Japan covering all aspects of quality compliance, from Good Tissue Practice (GTP) and product evaluation, to cGMP and Quality Management Systems (QMS) [26].

3.4 Development of Potency Assays

ATMP are unique drug entities composed of or derived from living cells as principle active ingredient [3, 27]. Apart from activation, expansion or other type of substantial manipulations, cells can be genetically edited or combined with scaffolds to shape complex 3D structures before implantation in patients [28]. In this context, the characterisation of such type of products is challenging, since they result from specific manufacturing processes that impact on their identity, purity and potency [29]. Multipotent Mesenchymal Stromal Cells (MSC) represent a good example of this [30]. Despite existing recommendations from the International Society for Cell and Gene Therapy (ISCT) [31, 32], developers may adapt the recommendations to their own products and intended application making it difficult to comprehend whether MSC from different laboratories are actually equivalent [33]. This in turn can compromise the relevance of systematic reviews and meta-analysis searching for efficacy of treatments [30, 34]. In this context, potency assessment may contribute to a better understanding of the pharmaceutical activity of such products, if standardised assays are implemented in different labs or if a centralised quality control (QC) laboratory can verify the reported potency of batches of cells produced elsewhere. However, this requires proper definition of the potency assay-specific target involved in the clinical indication according to the expected MoA. Ultimately, ATMP are complex, can present donor-variability, and can display more than one MoA. Frequent challenges faced in the development of potency assays for cell and gene therapy products are listed in Table 3.1.

3.4.1 Use of Surrogate Markers

When a suitable potency assay is not feasible or not sufficiently reliable, it may be necessary to identify a surrogate measurement of biological activity related to its specific ability to effect a clinical result [35]. This is independent of further improvements on existing potency assays, which can be modified and adapted with the development of the product and the scientific progress. Of note, developers can receive guidance from the competent regulatory authority to establish an acceptable potency assay using surrogate markers.

Surrogate markers must correlate to in vivo potency models. In the case of secreted factors, for instance, it is important to consider whether the biological effect is caused by a combination of factors rather than by a sole molecule [36]. Accordingly, some authors have proposed to identify multiple factors that may assist in predicting the therapeutic capacity of cell-based products before clinical transplantation (e.g. donor-dependency, gene expression profile, secretome) in a scorecard format that may complement deficient potency assays or even become a potency test itself [37–39]. It is believed that the combination of multiple parameters may assist to ensure minimal quality requirements for clinical use [40, 41].

Table 3.1 Challenges in the design of potency assays

Challenges	Examples
Inherent variability of starting materials	Autologous and allogeneic donor variability Cell line heterogeneity Error-prone replicating viruses
Limited batch size and limited material for testing	Single dose therapy using autologous cells suspended in a small volume
Limited stability	Viability of cellular products
Lack of appropriate reference standards	Autologous cellular material Novel gene therapy vectors
Multiple active ingredients	Multiple cell lines combined in final product Heterogeneous mixtures of peptide pulsed tumour and/or immune-modulatory cells Multiple vectors used in combination
The potential for interference or synergy between active ingredients	Multiple genes expressed by the same vector Multiple cell types present in cell preparations
Complex mechanism of action(s)	Multiple potential effector functions of cells Multiple steps required for function such as infection, integration, and expression of a transgene Vector containing multiple genes
In vivo fate of product	Migration from site of administration Cellular differentiation into the desired cell type Viral or cellular replication Viral vector infection, uncoating, and transgene expression

It is noteworthy that certain potency assays are time-consuming and can become a hurdle if used as release criteria. Therefore, for such cases, the regulatory agencies may accept a surrogate potency biomarker for product release, particularly for fresh products and if adequately validated. For instance, in a recent study, Thej and collaborators evaluated the in vitro angiogenic potency of Stempeucel®, which is an allogeneic product resulting from pooling human bone marrow-derived MSC used in critical limb ischaemia (CLI) patients [42]. Remarkably, a single angiogenic factor (VEGF) qualified as a surrogate potency marker through three in vitro functional assays to determine the angiogenic potency of Stempeucel®. Similarly, biological function may be based on either paracrine activity, e.g. use of apoptotic MSC in the management of GvHD [43], or cell-autonomous functions, as in the case of gene edited CD34 hematopoietic stem cells in Fanconi Anemia [44]. In all cases, adequate justification is needed and the assays should be revised regularly to revalidate their suitability.

3.4.2 Autologous and Allogeneic Products

At the earliest stage of designing potency tests, it is important to consider the nature of the ATMP under development and the formulation of the final product. Clearly, a fresh autologous product has different considerations in comparison to situations employing large batches of multiple-dose cryopreserved products for allogeneic use [30]. According to the situation, product release timing may be incompatible with the time required for the potency assay. Nonetheless, most regulatory authorities would accept assays performed post-administration in the patient, particularly in early phase clinical trials, if adequately justified and documented (Table 3.2). This information would certainly contribute to a better understanding of the ATMP under development.

When routine release testing is limited or not possible, the evaluation of process robustness through in-process controls (IPC) becomes more important in lieu of batch testing. This is compatible with the concept of Quality by Design (QbD) that quality of products can be planned and successfully achieved if processes involved are properly designed and developed [1]. Although release criteria may consider the performance of multiple tests, cumulative data gathered along the production process can support the decisions for product release (Table 3.3). For

Table 3.2 Summary of principal characteristics of ATMP according to their sourcing and use and considerations in the design of suitable potency assays

Characteristics of ATMP		Considerations on potency assays
Autologous	Allogeneic	
Immune compatibility	Non−/partial compatibility	Test system must take into consideration potential rejection of drug product
Time-consuming, not ready to use	If cryopreserved, ready-to-use	Assays must be fast for timely release of autologous products
One donor, one patient	Possibility to generate multiple doses from one single donor	Limited sample volume in autologous products. Need to define convenient sampling in multiple dose batches
QC need to be run immediately upon preparation (if fresh)	Products can be quarantined until QC results (if cryopreserved)	Time constraints for product release in fresh autologous products

Table 3.3 Potency testing at the release of fresh and cryopreserve products

Fresh product	Cryopreserved product
Limited QC panels for timely conditional release Reliance on surrogate markers Potency tests to be performed after treatment	QC performed on control samples from each batch, not on all vials Product release only when all tests result within specifications

instance, the MSC immunopotency assessment can involve the preparatory master and working cell banks in addition to the final DP [29, 45, 46].

3.4.3 Standardisation of Assays

The suitability of the analytical methods used in the characterisation of ATMP should be confirmed and preliminary acceptance limits defined (e.g. acceptance limits for the determined impurity content). The parameters for performing qualification of the analytical methods include: specificity, linearity, range, accuracy, precision, quantitation and limit of detection, as appropriate. Biological characterisation and potency assays are the most important parameters to perform comparability of ATMP on quality grounds [47]. Unfortunately, a lack of assay standardisation impedes further systematic reviews and meta-analysis aimed at a formal assessment of previous clinical research to derive robust conclusions on safety and efficacy of innovative therapies.

The reliability and robustness of results from potency assays need to be confirmed by including adequate controls with appropriate reference-standard reagents. For medicinal products, reference materials are normally utilised to ensure consistency between different batches but also to ensure the comparability of the product to be marketed with that used in clinical studies and to provide a link between process development and commercial manufacturing. However, definition of such controls may be challenging. On appropriate control tests, some authors have proposed optimised reference samples serving as a 'cell ruler' to compare final batches [48]. Therefore, it is recommended to establish a reference batch as soon as possible. Clearly, this field is still under development and therefore it is important to highlight a general lack of standardisation, yet to be attained.

Standardisation of potency assays is recognised as a key objective but little agreement in the design of assays and lack of inter-laboratory validated reagents and protocols makes it difficult to achieve [49]. The ICH promotes discussion of scientific and technical aspects of pharmaceuticals and has developed guidelines, working as a link between regulatory authorities and pharmaceutical industries. Moreover, scientific societies are also committed to provide guidance in the development of potency assays. This is the case of the ISCT that has published several white papers on this topic [15, 16, 50].

3.4.4 Further Considerations

Initial proposals for potency assays in early stages of product development may probably be qualitative instead of quantitative (e.g. pass/fail). If this is the case, they should be accompanied by quantitative assays. For instance, the osteogenic potential of MSC may be assessed by their capacity to become osteoblasts, typically characterised by in vitro assays such as alkaline phosphatase (ALP) activity or Alizarin Red Staining [7, 51]. However, these in vitro assays may not necessarily correlate with the in vivo bone-forming potential [52]. Moreover, human MSC derived from different tissue sources behaved differently in standard osteogenic conditions and this may lead to confusing results (e.g. partial differentiation outcomes in Wharton's jelly-derived MSC compared to bone marrow-derived MSC) [7]. In vitro potency assay conditions are often insufficient to recreate the clinical condition and may not include other cell types that can be involved in vivo, or consider concomitant treatments that can potentially interact with the activity of the ATMP. In the same way, the assay may not reflect all relevant biological properties (e.g. miscalculation of the impact of the DP on other cell types) or be non-specific due to the presence of impurities. Moreover, biodistribution as well as dose and schedule of the candidate medicine may greatly differ from the situation found in preclinical studies and therefore further efforts must be undergone to track ATMP in the patient [53, 54]. Of note, differences in engraftment, differentiation, persistence and immunogenicity between animals and humans may limit the predictive value of non-clinical dose-finding studies, as in the case of, e.g. genetically modified CD34 positive cells for treatment of severe immune deficiencies [47].

3.5 The Quality Target Product Profile

The target product profile (TPP) is a strategic document that summarises the key characteristics of a candidate medicine from multiple stakeholder perspectives [55]. The definition of a TPP according to the target-disease health requirements and user needs should drive the design of fundamental DP aspects such as stability requirements or logistics for product release and delivery to the patient. A deeper functional understanding of the DP results in improved convenience to the patients and regulatory compliance, improving focus on the unique CQA of the product under development. Furthermore, instituting the TPP will help achieve an integrated approach to product and process development contributing to clinical and commercial success [56].

The TPP is a dynamic, evolving, written document that organises all relevant information from multiple perspectives (i.e. medical, market, production, regulatory) in practice a focal compass reading for the entire duration of the product development programme. In addition to having a comprehensive TPP, it is encouraged to define in detail a Quality Target Product Profile (QTPP) of the cell- and/or gene-based medicine under development [56]. QTPP is defined as 'a prospective summary of the quality characteristics of a drug product that ideally will be achieved to ensure the desired quality, taking into account safety and efficacy of the drug product' [1].

The implementation of QbD concepts in the production process of ATMP and a properly defined QTTP based on sound science and quality risk management are tools to avoid variability in CQA [1]. Notably, CQA are defined based on the severity of harm to a patient (either safety or efficacy) resulting from failure to meet that quality attribute.

3.6 Final Remarks

Potency refers to what a cell-based medicine can do rather than what cells look like, provided that identity is only one attribute of the candidate medicine that, in most cases, does not preclude its biological functionality in the clinical setting. Rapid scientific advances bring new technologies to assess the properties of ATMP and may offer means for a better understanding of biological processes involved in the functionality of the new generation of cell and gene-based medicines. Miniaturisation of complex cell and tissue systems by means of 3D printing and organoid technologies are emerging trends that hold the potential to revolutionise the medicine of tomorrow.

References

1. Agency EM (2017) ICH guideline Q8 (R2) on pharmaceutical development, London
2. Guideline IHT (1999) ICH topic Q6B. Note for guidance on specifications: test procedures and acceptance criteria for biotechnological/biological products. European Medicines Agency, London
3. Schneider CK, Salmikangas P, Jilma B, Flamion B, Todorova LR, Paphitou A et al (2010) Challenges with advanced therapy medicinal products and how to meet them. Nat Rev Drug Discov 9(3):195–201
4. Guadix JA, López-Beas J, Clares B, Soriano-Ruiz JL, Zugaza JL, Gálvez-Martín P (2019) Principal criteria for evaluating the quality, safety and efficacy of hMSC-based products in clinical practice: current approaches and challenges. Pharmaceutics 11(11):552
5. Fang Y, Eglen RM (2017) Three-dimensional cell cultures in drug discovery and development. SLAS Discov 22(5):456–472
6. Khetani SR, Kanchagar C, Ukairo O, Krzyzewski S, Moore A, Shi J et al (2013) Use of micropatterned cocultures to detect compounds that cause drug-induced liver injury in humans. Toxicol Sci 132(1):107–117
7. Cabrera-Pérez R, Monguió-Tortajada M, Gámez-Valero A, Rojas-Márquez R, Borràs FE, Roura S et al (2019) Osteogenic commitment of Wharton's jelly mesenchymal stromal cells: mechanisms and implications for bioprocess development and clinical application. Stem Cell Res Ther 10(1):356
8. Kostadinova R, Boess F, Applegate D, Suter L, Weiser T, Singer T et al (2013) A long-term three dimensional liver co-culture system for improved prediction of clinically relevant drug-induced hepatotoxicity. Toxicol Appl Pharmacol 268(1):1–16
9. Messner S, Agarkova I, Moritz W, Kelm JM (2013) Multi-cell type human liver microtissues for hepatotoxicity testing. Arch Toxicol 87(1):209–213
10. Vives J, Batlle-Morera L (2020) The challenge of developing human 3D organoids into medicines. Stem Cell Res Ther 11(1):72
11. Nguyen DG, Funk J, Robbins JB, Crogan-Grundy C, Presnell SC, Singer T et al (2016) Bioprinted 3D primary liver tissues allow assessment of organ-level response to clinical drug induced toxicity in vitro. PLoS One 11(7):e0158674
12. Boráň T, Menezes-Ferreira M, Reischl I, Celis P, Ferry N, Gänsbacher B et al (2017) Clinical development and commercialization of advanced therapy medicinal products in the European Union: how are the product pipeline and regulatory framework evolving? Hum Gene Ther Clin Dev 28(3):126–135
13. Agency EM (2009) Assessment report for ChondroCelect, London. Contract No.: EMEA/724428/2009
14. Johnstone B, Hering TM, Caplan AI, Goldberg VM, Yoo JU (1998) In vitro chondrogenesis of bone marrow-derived mesenchymal progenitor cells. Exp Cell Res 238(1):265–272
15. Bravery CA, Carmen J, Fong T, Oprea W, Hoogendoorn KH, Woda J et al (2013) Potency assay development for cellular therapy products: an ISCT review of the requirements and experiences in the industry. Cytotherapy 15(1):9–19
16. Galipeau J, Krampera M, Barrett J, Dazzi F, Deans RJ, DeBruijn J et al (2016) International Society for Cellular Therapy perspective on immune functional assays for mesenchymal stromal cells as potency release criterion for advanced phase clinical trials. Cytotherapy 18(2):151–159
17. Salmikangas P, Schuessler-Lenz M, Ruiz S, Celis P, Reischl I, Menezes-Ferreira M et al (2015) Marketing regulatory oversight of Advanced Therapy Medicinal Products (ATMPs) in Europe: the EMA/CAT perspective. Adv Exp Med Biol 871:103–130
18. Agency EM (2016) Guideline on potency testing of cell based immunotherapy medicinal products for the treatment of cancer
19. Commission E. EudraLex (2017) The rules governing medicinal products in the European Union. Volume 4. Good manufacturing practice. Guidelines on good manufacturing practice specific to advanced therapy medicinal products, Brussels
20. Administration USDoHaHSFaD (2011) Guidance for industry potency tests for cellular and gene therapy products
21. Nagai S (2019) Flexible and expedited regulatory review processes for innovative medicines and regenerative medical products in the US, the EU, and Japan. Int J Mol Sci 20(15):3801
22. Nagata R, Rafizadeh-Kabe JD (2002) Japanese pharmaceutical and regulatory environment. Dialogues Clin Neurosci 4(4):470–474
23. Okita K, Ichisaka T, Yamanaka S (2007) Generation of germline-competent induced pluripotent stem cells. Nature 448(7151):313–317
24. Cyranoski D (2019) The potent effects of Japan's stem-cell policies. Nature 573(7775):482–485
25. Sipp D (2015) Conditional approval: Japan lowers the bar for regenerative medicine products. Cell Stem Cell 16(4):353–356
26. Kusakabe T (2015) Regulatory perspectives of Japan. Biologicals 43(5):422–424
27. Fischbach MA, Bluestone JA, Lim WA (2013) Cell-based therapeutics: the next pillar of medicine. Sci Transl Med 5(179):179ps7
28. Shukla V, Seoane-Vazquez E, Fawaz S, Brown L, Rodriguez-Monguio R (2019) The landscape of cellular and gene therapy products: authorization, discontinuations, and cost. Hum Gene Ther Clin Dev 30(3):102–113
29. Grau-Vorster M, Rodríguez L, Del Mazo-Barbara A, Mirabel C, Blanco M, Codinach M et al (2019) Compliance with good manufacturing practice in the assessment of immunomodulation potential of clinical grade multipotent mesenchymal stromal cells derived from Wharton's jelly. Cell 8(5):484

30. Vives J, Mirabel C (2019) Multipotent Mesenchymal stromal cells from bone marrow for current and potential clinical applications. In: Reis RL (ed) Encyclopedia of tissue engineering and regenerative medicine. Academic Press, Oxford, pp 503–512
31. Dominici M, Le Blanc K, Mueller I, Slaper-Cortenbach I, Marini F, Krause D et al (2006) Minimal criteria for defining multipotent mesenchymal stromal cells. The International Society for Cellular Therapy position statement. Cytotherapy 8(4):315–317
32. Viswanathan S, Shi Y, Galipeau J, Krampera M, Leblanc K, Martin I et al (2019) Mesenchymal stem versus stromal cells: International Society for Cell & Gene Therapy (ISCT®) Mesenchymal Stromal Cell committee position statement on nomenclature. Cytotherapy 21(10):1019–1024
33. Gastelurrutia P, Prat-Vidal C, Vives J, Coll R, Bayes-Genis A, Gálvez-Montón C (2021) Transitioning from preclinical evidence to advanced therapy medicinal product: a Spanish experience. Front Cardiovasc Med 8:604434
34. Mendicino M, Bailey AM, Wonnacott K, Puri RK, Bauer SR (2014) MSC-based product characterization for clinical trials: an FDA perspective. Cell Stem Cell 14(2):141–145
35. Agency EM (1999) ICH topic Q6B. Note for guidance on specifications: test procedures and acceptance criteria for biotechnological/biological products. European Medicines Agency, London
36. Porat Y, Abraham E, Karnieli O, Nahum S, Woda J, Zylberberg C (2015) Critical elements in the development of cell therapy potency assays for ischemic conditions. Cytotherapy 17(7):817–831
37. Janicki P, Boeuf S, Steck E, Egermann M, Kasten P, Richter W (2011) Prediction of in vivo bone forming potency of bone marrow-derived human mesenchymal stem cells. Eur Cell Mater 21:488–507
38. Sabbah N, Tamari T, Elimelech R, Doppelt O, Rudich U, Zigdon-Giladi H (2019) Predicting angiogenesis by endothelial progenitor cells relying on in-vitro function assays and VEGFR-2 expression levels. Biomol Ther 9(11):717
39. Cerignoli F, Abassi YA, Lamarche BJ, Guenther G, Santa Ana D, Guimet D et al (2018) In vitro immunotherapy potency assays using real-time cell analysis. PLoS One 13(3):e0193498
40. Torre ML, Lucarelli E, Guidi S, Ferrari M, Alessandri G, De Girolamo L et al (2015) Ex vivo expanded mesenchymal stromal cell minimal quality requirements for clinical application. Stem Cells Dev 24(6):677–685
41. Chinnadurai R, Rajan D, Qayed M, Arafat D, Garcia M, Liu Y et al (2018) Potency analysis of Mesenchymal stromal cells using a combinatorial assay matrix approach. Cell Rep 22(9):2504–2517
42. Thej C, Ramadasse B, Walvekar A, Majumdar AS, Balasubramanian S (2017) Development of a surrogate potency assay to determine the angiogenic activity of Stempeucel®, a pooled, ex-vivo expanded, allogeneic human bone marrow mesenchymal stromal cell product. Stem Cell Res Ther 8(1):47
43. Galleu A, Riffo-Vasquez Y, Trento C, Lomas C, Dolcetti L, Cheung TS et al (2017) Apoptosis in mesenchymal stromal cells induces in vivo recipient-mediated immunomodulation. Sci Transl Med 9(416):eaam7828
44. Río P, Navarro S, Wang W, Sánchez-Domínguez R, Pujol RM, Segovia JC et al (2019) Successful engraftment of gene-corrected hematopoietic stem cells in non-conditioned patients with Fanconi anemia. Nat Med 25(9):1396–1401
45. Oliver-Vila I, Ramírez-Moncayo C, Grau-Vorster M, Marín-Gallén S, Caminal M, Vives J (2018) Optimisation of a potency assay for the assessment of immunomodulative potential of clinical grade multipotent mesenchymal stromal cells. Cytotechnology 70(1):31–44
46. Oliver-Vila I, Coca MI, Grau-Vorster M, Pujals-Fonts N, Caminal M, Casamayor-Genescà A et al (2016) Evaluation of a cell-banking strategy for the production of clinical grade mesenchymal stromal cells from Wharton's jelly. Cytotherapy 18(1):25–35
47. Agency EM (2019) Guideline on quality, non -clinical and clinical requirements 4 for investigational advanced therapy medicinal products 5 in clinical trials, London
48. Deans R (2015) Towards the creation of a standard MSC line as a calibration tool. Cytotherapy 17(9):1167–1168
49. McGowan NWA, Campbell JDM, Mountford JC (2018) Good Manufacturing Practice (GMP) translation of advanced cellular therapeutics: lessons for the manufacture of erythrocytes as medicinal products. Methods Mol Biol 1698:285–292
50. Lehman N, Cutrone R, Raber A, Perry R, Van't Hof W, Deans R et al (2012) Development of a surrogate angiogenic potency assay for clinical-grade stem cell production. Cytotherapy 14(8):994–1004
51. Prins HJ, Braat AK, Gawlitta D, Dhert WJ, Egan DA, Tijssen-Slump E et al (2014) In vitro induction of alkaline phosphatase levels predicts in vivo bone forming capacity of human bone marrow stromal cells. Stem Cell Res 12(2):428–440
52. Larsen KH, Frederiksen CM, Burns JS, Abdallah BM, Kassem M (2010) Identifying a molecular phenotype for bone marrow stromal cells with in vivo bone-forming capacity. J Bone Miner Res 25(4):796–808
53. Vives J, Casademont-Roca A, Martorell L, Nogués N (2020) Beyond chimerism analysis: methods for tracking a new generation of cell-based medicines. Bone Marrow Transplant 55:1229
54. Reyes B, Coca MI, Codinach M, López-Lucas MD, Del Mazo-Barbara A, Caminal M et al (2017) Assessment of biodistribution using mesenchymal stromal cells: algorithm for study design and challenges in detection methodologies. Cytotherapy 19(9):1060–1069
55. Curry S, Brown R (2003) The target product profile as a planning tool in drug discovery research. Int J Pharm Technol 6:7867
56. Vives J, Carmona G, Vives J, Carmona G (2015) Guide to cell therapy GxP, 1st edn. Academic Press (Elsevier), p 266

Identifying Biomarkers for Osteogenic Potency Assay Development

4

Jorge S. Burns and Moustapha Kassem

4.1 Skeletal Stem Cell for Conservative Bone Healing

The landmark demonstration that it was possible to transplant whole bone marrow tissue pieces autologously to extramedullary sites [51] and histologically observe reticular and spindle shaped cells with a developmental capacity that showed distinct morphological features of osteoclastic and osteoblastic elements [147], paved the way for an extensive exploration of how cells may serve as advanced therapy medicinal products to treat skeletal pathologies. The early speculation that osteoblasts and osteoclasts, the principal cell types responsible for bone remodelling, might be derived from a common osteoprogenitor cell was refuted by experiments showing that instead, osteoclasts were a product of the mononuclear phagocyte system with no evidence for a common stem cell that could give rise to both [104]. Regarding osteoblast progenitors, pioneering experiments by Friedenstein et al. [3] characterised the rare bone marrow subpopulation of plastic-adherent cells that could proliferate to form single cell derived colonies consisting of fibroblastoid cells, termed colony-forming unit-fibroblasts (CFU-F). Notably, these cells could differentiate to aggregates resembling small areas of bone or cartilage [105], leading to an eventual definition as multipotent stromal cells (MSC) that could differentiate into osteoblasts, chondrocytes and adipocytes. Solid experimental evidence based on the use of in vivo transplantation assays substantiated the proposed differentiation potential [50] and this included the capacity for human bone marrow derived MSC to support haematopoiesis in culture [92] and long-term haematopoietic stem cell engraftment in vivo [1]. Most significantly, autologous human bone marrow derived osteoprogenitor cells could be isolated for ex vivo expansion and subsequently implanted with a hydroxyapatite scaffold at a large bone defect site, leading to favourable radiographic evidence of abundant callus formation indicative of accelerated healing [111]. This spurred excitement that cell-based tissue engineering approaches could be used to treat patients, yet the complexity of MSC biology and bone formation has made the routine realisation of this objective a most challenging quest.

Conceptualisations and assumptions have preceded scientific discovery, yet compliance with

J. S. Burns (✉)
Department of Environmental and Prevention Sciences, University of Ferrara, Ferrara, Italy
e-mail: js.burns@unife.it

M. Kassem
University Hospital of Odense, University of Southern Denmark, Odense, Denmark

Danish Stem Cell Center, University of Copenhagen, Copenhagen, Denmark

College of Medicine, King Saud University, Riyadh, Saudi Arabia

J. S. Burns (ed.), *Potency Assays for Advanced Stem Cell Therapy Medicinal Products*, Advances in Experimental Medicine and Biology 1420, https://doi.org/10.1007/978-3-031-30040-0_4

the Hippocratic oath of doing no harm requires bridging a significant knowledge gap between demonstrations of feasibility and establishment of a rational and validated stem cell therapeutic approach. Ironically, research approaches have to some extent contributed to a confusion that risked undermining expectations and safe practice [80, 133]. Raising concern [16], the reality of a bone marrow CFU-F as the putative progenitor cell of skeletal tissues was misconceptually extrapolated to the idea that such cells may represent stem cells for 'mesenchymal' tissues in general, to be found broadly in postnatal tissues [24]. Lapse use of the term 'mesenchymal stem cell' contradicts what is known about cell lineages from developmental biology, yet readily arose because ex vivo assessment of the cellular phenotype does not have the same stringency as tissue specific differentiation. The popular terminology 'MSC' [92] was maintained by having 'multipotent stromal' replace 'mesenchymal stem' preserving the acronym's semantic continuity, yet perhaps not avoiding confusion. The extent to which there can be genuine commonality between 'MSC' sourced from various tissues remains a long-debated topic [4, 68]. Apropos osteogenic potency, avoidance of the term mesenchymal and conservative use of the term multipotent stromal cells or skeletal stem cells (SSC) is advised [17]. More specific terms and acronyms such human Bone Marrow Stromal Cell (hBMSC), human Bone Marrow Multipotent Stromal Cell (hBM-MSC), Adipose Derived Stem Cell (ADSC) or human Adipose Derived Multipotent Stromal Cells (AD-MSC) can help avoid confusion between similar 'MSC' stromal cells derived from different tissues or species. These alternatively sourced cells may exert different therapeutic effects or introduce different responses [148], diminishing predictability of their usefulness in the context of bone healing [117].

The most straightforward concept is that bone marrow stroma includes self-renewing, multipotent progenitors termed SSC that give rise to skeletal lineages (bone, cartilage, fibroblasts and potentially marrow adipocytes) [70]. This reservoir of bone-forming cells is dedicated to bone growth during development and bone remodelling in the adult where they regulate the differentiation of bone-resorbing osteoclasts and maintain the haemopoietic microenvironment necessary for blood cell maturation and growth. The conservative principle of osteogenic regenerative medicine is that these cells can be isolated from the bone marrow, expanded to a critical therapeutic dose ex vivo and then re-introduced into the patient at a site requiring bone fracture healing to accelerate the process of new bone formation. This was a primary concept, but now there is scope for considering skeletal stem cells, or even umbilical cord MSC [162] as a 'humoral factory', releasing regeneration-stimulating factors that in turn benefit from a more multifaceted approach for estimating 'cell potency' [31, 162].

4.2 The Challenge of hBM-MSC Donor-Specific Heterogeneity

Skeletal stem cells, may themselves be further refined to have specific cell subsets with dedicated functions, since the ordered development of cartilage, bone, stroma and marrow adipocytes occurs at different times and via different embryonic lineages. Facial bones are derived from neural crest ectoderm, the axial skeleton has derivation from paraxial mesoderm and the limb skeleton is formed from lateral plate mesodermal cells [102]. Murine models allowing cell fate tracking suggest that metaphysis and diaphysis bone-forming osteoblast lineage cells are fundamentally distinct [134]. Much of what is known about the cellular and molecular basis of skeletal development comes from the study of human bone disorders [135] and animal models [25], but for the purposes of an osteogenic potency assay, one needs to understand how osteogenic biomarkers or genetic signatures in a cell culture context retain relevance for osteogenic behaviour in vivo.

Isolation of skeletal progenitor cells, estimated to comprise only 0.01% of the mononuclear cells in the bone marrow, is complicated by the lack of any robust biomarkers unique to skeletal stem cells that would allow convenient prompt isolation from bone marrow tissue. Nonetheless, seeking to harmonise the situation, a minimal set of criteria for defining multipotent mesenchymal stromal cells (MSC) was derived, whereby the cells should be plastic-adherent under standard culture conditions and express Cluster of Differentiation (CD) surface antigens CD73, CD90 CD105 but not CD11b, CD14, CD19, CD34, CD45, CD79α or Human Leukocyte Antigen – DR isotype (HLA-DR) surface molecules. They should also be able to show multipotent differentiation to osteoblasts, adipocytes and chondroblasts in vitro [36]. Although these represent sensitive MSC markers, they fail to be unique or specific as they are expressed by variety of cells and they cannot be used to predict the differentiation potential of the cells [74, 75, 122]. These minimal criteria became very popular as a means of bringing greater uniformity when exchanging data between laboratories, yet since their description, cell culture conditions and descriptive features have changed to meet clinical grade circumstances, contributing to an evolved understanding of cell features [37]. Encouragingly, the multipotent differentiation potential persisted when fetal bovine serum (FBS) was replaced by serum-free formulations to comply with current good manufacturing practice (cGMP), requiring xeno-free medicinal products [46, 47]. Alternative culture methods have included 3D culture and use of hypoxic conditions, devised to help preserve potency [63]. Variability among clinical trial outcomes promoted calls for more detailed standards [145] including a well-characterised reference cell material for calibration and improved comparability among different laboratories [154]. However, this remains a very challenging proposition, given the need to show that such a reference material would function in an in vivo therapeutic context.

Such calls for greater consistency were indicative of the very challenging situation presented by primary culture of bone marrow derived skeletal stem cells. There can be considerable donor variation in the growth properties and dramatic differences in the expression levels of osteogenic genes when the cells were exposed to osteoinductive medium [109]. The heterogeneity in trilineage differentiation potential in bone marrow derived MSC was found to be more complex than presumed; the colony-forming efficiency of attached MSC preparations was about 50–60% and tripotent MSC accounted for about 50% of the colony-forming cells. The biomarker CD146 showed greater mean fluorescence intensity in bipotent and tripotent CFU-F derived clones [114]. Multipotent (tripotent) CFU-F derived MSC were found to have significantly higher proliferative potential than CFU-F that had a more restricted lineage commitment. Populations of MSC composed of pooled CFU-F represented a heterogenous mixture of cells with different lineage-commitments and proliferation rates [115]. Although these observations suggested a tripotent MSC may end up predominating the primary culture as cells are passaged, other factors influencing the cell phenotype need to also be considered. Compounding variability, cell culture seeding density influenced the MSC metabolism [88] and expansion of primary MSC as monolayer cultures resulted in phenotypes that varied with cell doublings over time [11] with a dramatically decreased in vitro osteogenic potential as cells exited the cell cycle, reaching senescence [58, 140, 159]. This was consistent with a 36-fold reduced in vivo ectopic bone-forming potential in nude mice when using cells had been expanded to first confluence compared to using fresh bone marrow [10]. Clearly, the quest for finding biomarkers suitable for osteogenic potency assays would be complicated by these many changing variables. Heterogeneity at multiple levels; the cell populations between individual donors [130], differences according to tissue source [161], between individual cells during culture

[156, 159] and according to cell metabolism [88]; collectively presenting numerous confounding factors for the practical application of cell-based osteogenic therapy [108].

4.3 Telomerised MSC; Scalable Clonal Populations with Consistent Bone-Forming Potential

Given the relatively large number of ex vivo expanded BM-MSC required to repair bone fractures in animal models and the limited culturable life-span of primary hBM-MSC, with progressive loss of osteogenic potential after extensive propagation [76, 139], two independent research groups explored ectopic expression of human telomerase reverse transcriptase (hTERT) as a means of extending proliferative potential in what were termed human bone marrow stromal cells (hMSC-TERT) [132] or human bone marrow stromal stem cells (BMSSC-Ts) [127]. Both research groups found that greatly enhanced ex vivo cell expansion was accompanied by maintenance of an osteogenic stromal cell pool and when transplanted subcutaneously in immunodeficient mice, hTERT transduced cells formed more bone than equivalent xenografts using primary hBM-MSC. While Shi et al. noted that the bone-forming capacity of their BMSSC-Ts decreased markedly after prolonged culture of over 80 population doublings, Simonsen et al. observed excellent osteogenic potential even though their hMSC-TERT cells had undergone 260 population doublings, a phenotype that may have involved serendipitous use of cells bearing a polymorphic biomarker glutathione S-transferase theta 1 (GSTT1) correlated with enhanced culturability [21, 121]. Most significantly, hMSC-TERT cells overcame critical technical barriers for molecular analysis and when grown at a 1:4 passage ratio, generated cell populations termed hBMSC-TERT that could provide single-cell derived clones ideally suited for comparative analysis to derive biomarkers associated with bone-forming capacity [81] or for identifying genetic [113] or proteomic signatures [49, 77] (Fig. 4.1).

4.4 Comparative Analysis of Gene Expression, microRNA, Morphological Phenotypes and Cell Membrane or Secreted Proteins

Consistent with the requirements of a continuous bone remodeling process throughout life, osteoblast proliferation and differentiation are coupled events [136, 137]. Ex vivo stromal cell responsiveness to physiological chemical and physical mediators of osteoblast differentiation, have introduced a means of controllably guiding their fate decision [2, 29] towards osteogenic differentiation via extracellular signals [131]. A sequential expression of cell-growth regulated genes and genes associated with progressive development of the osteogenic phenotype was identified at both the level of transcription and mRNA stability [138]. Three principal steps of an osteoblast development sequence timeline could be described, namely, proliferation, matrix maturation and mineralisation. Experimentally established transition points in the developing sequence included a completion of proliferation with upregulation of genes associated with extracellular matrix synthesis and maturation involving upregulation of genes such as alkaline phosphatase that peaked before the onset of matrix mineralisation when osteopontin and osteocalcin mRNA levels reached their peak. Diverse ways of modulating transcriptional control and the emergence of options for redundancy in signalling pathways support need for prompt physiological responsiveness, yet make identification of appropriate potency assay biomarkers more challenging.

Notably, single-cell derived clones of hBMSC-TERT with markedly different bone forming capacity expanded as cell monolayers, provided a comparative research platform that indicated traditional osteoblastic biomarkers e.g., Alkaline phosphatase, collagen type I, osteopontin and

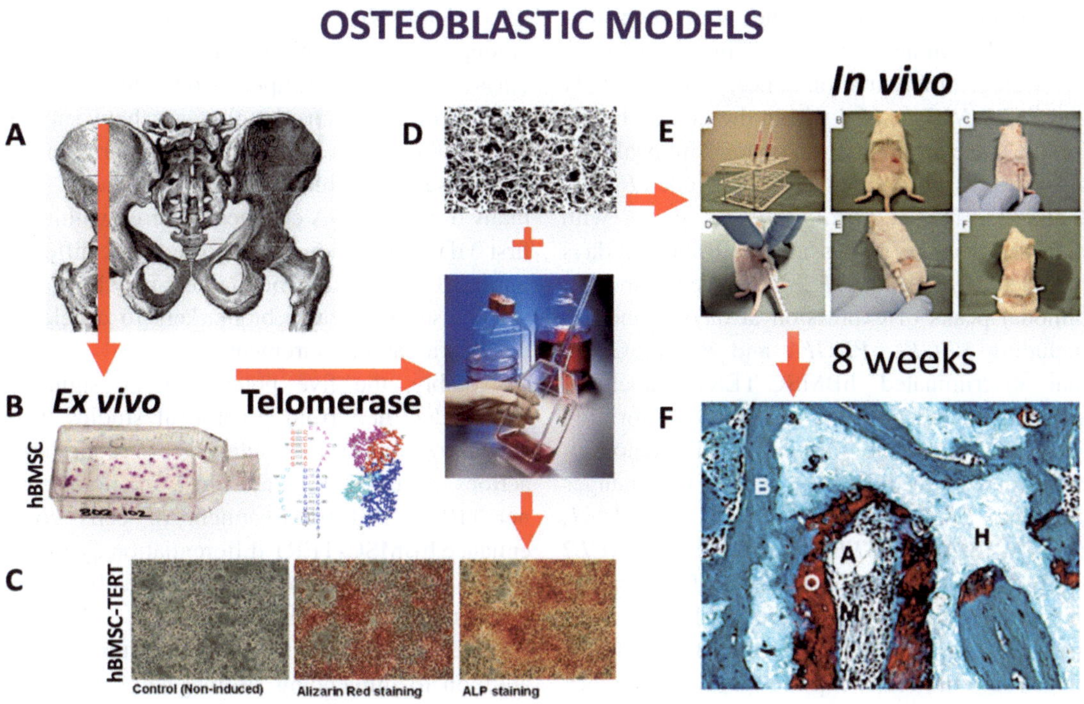

Fig. 4.1 Osteoblastic Models from telomerase immortalised human bone marrow stromal cells. (**a**) Cells harvested from the bone marrow of the iliac crest were placed in culture and (**b**) treated with retroviral vectors conferring ectopic expression of the hTERT gene. (**c**) Consequently, immortalised hBMSC-TERT cells could be expanded in culture and osteogenically induced ex vivo with induction factors to express alizarin red and alkaline phosphatase. (**d**) Osteoconductive scaffold biomaterial combined with expanded cultures of hTERT transduced cells were (**e**) implanted subcutaneously in immune deficient mice resulting in (**f**) histologically demonstrable bone formation after 8 weeks

Runx2 were not necessarily directly correlated with bone forming potential. Instead, significant correlation was found for enhanced gene expression of decorin, lysyl oxidase-like 4, natriuretic peptide receptor C and tetranectin [81], all genes associated with development of the osteogenic extracellular matrix [85] and its subsequent mineralisation [66, 158]. In addition, hBMSC-TERT grown as three-dimensional osteospheroids and similarly compared for correlations with bone-forming potential, revealed that also for the ex vivo 3D context, genes for matrix proteins served as good correlates for predicting in vivo bone forming potential. Quantification of collagen birefringence, a characteristic of osteogenic matrix maturation, measured using polarised light in histochemically processed sections of 3D cultures treated with osteogenic medium for 2 weeks, was also found to correlate with bone formation, providing independent corroborative evidence to support the gene expression studies [22]. These studies highlighted that biomarkers with correlations between ex vivo and in vivo contexts could be found, yet since the telomerised cells demonstrated an enhanced bone forming potential greater than that seen with primary cells, the relevance of these biomarkers for the context of clinical grade primary human BMSC remained to be determined. Nevertheless, the telomerised hBMSC model provided a uniquely powerful means of comparing appropriately uniform populations of cells that differed in osteogenic potential, allowing extension to other measurement platforms. RNA sequencing examining skeletally-related genes across 8 time points between 0 to 12 days of ex vivo osteoblastic differentiation identified 123 genes with chronologically determined changes in expression. Early-stage differentiation genes, *COL1A1*, *LOX* and *SERPINH1* peaked within the first 24 hours, middle stage differentiation genes with peak levels of expression at 3 and 6 days included *BMP4*, *CYP24A1* and *TGFBR2*, whereas genes with highest levels of expression at 9 and 12 days included *BMP2* and *IGF2*. Other genes showed bimodal peaks of expression at days 0 and 12, including *VEGFA*, *PDGFA* and *FGF2*. Genes that discriminated hBMSC-TERT subclones selected on the basis of a high bone forming (HBF) or low bone forming (LBF) xenograft behaviour, with relatively high fold changes included *ELN*, *COL1A1*, *BMP4*, *COL16A1*, *POSTN*, *SMAD6*, *TGFB2*, *ALPL*, *IL8* and *CXCL2* [152]. Not all of the potentially useful biomarkers obtained by this systematic and largely unbiased genetic explorations of hBMSC-TERT clones have been fully explored for use in potency assays, but *ALPL*, *COL1A2*, *DCN*, *ELN* and *RUNX2* provided a set of signature genes in potency assays using cGMP cultured primary hBM-MSC [100] and *TGFB2* has also proved to be highly relevant [101].

Critical regulators of gene expression, microRNA (miRNA), help regulate osteogenesis [55, 82, 83, 86, 95] and are increasingly recognised as highly influential non-coding RNA family members that influence the outcomes of diverse biological processes including fracture healing [73]. The significant miRNA impact on osteoblastic differentiation has been demonstrated using antimiR and miRNA overexpression in suitable target cells, exerting significant phenotypic effects, targeting genes very relevant to the process of osteoblast differentiation, e.g. miR-138 can inhibit osteoblastic differentiation, moreover functional inhibition of miR-138 can accelerate osteogenic differentiation of hBMSC ex vivo and increase bone formation in vivo [45] and a similar impact was demonstrated by miR-34a [27]. The notable pivotal aspect of hBMSC on the predominant commitment choice of osteogenic or adipogenic pathways, may also be influenced by microRNA, e.g., miRNA-4739 [40]. Global microRNA profiling of hBMSC has identified 15 miRNAs, with miR-222 and miR-423 as among as most significant regulators of osteoblastogenesis [26]. Emphasis on changes occurring during the transitional stages between cell proliferation, extracellular matrix maturation and its subsequent mineralisation, indicated that many of the miRNA changes occurred within the first 3 days after induction of osteogenic differentiation. Thus, miRNA biomarkers may be particularly useful as early biomarkers to accelerate potency assay measurement.

Notably, the five potency assay signature genes identified from preclinical studies were shown by bioinformatic analysis of protein interactions to have TGF-ß1 as a close functional partner [100]. TGF-ß1 treatment of cells could enhance hBMSC-TERT differentiation by induction of genes in the skeletal and extracellular categories e.g. transgelin (*TAGLN*) [42], that together regulate the actin cytoskeleton [41]. In addition to influencing stromal cell differentiation and commitment, the actin cytoskeletal control of the morphology and mechanical properties of the nucleus can play an important role in BMSC migration [87], cellular mobilisation being itself an important factor for recruitment to sites of injury and in vivo bone forming ability [6]. Collectively, these results implied that cytoskeletally-regulated morphological features of the cells may also be indicative of osteogenic potency. Indeed, nuclear morphology and geometry of cultured primary hBM-MSC measured by a high-content imaging system with multivariable analysis demonstrated that nuclear geometry and texture could stably predict hBM-MSC differentiation potential to osteoblasts or adipocytes [75]. Most pragmatically, analysis of native morphological features of primary hBM-MSC cultures, without treatment using osteoblastic inductive media, may contribute to effective early quality screening tests of hBM-MSC prior to clinical use (Fig. 4.2).

Prompt potency assay tests that minimally interfere with hBM-MSC cell expansion would be advantageous. Increased understanding of how secreted cell products influence hBM-MSC bone forming potential introduces prospects for monitoring novel biomarkers in cell supernatant

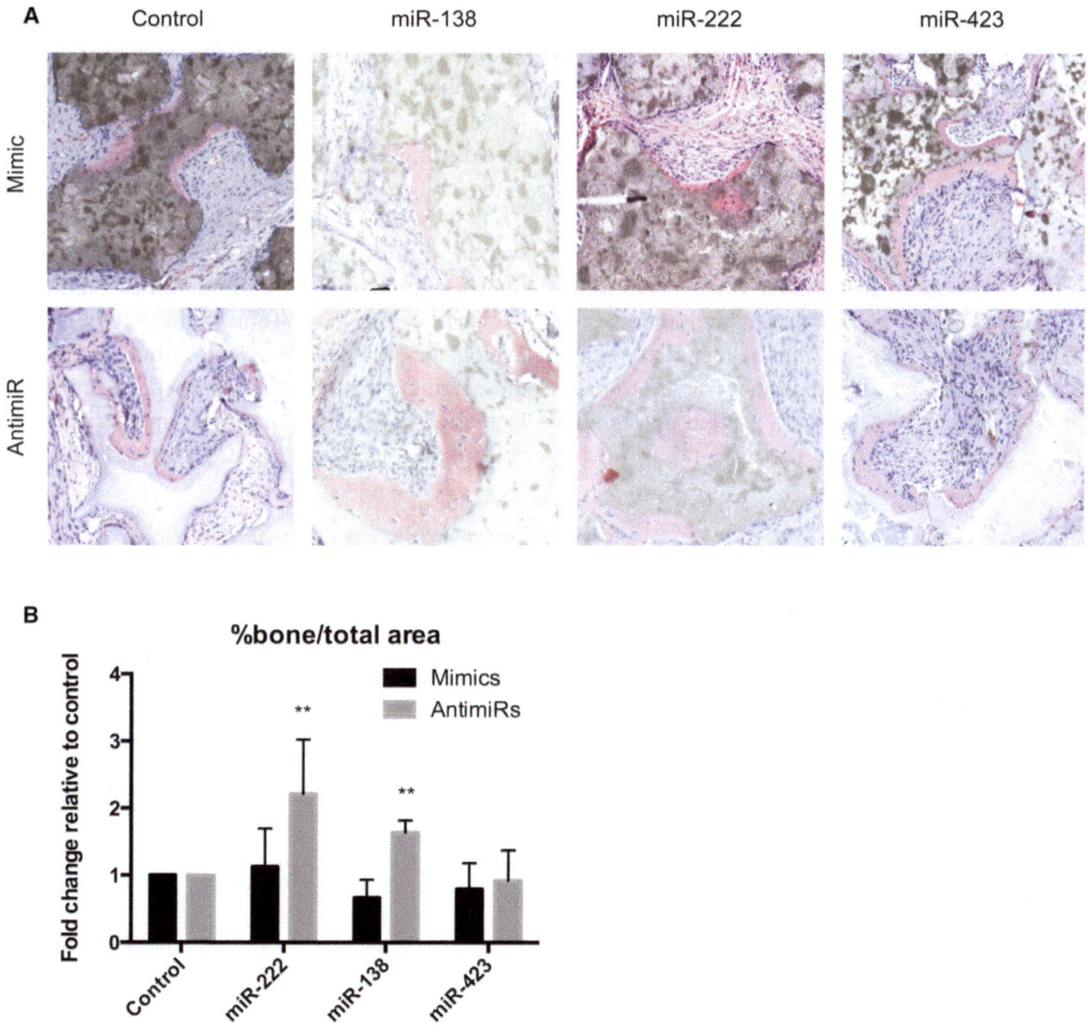

Fig. 4.2 Anti-miR-138 and Anti-miR-222 Increases Heterotrophic Bone Formation In Vivo. The hBMSC-TERT cells were transfected with 25 nM control, miR-138, miR-222, or miR-423 mimics or anti-miRs, seeded onto TCP/HA scaffolds, and transplanted into 8-week- old female NOD/SCID mice. Scaffolds and cells were stained with Haematoxylin and Eosin (H & E) 8 weeks post-implantation. Bone formation was quantified as the bone volume/total volume and normalised to their respective miRNA controls ($n = 6$ per treatment). (**a**) Representative images of H & E-stained scaffolds showing bone formation. (**b**) Quantified bone volume (% bone/total area). $^{**}p < 0.01$ (Reprinted from Chang et al., 2018 [26])

during cell expansion [96]. For example, both secreted frizzled-related protein 1 (sFRP-1) and delta-like 1/fetal antigen 1 (Dlk1/FA1) exert regulatory effects on adipogenesis and osteoblast differentiation [2]. A link between telomerase expression and upregulated insulin-like growth factor (IGF-1) signaling, a secreted protein showing reduced serum levels in telomerase deficient (Terc-/-) transgenic mice of low bone mass, may also help explain the enhanced bone-forming potential of hBMSC-TERT cells [118]. The secretome of hBMSC during osteoblastic differentiation has been analysed by sensitive quantitative proteomic techniques and is certainly complex [77]. The hormone stanniocalcin 2 (STC2) has autocrine effects enhancing osteoblastic differentiation and cell migration induced by hyaluronan binding protein (KIAA1199/CEMIP), a mobilising factor interacting with Wnt-signaling that can induce changes in the actin skeleton [28] required for the important phenotype of cell migration to bone formation sites. Secreted factors act as coupling factors for interaction with other cell types including osteoclasts and can reflect the broader metabolic status of the individual [157]. In this respect they may provide excellent additional information about host status when performing the potency assay for autologous cell therapy; at the same time this highlights the challenging complexity for an ex vivo potency assay to have in vivo relevance. A secreted protein that is also found on the cell surface that may be useful for BM-MSC characterisation is Meflin, expressed by hBM-MSC in their undifferentiated state and downregulated upon their differentiation. Notably, Meflin is found on stromal cells distributed throughout the bone marrow and on pericytes and perivascular cells in various organs [91]. It is absent in epithelial, endothelial and smooth muscle cells, and my serve as a potentially useful marker for cultured hMSC, although to what extent it correlates directly with the osteogenic potency of cGMP cultured cells for therapeutic purposes has yet to be determined. Detection of Meflin on pericytes and perivascular cells of various organs, agrees with early descrip-

a

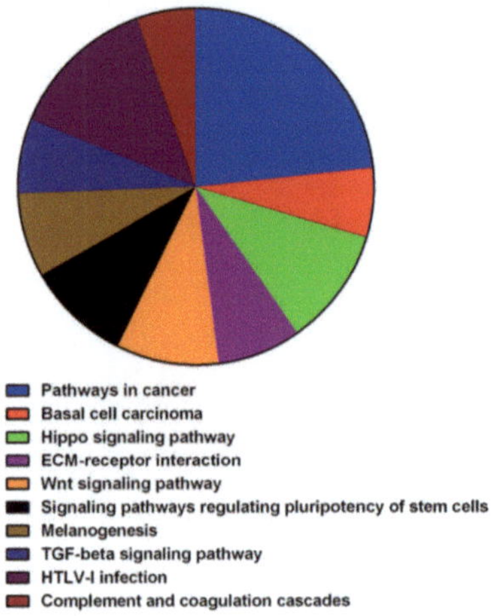

Gene Symbol	FC (hBMSC^{-Bone} vs. hBMSC^{+Bone})
RUNX2	-2.6
BMP2	-7.8
BMP4	-9.2
SMAD6	-8.1
SMAD9	-8.9
TGFB2	-8.7
TAGLN	-20.9
TPM1	-2.3
ACTA2	-2.1
COL1A1	-2.8
SERPINB2	3.2
NOGGIN	12.4

c

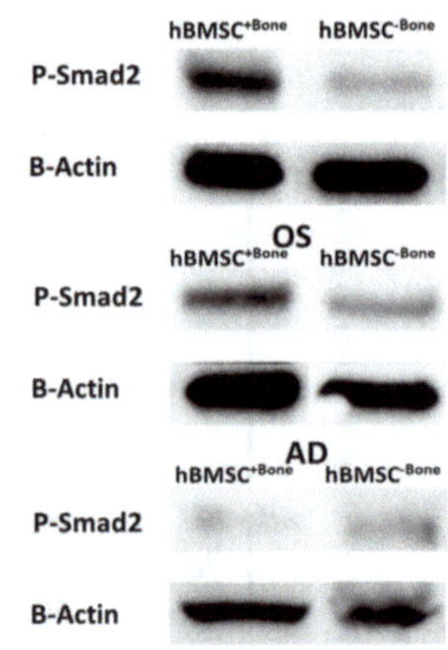

b

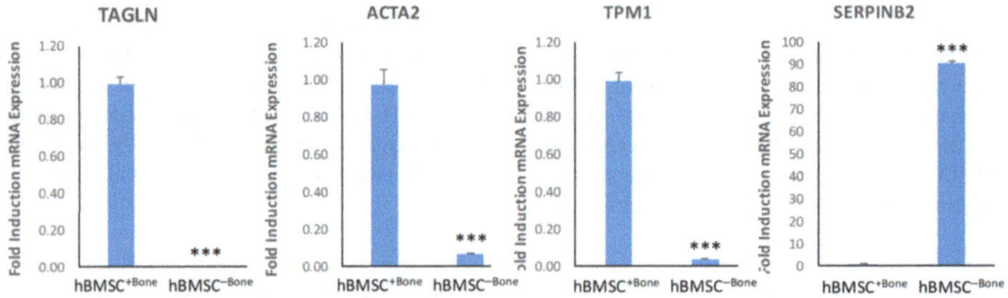

Fig. 4.3 Impaired TGFβ signalling in hBMSC−Bone cells. (**a**) Pie chart illustrating the distribution of the top 10 KEGG pathways in the down-regulated genes. The pie size corresponds to the number of matched entities. List of TGFβ-responsive genes, which were differentially expressed in hBMSC−Bone vs. hBMSC+Bone as revealed by whole genome microarray profiling is shown. (**b**) qRT-PCR validation for the expression of a panel of TGFβ responsive genes (*TAGLN*, *ACTA2*, *TPM1*, and *SERPINB2*) in hBMSC−Bone compared to hBMSC+Bone cells. Expression of each target gene was normalised to *GAPDH*. Data is shown as the mean ± SD from three independent experiments, ***$p < 0.0005$. (**c**) Western blotting for P-SMAD2 in hBMSC−Bone compared to hBMSC+Bone cells (upper panel), whereas B-Actin (ACTB, lower panel) was used as a loading control. Phosphorylation of SMAD2 is also shown during the osteogenic and adipogenic differentiation of both cell lines (Reprinted from Elsafadi et al., [43]).

tions for the in situ identity of archetypal MSC in the bone marrow [15], but pericytes do not form a discrete lineage [67] and this observation should not encourage an anatomical-based credence that MSC are ubiquitous functionally equivalent cells throughout the body [14] (Fig. 4.3).

A cell surface membrane biomarker that has been rigorously tested as pertaining to MSC, emerging to be useful for prospective characterisation with regard to therapeutic efficacy is melanoma-associated cell adhesion molecule CD146/MCAM [129]. More stably expressed when cultured MSC were fed serum-free platelet lysate rather than fetal bovine serum [99], CD146 has been found to be present in stromal cell populations derived from adipose tissue vascular fractions [9] and in what were termed mesenchymal stem-like cells from human endometrium [124]. The anatomical distribution of CD146^{+} stromal cells in situ correlated with observations that CD146 expression can be elevated in normoxia and downregulated in hypoxia [151]. In bone marrow derived stromal cells, the most salient feature of CD146^{+} osteoprogenitor cells was a clonal self-renewal and a capacity to form bone at heterotopic sites with the more stringent quality of supporting a hematopoietic environment upon cell transplantation to heterotopic sites. In contrast CD146^{-} stromal cells could generate osteoblasts and bone in vivo, but were unable to establish a sinusoidal system and establish a hematopoietic environment [116]. The enhanced functionalities of CD146^{+} cells include trans-endothelial migration and recruitment to bone surfaces, whereupon committed osteoblastic cells on active bone-forming surfaces were CD146^{-} [60]. CD146^{+} hBM-MSC showed higher secretory capacity, plus immunomodulatory and anti-inflammatory protein production in comparison to CD146^{-} counterparts, properties consistent with a greater therapeutic potency [18]. An elevated fraction of CD146^{+} hBMSC contributed to a clinical signature predicting osteogenic potency [74] and CD146^{+} dental pulp derived hDP-MSC were also deemed to have good therapeutic potency [90] (Fig. 4.4).

Paracrine activity is increasingly appreciated to be a principal mediator of pathological processes [7, 125]. Extracellular vesicles (EV) are key components of the mineralisation [32] and regenerative processes [150], recognised as

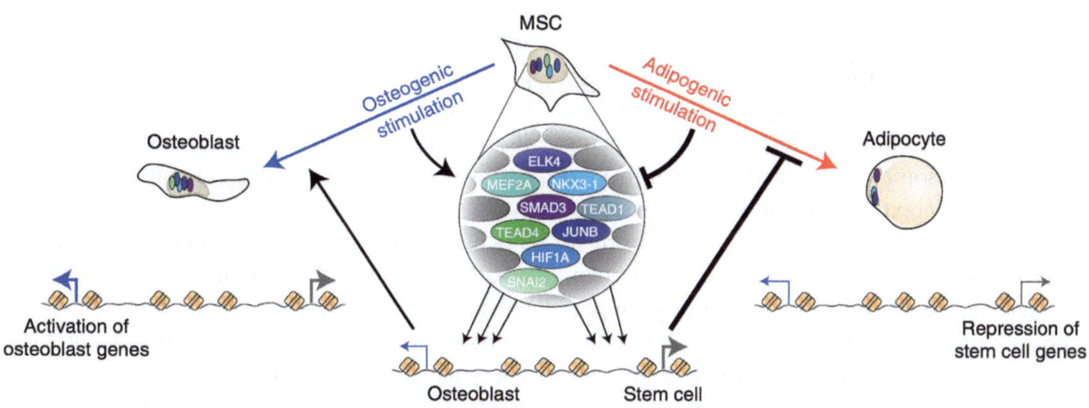

Fig. 4.4 Model for osteoblast and adipocyte differentiation. The model illustrates the action of a group of transcription factors that are present in stem cells and are pivotal for osteoblast differentiation, thereby controlling the balance between osteoblast and adipocyte differentiation. These osteogenic stem cell factors are implicated in transcriptional control of both stem cell genes and genes that are activated upon osteogenic differentiation. In contrast, adipogenic stimulation decreases the activity of the osteogenic stem cell factors, leading to downregulation of stem cell genes during adipocyte differentiation. In this case, inactivation of stem cell genes represents an important part of the remodelling of the transcriptional networks during adipocyte differentiation. Of note, adipocyte differentiation in particular is dependent on de novo activation of transcription factors that are not expressed in undifferentiated stem cells. (Reprinted from Rauch et al., [113]).

mediators for the therapeutic activity of hBMSC [33, 103]. Whether a cell-free secretome-based therapeutic modality may complement cell-based approaches [23] and supplant use of hBMSC is a debated future perspective [39, 89, 94]. Advantageously, EV may more readily help with epigenetic preconditioning or reprogramming [84, 93] yet for potency tests the challenge remains to improve upon the characterisation and classification of EV [160], measure the key components responsible for the intended biological activity and determine how to enhance manufacturing efficiency and consistency [53].

4.5 Compelling Pathways for Functional Attributes in Osteogenic Potency Assays

A number of confounding factors have made exploration of the key functional pathways to be measured in potency assays for osteoblastic ATMP particularly difficult.

Studies of primary hBMSC from individual donors have demonstrated cell product functional heterogeneity can reflect different procurement methods, donor age, gender, in vitro replicative senescence, details of in vitro manufacture approaches and measurement technologies. Key signaling pathways necessary for cell function are nonetheless emerging and there is growing evidence that some congruence between in vivo and ex vivo contexts can be found. It may be prudent to not limit the potency assay approach to ex vivo assays, but to adopt a more comprehensive approach considering the possibility for preemptive screening of the donors undergoing surgery for bone fracture, given that a growing number of indicative biomarkers can now be brought into consideration. Signature characteristics that may be positively correlated with clinical efficacy include donor sex (male), absence of any pre-diagnosed osteoporosis, intake of vitamin D supplements and a higher fraction of CD146+ and ALPL+ cells [74]. Thus, selection of donors for bone regeneration clinical trials may be guided by prospectively testable relevant variables, favouring prudent choices to maximise the likelihood of a successful therapeutic outcome.

Returning to quantitively relevant measurements that may be performed on the cGMP cultured cells being expanded to a clinical dose, the powerful approach of global direct comparison of hBMSC cell lines showing consistently different bone forming potential, highlighted the significance of microRNA regulation [143] and TGF-ß plus BMP signaling for regulating hBMSC lineage commitment and differentiation [43]. These observations have been broadly confirmed to be of clinical relevance [38, 142]. Wnt (Wingless-related integration site) signaling pathways that pass signals via Wnt-protein ligand acting on cell surface Frizzled family receptors are crucial to bone formation. High or low bone mass abnormalities can result from mutations in a large family of proteins constituting either canonical Wnt-ß-catenin or non-canonical Wnt-planar cell polarity or Wnt-calcium pathways [12]. Both non-canonical Wnt3a [110] and canonical Wnt7a protein signals [163] are integrated in the differentiation commitment of hBMSC to favour osteoblasts as opposed to adipocytes.

Wnt signaling can be influenced by mechano-responsive mechanisms involved in exercise-stimulated skeletal integrity [30]. Additional key players in the molecular mechanisms underlying bone fracture healing include Connexin-43 (Cx43) gap junction protein dependent signaling pathways, plus a signaling axis involving membrane anchored metalloproteinase (MT1-MMP), Yes associated protein (YAP) and transcriptional co-activator with PDZ binding motif (TAZ) that galvanises osteoblastic commitment [146] and promotes osteoblast precursor expansion [71]. The small GTPase protein Ras homolog family member A (RhoA) and its effector Rho-associated coiled-coil kinase (ROCK), also responsive to mechanical stimuli and extracellular matrix cues, can antagonise Wnt/ß-catenin signaling [128]. RhoA loss of function in preosteoblasts and inhibition of ROCK signaling can increase osteoblast differentiation and bone formation in a topography-related manner [141]. Responsive to numerous stimulatory cues, hBMSC lineage commitment is a complex process, involving a complex set of signaling cascades with two distinct waves of phosphoproteomic signalling [13]. The TAM family of receptor tyrosine kinases, including TYRO3 and MERTK trigger phosphorylation and activation of multiple downstream signalling proteins and represent potent positive and negative regulators of bone homeostasis, whereby blockade of MERTK function has osteoanabolic consequences [44].

Despite extensive insights into the molecular mechanisms mentioned above and known involvement of Notch, Hedgehog and NELL pathways and their crosstalk with hormone signaling networks [149], the challenge to development of a potency assay concerns the convenience and relevance with which a parameter responsible for a functional attribute can be measured. The cells in question are exquisitely responsive to their contextual microenvironment and with limitations for fully reproducing the fracture site context ex vivo, it is understandable that derivation of osteogenic potency assays has not been straightforward. Many of the molecular regulators act at the transcriptional level, where relatively small-fold changes in expression may be significant, but challenging to measure with predictive precision in the real-world context of a heterogeneous population of primary hBMSC undergoing expansion under cGMP culture conditions. Nonetheless, a subset of gene expression markers in monolayer cultures of hBMSC lines were found to be well correlated with an in-vivo bone forming phenotype, prominently, genes involved in extracellular matrix regulation [81]. Of these, decorin (*DCN*) expression serves as an informative example, although it may be considered to have some counterintuitive aspects regarding its choice as a useful biomarker for functional attributes in osteogenic potency assays.

Decorin was named from its ability to bind and 'decorate' collagen fibrils, a property shared by other Small Leucine Rich Proteoglycans (SLRPs) involved in matrix maturation, yet only decorin could faithfully recapitulate the native organisation of type I collagen in vitro, organising collagen fibrils into fibers compact enough to mimic the superfibrillar organisation of natural tissues [112]. Decorin gene expression was consistently upregulated to a significant extent upon osteogenic induction of hBM-MSC despite different contexts; (i) when hBMSC-TERT cells were cultured as 3D osteospheroids [22], (ii) in primary hBM-MSC cultured in either Fetal Bovine Serum or Platelet lysate [100] and (iii) when the primary hBM-MSC were tested in another laboratory using different batches of growth medium and a different osteoinduction medium formulation [101]. Thus, decorin represents a biomarker that could qualify as being 'robustly expressed' across different cell expansion platforms. The fold-change in hBMSC-TERT *DCN* mRNA expression was >3-fold for monolayer cultures or >2-fold for 3D cultures osteogenically induced in FBS containing medium. For monolayer cultures of primary hBM-MSC grown with platelet lysate instead of FBS, the *DCN*- specific transcript expression typically increased 20 to 30-fold after just 1 week of osteogenic induction and this was reproducible despite use of alternative osteogenic induction agents across different laboratories [101]. Good for potency biosensing purposes, significant changes in decorin expression occurred early in the induced transition from precursor cell to osteoblast [34, 35, 65, 97, 98, 155]. Beyond influencing collagen fibril organisation, preventing aberrant premature osteoid mineralisation, decorin is likely to have an important role in blocking excess TGF-ß signaling from inhibiting osteoblast maturation [19]. Although TGF-ß1 is required for optimal bone formation [19, 20, 52, 61, 144, 164], high doses of TGF-ß1 could suppress mineralisation in an orthotopic implant model [20]. Decorin can interact directly with all three TGF-ß protein isoforms [62, 72]. When immobilised on collagen fibrils, decorin could antagonise TGF-ß1 mediated stimulation of collagen gel retraction and biglycan induction, presumably by sequestering TGF-ß1 in the extracellular matrix [165] With a specific leucin-rich collagen binding region [69] the periodic binding pattern of decorin on collagen may lead to multiple interaction patterns in vivo [112], nonetheless it provides an elegant stoichiometric means of coupling the intensity of TGF-ß signalling pathways to the physical quantity of extracellular matrix produced and is likely to be deterministic in bone formation, a tissue in which physical density is directly related to functional performance. As such, decorin constitutes an excellent osteogenic potency biomarker, a functional attribute that is expressed at readily measurable levels, consistently expressed ex vivo and in vivo.

4.6 Lessons Learned for Bone Repair ATMP Development

The striking demonstration of a cell-therapy based benefit for skeletal pathologies [111] has occurred decades before availability of any accredited cell therapy products, reflecting the complexity of bringing technologies to scale. From the outset a sophisticated level of understanding is needed to appreciate the dosing required and best methods of harvest, expansion and deployment of the therapeutic cells. In osteogenic therapy, overcoming complexity of the bone marrow to gain an understanding the role of stem cells in the identity, nature, origin and function of the differentiated progenitor cells has been an enormous accomplishment, still in progress, crucial for making the prospect of stem cell therapy possible. A five-year follow-up of a European multicentric clinical trial has confirmed safety and early efficacy in 80% of 21 cases of early femoral head osteonecrosis treated through minimally invasive surgical implantation of autologous hBM-MSC expanded from bone marrow under cGMP protocols [56]. This notably positive outcome invites further development and emphasis on provision of potency assays to discriminate the most significant mechanism of action, whether the transplanted cells differentiate into osteogenic cells or whether they modulate the healing process by their secreted factors.

Such therapy inevitably incurs relatively expensive procedures, encouraging consideration that wherever possible, it would be helpful to mitigate potency assay costs. This can involve alternative creative approaches to traditional methods. The use of exogenous osteogenic induction growth factors beyond those used to expand the cells in culture, required to perform some osteogenic potency assay tests, already introduces additional need for quality and safety controls for the relevant reagents. Noteworthy alternative differentiation methods include use of osteoinductive materials such as graphene oxide composites [78] or introduction of nano vibrations [64] as procedures that can stimulate innate potent bioactive metabolites that specifically potentiate osteogenesis, without incurring the potential artefact of an arbitrarily derived ex vivo differentiation factor. Furthermore, experimental models have demonstrated that both siRNA [5] and microRNA [106] can be used to functionalise scaffolds to influence multilineage differentiation and accelerate bone regeneration, although potency assays that incorporate the contribution of scaffold dynamics are at an early stage of development. Notably, for the compromised circumstances found in large bone fractures, the cellular component combined with the scaffold was significant for repair [79]. Additional approaches, exploring whether scaffolds can be engineered to encourage activation of endogenous cells to regenerate the appropriate skeletal tissue healing are under development [8].

Although there may be concerns that immortalised cell models fail to mimic cells directly obtained and cultured from the bone marrow, the telomerised hBM-MSC-TERT cells have proved extremely informative for identifying relevant biomarkers indicative of osteogenic potency. This likely reflects the extensive number of molecular phenotype similarities found when directly comparing primary and hBM-MSC-TERT cells [153]. The ability to test clonal derivatives of different in vivo bone forming potential in a reproducible manner has allowed Omics-scale exploration of mRNA, microRNA and proteins. The important deterministic aspect of extracellular matrix proteins as early differentiation stage biomarkers, have highlighted biomarkers that would not necessarily be anticipated, given that decorin expression is more prevalent in adult skin rather than bone and prior transgenic mouse studies indicating its dispensability for bone formation. In addition to appropriate collagen architecture requisites for mineralisation, the extracellular matrix is important for establishing the bone vasculature that is essential for therapeutic efficacy [123] and appropriate bone development [48]. Raman spectroscopy can be used as a non-invasive label-free technique to assess osteoblast matrix maturation with relevance for clinical application [59, 126]. Novel advancements in sensor and probe technology are enhancing the anticipated prospects [63] of online monitoring of proliferation and multipotency, with a more holistic quantitative evaluation of hBM-MSC that genuinely reflects their therapeutic potential.

What is the ultimate osteogenic potency assay? Large animal models that demonstrate bone regenerative potential of autologous bone marrow derived MSC can have greater relevance than other experimental animal models of mice and rabbits [54], but it is debatable as to whether evidence for in vivo bone formation defines a 'gold standard' potency assay for osteogenic ATMP. Certainly, it demonstrates a desired functional outcome, but does not necessarily reveal the fundamental mechanism of action; whether there is real integration of the transplanted cells at the therapeutically relevant site or whether the transplanted cells mediate bone formation via secreted factors acting on host cells. The principal mechanism of action is likely to differ according to the diverse requirements of different types of bone lesion [57, 107, 120]. Further research is needed to more precisely define the molecular mechanisms underpinning bone development [119], to derive potency assays that accurately, promptly and conveniently measure the therapeutic capacity of ATMP to guide cell therapy for bone repair.

References

1. Abbuehl JP, Tatarova Z, Held W, Huelsken J (2017) Long-term engraftment of primary bone marrow stromal cells repairs niche damage and improves hematopoietic stem cell transplantation. Cell Stem Cell 21:241–255.e6
2. Abdallah BM, Kassem M (2012) New factors controlling the balance between osteoblastogenesis and adipogenesis. Bone 50:540–545
3. Afanasyev BV, Elstner EE, AR Z (2009) AJ Friedenstein, founder of the mesenchymal stem cell concept. Cell Ther Transpl 1:35
4. Al-Nbaheen M, Vishnubalaji R, Ali D, Bouslimi A, Al-Jassir F, Megges M, Prigione A, Adjaye J, Kassem M, Aldahmash A (2013) Human stromal (mesenchymal) stem cells from bone marrow, adipose tissue and skin exhibit differences in molecular phenotype and differentiation potential. Stem Cell Rev Rep 9:32–43
5. Andersen MØ, Nygaard JV, Burns JS, Raarup MK, Nyengaard JR, Bünger C, Besenbacher F, Howard KA, Kassem M, Kjems J (2010) siRNA nanoparticle functionalization of nanostructured scaffolds enables controlled multilineage differentiation of stem cells. Mol Ther 18:2018–2027
6. Andersen RK, Zaher W, Larsen KH, Ditzel N, Drews K, Wruck W, Adjaye J, Abdallah BM, Kassem M (2015) Association between in vivo bone formation and ex vivo migratory capacity of human bone marrow stromal cells. Stem Cell Res Ther 6:196
7. Anderson HC, Mulhall D, Garimella R (2010) Role of extracellular membrane vesicles in the pathogenesis of various diseases, including cancer, renal diseases, atherosclerosis, and arthritis. Lab Investig 90:1549–1557
8. Arthur A, Gronthos S (2020) Clinical application of bone marrow mesenchymal stem/stromal cells to repair skeletal tissue. Int J Mol Sci 21:E9759
9. Astori G, Vignati F, Bardelli S, Tubio M, Gola M, Albertini V, Bambi F, Scali G, Castelli D, Rasini V, Soldati G, Moccetti T (2007) "In vitro" and multicolor phenotypic characterization of cell subpopulations identified in fresh human adipose tissue stromal vascular fraction and in the derived mesenchymal stem cells. J Transl Med 5:55
10. Banfi A, Muraglia A, Dozin B, Mastrogiacomo M, Cancedda R, Quarto R (2000) Proliferation kinetics and differentiation potential of ex vivo expanded human bone marrow stromal cells: implications for their use in cell therapy. Exp Hematol 28:707–715
11. Bara JJ, Richards RG, Alini M, Stoddart MJ (2014) Concise review: bone marrow-derived mesenchymal stem cells change phenotype following in vitro culture: implications for basic research and the clinic. Stem Cells 32:1713–1723
12. Baron R, Kneissel M (2013) WNT signaling in bone homeostasis and disease: from human mutations to treatments. Nat Med 19:179–192
13. Barrio-Hernandez I, Jafari A, Rigbolt KTG, Hallenborg P, Sanchez-Quiles V, Skovrind I, Akimov V, Kratchmarova I, Dengjel J, Kassem M, Blagoev B (2020) Phosphoproteomic profiling reveals a defined genetic program for osteoblastic lineage commitment of human bone marrow-derived stromal stem cells. Genome Res 30:127–137
14. Bianco P (2014) "Mesenchymal" stem cells. Annu Rev Cell Dev Biol 30:677–704

15. Bianco P, Boyde A (1993) Confocal images of marrow stromal (Westen-Bainton) cells. Histochemistry 100:93–99
16. Bianco P, Cao X, Frenette PS, Mao JJ, Robey PG, Simmons PJ, Wang CY (2013) The meaning, the sense and the significance: translating the science of mesenchymal stem cells into medicine. Nat Med 19:35–42
17. Bianco P, Robey PG (2015) Skeletal stem cells. Development 142:1023–1027
18. Bowles AC, Kouroupis D, Willman MA, Perucca Orfei C, Agarwal A, Correa D (2020) Signature quality attributes of CD146+ mesenchymal stem/stromal cells correlate with high therapeutic and secretory potency. Stem Cells 38:1034–1049
19. Breen EC, Ignotz RA, McCabe L, Stein JL, Stein GS, Lian JB (1994) TGF beta alters growth and differentiation related gene expression in proliferating osteoblasts in vitro, preventing development of the mature bone phenotype. J Cell Physiol 160:323–335
20. Broderick E, Infanger S, Turner TM, Sumner DR (2005) Depressed bone mineralization following high dose TGF-beta1 application in an orthopedic implant model. Calcif Tissue Int 76:379–384
21. Burns JS, Kassem M (2020) Less is more: corroborating a genomic biomarker identifying human bone marrow multipotent stromal cells with high scalability. Stem Cells 38:E5–E6
22. Burns JS, Rasmussen PL, Larsen KH, Schrøder HD, Kassem M (2010) Parameters in three-dimensional osteospheroids of telomerized human mesenchymal (stromal) stem cells grown on osteoconductive scaffolds that predict in vivo bone-forming potential. Tissue Eng Part A 16:2331–2342
23. Cabrera-Pérez R, Monguió-Tortajada M, Gámez-Valero A, Rojas-Márquez R, Borràs FE, Roura S, Vives J (2019) Osteogenic commitment of Wharton's jelly mesenchymal stromal cells: mechanisms and implications for bioprocess development and clinical application. Stem Cell Res Ther 10:356
24. Caplan AI (1991) Mesenchymal stem cells. J Orthop Res 9:641–650
25. Chan CK, Seo EY, Chen JY, Lo D, McArdle A, Sinha R, Tevlin R, Seita J, Vincent-Tompkins J, Wearda T, Lu WJ, Senarath-Yapa K, Chung MT, Marecic O, Tran M, Yan KS, Upton R, Walmsley GG, Lee AS, Sahoo D, Kuo CJ, Weissman IL, Longaker MT (2015) Identification and specification of the mouse skeletal stem cell. Cell 160:285–298
26. Chang CC, Venø MT, Chen L, Ditzel N, Le DQS, Dillschneider P, Kassem M, Kjems J (2018) Global MicroRNA profiling in human bone marrow skeletal-stromal or mesenchymal-stem cells identified candidates for bone regeneration. Mol Ther 26:593–605
27. Chen L, Holmstrøm K, Qiu W, Ditzel N, Shi K, Hokland L, Kassem M (2014) MicroRNA-34a inhibits osteoblast differentiation and in vivo bone formation of human stromal stem cells. Stem Cells 32:902–912
28. Chen L, Shi K, Andersen TL, Qiu W, Kassem M (2019) KIAA1199 is a secreted molecule that enhances osteoblastic stem cell migration and recruitment. Cell Death Dis 10:126
29. Chen Q, Shou P, Zheng C, Jiang M, Cao G, Yang Q, Cao J, Xie N, Velletri T, Zhang X, Xu C, Zhang L, Yang H, Hou J, Wang Y, Shi Y (2016) Fate decision of mesenchymal stem cells: adipocytes or osteoblasts. Cell Death Differ 23:1128–1139
30. Choi RB, Robling AG (2021) The Wnt pathway: An important control mechanism in bone's response to mechanical loading. Bone 153:116087
31. Christy BA, Herzig MC, Delavan CP, Abaasah I, Cantu C, Salgado C, Lovelace S, Garcia L, Jensen K, Montgomery R, Cap AP, Bynum JA (2020) Use of multiple potency assays to evaluate human mesenchymal stromal cells. J Trauma Acute Care Surg 89:S109–S117
32. Cmoch A, Strzelecka-Kiliszek A, Palczewska M, Groves P, Pikula S (2011) Matrix vesicles isolated from mineralization-competent Saos-2 cells are selectively enriched with annexins and S100 proteins. Biochem Biophys Res Commun 412:683–687
33. Cooper LF, Ravindran S, Huang CC, Kang M (2019) A role for exosomes in craniofacial tissue engineering and regeneration. Front Physiol 10:1569
34. Corsi A, Xu T, Chen XD, Boyde A, Liang J, Mankani M, Sommer B, Iozzo RV, Eichstetter I, Robey PG, Bianco P, Young MF (2002) Phenotypic effects of biglycan deficiency are linked to collagen fibril abnormalities, are synergized by decorin deficiency, and mimic Ehlers-Danlos-like changes in bone and other connective tissues. J Bone Miner Res 17:1180–1189
35. Danielson KG, Baribault H, Holmes DF, Graham H, Kadler KE, Iozzo RV (1997) Targeted disruption of decorin leads to abnormal collagen fibril morphology and skin fragility. J Cell Biol 136:729–743
36. Dominici M, Le Blanc K, Mueller I, Slaper-Cortenbach I, Marini F, Krause D, Deans R, Keating A, Prockop D, Horwitz E (2006) Minimal criteria for defining multipotent mesenchymal stromal cells. The International Society for Cellular Therapy position statement. Cytotherapy 8:315–317
37. Dominici M, Paolucci P, Conte P, Horwitz EM (2009) Heterogeneity of multipotent mesenchymal stromal cells: from stromal cells to stem cells and vice versa. Transplantation 87:S36–S42
38. Du G, Cheng X, Zhang Z, Han L, Wu K, Li Y, Lin X (2021) TGF-Beta induced key genes of osteogenic and adipogenic differentiation in human mesenchymal stem cells and MiRNA-mRNA regulatory networks. Front Genet 12:759596
39. Eichholz KF, Woods I, Riffault M, Johnson GP, Corrigan M, Lowry MC, Shen N, Labour MN, Wynne K, O'Driscoll L, Hoey DA (2020) Human bone marrow stem/stromal cell osteogenesis is regulated via mechanically activated osteocyte-derived extracellular vesicles. Stem Cells Transl Med 9:1431–1447

40. Elsafadi M, Manikandan M, Alajez NM, Hamam R, Dawud RA, Aldahmash A, Iqbal Z, Alfayez M, Kassem M, Mahmood A (2017) MicroRNA-4739 regulates osteogenic and adipocytic differentiation of immortalized human bone marrow stromal cells via targeting LRP3. Stem Cell Res 20:94–104
41. Elsafadi M, Manikandan M, Almalki S, Mobarak M, Atteya M, Iqbal Z, Hashmi JA, Shaheen S, Alajez N, Alfayez M, Kassem M, Dawud RA, Mahmood A (2018) TGFβ1-induced differentiation of human bone marrow-derived MSCs is mediated by changes to the actin cytoskeleton. Stem Cells Int 2018:6913594
42. Elsafadi M, Manikandan M, Dawud RA, Alajez NM, Hamam R, Alfayez M, Kassem M, Aldahmash A, Mahmood A (2016) Transgelin is a TGFβ-inducible gene that regulates osteoblastic and adipogenic differentiation of human skeletal stem cells through actin cytoskeleston organization. Cell Death Dis 7:e2321
43. Elsafadi M, Shinwari T, Al-Malki S, Manikandan M, Mahmood A, Aldahmash A, Alfayez M, Kassem M, Alajez NM (2019) Convergence of TGFβ and BMP signaling in regulating human bone marrow stromal cell differentiation. Sci Rep 9:4977
44. Engelmann J, Zarrer J, Gensch V, Riecken K, Berenbrok N, Luu TV, Beitzen-Heineke A, Vargas-Delgado ME, Pantel K, Bokemeyer C, Bhamidipati S, Darwish IS, Masuda E, Burstyn-Cohen T, Alberto EJ, Ghosh S, Rothlin C, Hesse E, Taipaleenmäki H, Ben-Batalla I, Loges S (2022) Regulation of bone homeostasis by MERTK and TYRO3. Nat Commun 13:7689
45. Eskildsen T, Taipaleenmäki H, Stenvang J, Abdallah BM, Ditzel N, Nossent AY, Bak M, Kauppinen S, Kassem M (2011) MicroRNA-138 regulates osteogenic differentiation of human stromal (mesenchymal) stem cells in vivo. Proc Natl Acad Sci U S A 108:6139–6144
46. Fekete N, Rojewski MT, Fürst D, Kreja L, Ignatius A, Dausend J, Schrezenmeier H (2012) GMP-compliant isolation and large-scale expansion of bone marrow-derived MSC. PLoS One 7:e43255
47. Felka T, Schäfer R, De Zwart P, Aicher WK (2010) Animal serum-free expansion and differentiation of human mesenchymal stromal cells. Cytotherapy 12:143–153
48. Filipowska J, Tomaszewski KA, Niedźwiedzki Ł, Walocha JA, Niedźwiedzki T (2017) The role of vasculature in bone development, regeneration and proper systemic functioning. Angiogenesis 20:291–302
49. Foster LJ, Zeemann PA, Li C, Mann M, Jensen ON, Kassem M (2005) Differential expression profiling of membrane proteins by quantitative proteomics in a human mesenchymal stem cell line undergoing osteoblast differentiation. Stem Cells 23:1367–1377
50. Friedenstein AJ (1980) Stromal mechanisms of bone marrow: cloning in vitro and retransplantation in vivo. Haematol Blood Transfus 25:19–29
51. Friedenstein AJ, Piatetzky-Shapiro II, Petrakova KV (1966) Osteogenesis in transplants of bone marrow cells. J Embryol Exp Morphol 16:381–390
52. Geiser AG, Zeng QQ, Sato M, Helvering LM, Hirano T, Turner CH (1998) Decreased bone mass and bone elasticity in mice lacking the transforming growth factor-β1 gene. Bone 23:87–93
53. Gimona M, Brizzi MF, Choo ABH, Dominici M, Davidson SM, Grillari J, Hermann DM, Hill AF, de Kleijn D, Lai RC, Lai CP, Lim R, Monguió-Tortajada M, Muraca M, Ochiya T, Ortiz LA, Toh WS, Yi YW, Witwer KW, Giebel B, Lim SK (2021) Critical considerations for the development of potency tests for therapeutic applications of mesenchymal stromal cell-derived small extracellular vesicles. Cytotherapy 23:373–380
54. Gjerde CG, Santis DD, Dominic M, Zanotti G, Hellem S, Piccinno S, Burns JS, Murgia A, Candini O, Krampera M, Nocini P, Addis A, Amiaud J, Layrolle P, Mustafa K, Veronesi E (2017) Autologous porcine bone marrow mesenchymal cells for reconstruction of a resorbed alveolar bone: a preclinical model in mini-pigs. Int J Stem Cell Res 4:50
55. Goff LA, Boucher S, Ricupero CL, Fenstermacher S, Swerdel M, Chase LG, Adams CC, Chesnut J, Lakshmipathy U, Hart RP (2008) Differentiating human multipotent mesenchymal stromal cells regulate microRNAs: prediction of microRNA regulation by PDGF during osteogenesis. Exp Hematol 36:1354–1369
56. Gómez-Barrena E, Padilla-Eguiluz NG, Rosset P, Hernigou P, Baldini N, Ciapetti G, Gonzalo-Daganzo RM, Avendaño-Solá C, Rouard H, Giordano R, Dominici M, Schrezenmeier H, Layrolle P, On BOTRC (2021) Osteonecrosis of the femoral head safely healed with autologous, expanded, bone marrow-derived mesenchymal stromal cells in a multicentric trial with minimum 5 years follow-up. J Clin Med 10:508
57. Goodnough LH, Goodman SB (2022) Relationship of aging, inflammation, and skeletal stem cells and their effects on fracture repair. Curr Osteoporos Rep 20:320–325
58. Grotheer V, Skrynecki N, Oezel L, Grassmann J, Windolf J, Suschek CV (2021) Osteogenic differentiation of human mesenchymal stromal cells and fibroblasts differs depending on tissue origin and replicative senescence. Sci Rep 11:1–17
59. Harkness L, Novikov SM, Beermann J, Bozhevolnyi SI, Kassem M (2012) Identification of abnormal stem cells using Raman spectroscopy. Stem Cells Dev 21:2152–2159
60. Harkness L, Zaher W, Ditzel N, Isa A, Kassem M (2016) CD146/MCAM defines functionality of human bone marrow stromal stem cell populations. Stem Cell Res Ther 7:4
61. He R, Lu Y, Ren J, Wang Z, Huang J, Zhu L, Wang K (2017) Decreased fibrous encapsulation and enhanced osseointegration in vitro by

decorin-modified titanium surface. Colloids Surf B Biointerfaces 155:17–24
62. Hildebrand A, Romarís M, Rasmussen LM, Heinegård D, Twardzik DR, Border WA, Ruoslahti E (1994) Interaction of the small interstitial proteoglycans biglycan, decorin and fibromodulin with transforming growth factor beta. Biochem J 302:527–534
63. Hoch AI, Leach JK (2014) Concise review: optimizing expansion of bone marrow mesenchymal stem/stromal cells for clinical applications. Stem Cells Transl Med 3:643–652
64. Hodgkinson T, Tsimbouri PM, Llopis-Hernandez V, Campsie P, Scurr D, Childs PG, Phillips D, Donnelly S, Wells JA, O'Brien FJ, Salmeron-Sanchez M, Burgess K, Alexander M, Vassalli M, Oreffo ROC, Reid S, France DJ, Dalby MJ (2021) The use of nanovibration to discover specific and potent bioactive metabolites that stimulate osteogenic differentiation in mesenchymal stem cells. Sci Adv 7:eabb7921
65. Hoshi K, Kemmotsu S, Takeuchi Y, Amizuka N, Ozawa H (1999) The primary calcification in bones follows removal of decorin and fusion of collagen fibrils. J Bone Miner Res 14:273–280
66. Iba K, Chiba H, Yamashita T, Ishii S, Sawada N (2001) Phase-independent inhibition by retinoic acid of mineralization correlated with loss of tetranectin expression in a human osteoblastic cell line. Cell Struct Funct 26:227–233
67. Jain RK (2003) Molecular regulation of vessel maturation. Nat Med 9:685–693
68. Jones E, Schäfer R (2015) Where is the common ground between bone marrow mesenchymal stem/stromal cells from different donors and species. Stem Cell Res Ther 6:143
69. Kalamajski S, Aspberg A, Oldberg Å (2007) The decorin sequence SYIRIADTNIT binds collagen type I. J Biol Chem 282:16062–16067
70. Kassem M, Bianco P (2015) Skeletal stem cells in space and time. Cell 160:17–19
71. Kegelman CD, Nijsure MP, Moharrer Y, Pearson HB, Dawahare JH, Jordan KM, Qin L, Boerckel JD (2021) YAP and TAZ promote periosteal osteoblast precursor expansion and differentiation for fracture repair. J Bone Miner Res 36:143–157
72. Kolb M, Margetts PJ, Sime PJ, Gauldie J (2001) Proteoglycans decorin and biglycan differentially modulate TGF-beta-mediated fibrotic responses in the lung. Am J Physiol Lung Cell Mol Physiol 280:L1327–L1334
73. Komatsu DE, Duque E, Hadjiargyrou M (2021) MicroRNAs and fracture healing: preclinical studies. Bone 143:115758
74. Kowal JM, Möller S, Ali D, Figeac F, Barington T, Schmal H, Kassem M (2021) Identification of a clinical signature predictive of differentiation fate of human bone marrow stromal cells. Stem Cell Res Ther 12:265
75. Kowal JM, Schmal H, Halekoh U, Hjelmborg JB, Kassem M (2020) Single-cell high-content imaging parameters predict functional phenotype of cultured human bone marrow stromal stem cells. Stem Cells Transl Med 9:189–202
76. Krebsbach PH, Kuznetsov SA, Satomura K, Emmons RVB, Rowe DW, Robey PG (1997) Bone formation in vivo: comparison of osteogenesis by transplanted mouse and human marrow stromal fibroblasts. Transplantation 63:1059–1069
77. Kristensen LP, Chen L, Nielsen MO, Qanie DW, Kratchmarova I, Kassem M, Andersen JS (2012) Temporal profiling and pulsed SILAC labeling identify novel secreted proteins during ex vivo osteoblast differentiation of human stromal stem cells. Mol Cell Proteomics 11:989–1007
78. Krukiewicz K, Putzer D, Stuendl N, Lohberger B, Awaja F (2020) Enhanced osteogenic differentiation of human primary mesenchymal stem and progenitor cultures on graphene oxide/poly(methyl methacrylate) composite scaffolds. Materials (Basel) 13:E2991
79. Lammens J, Maréchal M, Delport H, Geris L, Oppermann H, Vukicevic S, Luyten FP (2020) A cell-based combination product for the repair of large bone defects. Bone 138:115511
80. Langrzyk A, Nowak WN, Stępniewski J, Jaźwa A, Florczyk-Soluch U, Józkowicz A, Dulak J (2018) Critical view on mesenchymal stromal cells in regenerative medicine. Antioxid Redox Signal 29:169–190
81. Larsen KH, Frederiksen CM, Burns JS, Abdallah BM, Kassem M (2010) Identifying a molecular phenotype for bone marrow stromal cells with in vivo bone-forming capacity. J Bone Miner Res 25:796–808
82. Li Z, Hassan MQ, Volinia S, van Wijnen AJ, Stein JL, Croce CM, Lian JB, Stein GS (2008) A microRNA signature for a BMP2-induced osteoblast lineage commitment program. Proc Natl Acad Sci U S A 105:13906–13911
83. Lian JB, Stein GS, van Wijnen AJ, Stein JL, Hassan MQ, Gaur T, Zhang Y (2012) MicroRNA control of bone formation and homeostasis. Nat Rev Endocrinol 8:212–227
84. Liang B, Liang JM, Ding JN, Xu J, Xu JG, Chai YM (2019) Dimethyloxaloylglycine-stimulated human bone marrow mesenchymal stem cell-derived exosomes enhance bone regeneration through angiogenesis by targeting the AKT/mTOR pathway. Stem Cell Res Ther 10:335
85. Lin X, Patil S, Gao YG, Qian A (2020) The bone extracellular matrix in bone formation and regeneration. Front Pharmacol 11:757
86. Liu CJ (2009) MicroRNAs in skeletogenesis. Front Biosci (Landmark Ed) 14:2757–2764
87. Liu L, Luo Q, Sun J, Song G (2019) Cytoskeletal control of nuclear morphology and stiffness are required for OPN-induced bone-marrow-derived

mesenchymal stem cell migration. Biochem Cell Biol 97:463–470
88. Liu Y, Muñoz N, Bunnell BA, Logan TM, Ma T (2015) Density-dependent metabolic heterogeneity in human mesenchymal stem cells. Stem Cells 33:3368–3381
89. Lu CH, Chen YA, Ke CC, Liu RS (2021) Mesenchymal stem cell-derived extracellular vesicle: a promising alternative therapy for osteoporosis. Int J Mol Sci 22:12750
90. Ma L, Huang Z, Wu D, Kou X, Mao X, Shi S (2021) CD146 controls the quality of clinical grade mesenchymal stem cells from human dental pulp. Stem Cell Res Ther 12:488
91. Maeda K, Enomoto A, Hara A, Asai N, Kobayashi T, Horinouchi A, Maruyama S, Ishikawa Y, Nishiyama T, Kiyoi H, Kato T, Ando K, Weng L, Mii S, Asai M, Mizutani Y, Watanabe O, Hirooka Y, Goto H, Takahashi M (2016) Identification of Meflin as a potential marker for mesenchymal stromal cells. Sci Rep 6:22288
92. Majumdar MK, Thiede MA, Haynesworth SE, Bruder SP, Gerson SL (2000) Human marrow-derived mesenchymal stem cells (MSCs) express hematopoietic cytokines and support long-term hematopoiesis when differentiated toward stromal and osteogenic lineages. J Hematother Stem Cell Res 9:841–848
93. Man K, Brunet MY, Fernandez-Rhodes M, Williams S, Heaney LM, Gethings LA, Federici A, Davies OG, Hoey D, Cox SC (2021) Epigenetic reprogramming enhances the therapeutic efficacy of osteoblast-derived extracellular vesicles to promote human bone marrow stem cell osteogenic differentiation. J Extracell Vesicles 10:e12118
94. Marolt Presen D, Traweger A, Gimona M, Redl H (2019) Mesenchymal stromal cell-based bone regeneration therapies: from cell transplantation and tissue engineering to therapeutic secretomes and extracellular vesicles. Front Bioeng Biotechnol 7:352
95. Mazziotta C, Lanzillotti C, Iaquinta MR, Taraballi F, Torreggiani E, Rotondo JC, Otòn-Gonzalez L, Mazzoni E, Frontini F, Bononi I, De Mattei M, Tognon M, Martini F (2021) MicroRNAs modulate signaling pathways in osteogenic differentiation of mesenchymal stem cells. Int J Mol Sci 22:2362
96. Mizukami A, Thomé CH, Ferreira GA, Lanfredi GP, Covas DT, Pitteri SJ, Swiech K, Faça VM (2019) Proteomic identification and time-course monitoring of secreted proteins during expansion of human mesenchymal stem/stromal in stirred-tank bioreactor. Front Bioeng Biotechnol 7:154
97. Mochida Y, Duarte WR, Tanzawa H, Paschalis EP, Yamauchi M (2003) Decorin modulates matrix mineralization in vitro. Biochem Biophys Res Commun 305:6–9
98. Mochida Y, Parisuthiman D, Pornprasertsuk-Damrongsri S, Atsawasuwan P, Sricholpech M, Boskey AL, Yamauchi M (2009) Decorin modulates collagen matrix assembly and mineralization. Matrix Biol 28:44–52
99. Müller I, Kordowich S, Holzwarth C, Spano C, Isensee G, Staiber A, Viebahn S, Gieseke F, Langer H, Gawaz MP, Horwitz EM, Conte P, Handgretinger R, Dominici M (2006) Animal serum-free culture conditions for isolation and expansion of multipotent mesenchymal stromal cells from human BM. Cytotherapy 8:437–444
100. Murgia A, Veronesi E, Candini O, Caselli A, D'souza N, Rasini V, Giorgini A, Catani F, Iughetti L, Dominici M, Burns JS (2016) Potency biomarker signature genes from multiparametric osteogenesis assays: will cGMP human bone marrow mesenchymal stromal cells make bone. PLoS One 11:e0163629
101. Ofiteru AM, Becheru DF, Gharbia S, Balta C, Herman H, Mladin B, Ionita M, Hermenean A, Burns JS (2020) Qualifying osteogenic potency assay metrics for human multipotent stromal cells: TGF-β2 a telling eligible biomarker. Cell 9:E2559
102. Olsen BR, Reginato AM, Wang W (2000) Bone development. Annu Rev Cell Dev Biol 16:191–220
103. Otsuru S, Desbourdes L, Guess AJ, Hofmann TJ, Relation T, Kaito T, Dominici M, Iwamoto M, Horwitz EM (2018) Extracellular vesicles released from mesenchymal stromal cells stimulate bone growth in osteogenesis imperfecta. Cytotherapy 20:62–73
104. Owen M (1980) The origin of bone cells in the postnatal organism. Arthritis Rheum 23:1073–1080
105. Owen M, Friedenstein AJ (1988) Stromal stem cells: marrow-derived osteogenic precursors. Cell Mol Biol Verteb Hard Tiss 136:42–60
106. Pan T, Song W, Xin H, Yu H, Wang H, Ma D, Cao X, Wang Y (2022) MicroRNA-activated hydrogel scaffold generated by 3D printing accelerates bone regeneration. Bioact Mater 10:1–14
107. Perez JR, Kouroupis D, Li DJ, Best TM, Kaplan L, Correa D (2018) Tissue engineering and cell-based therapies for fractures and bone defects. Front Bioeng Biotechnol 6:105
108. Phinney DG (2012) Functional heterogeneity of mesenchymal stem cells: implications for cell therapy. J Cell Biochem 113:2806–2812
109. Phinney DG, Kopen G, Righter W, Webster S, Tremain N, Prockop DJ (1999) Donor variation in the growth properties and osteogenic potential of human marrow stromal cells. J Cell Biochem 75:424–436
110. Qiu W, Chen L, Kassem M (2011) Activation of non-canonical Wnt/JNK pathway by Wnt3a is associated with differentiation fate determination of human bone marrow stromal (mesenchymal) stem cells. Biochem Biophys Res Commun 413:98–104
111. Quarto R, Mastrogiacomo M, Cancedda R, Kutepov SM, Mukhachev V, Lavroukov A, Kon E, Marcacci M (2001) Repair of large bone defects with the use of autologous bone marrow stromal cells. N Engl J Med 344:385–386

112. Raspanti M, Viola M, Forlino A, Tenni R, Gruppi C, Tira ME (2008) Glycosaminoglycans show a specific periodic interaction with type I collagen fibrils. J Struct Biol 164:134–139
113. Rauch A, Haakonsson AK, Madsen JGS, Larsen M, Forss I, Madsen MR, Van Hauwaert EL, Wiwie C, Jespersen NZ, Tencerova M, Nielsen R, Larsen BD, Röttger R, Baumbach J, Scheele C, Kassem M, Mandrup S (2019) Osteogenesis depends on commissioning of a network of stem cell transcription factors that act as repressors of adipogenesis. Nat Genet 51:716–727
114. Russell KC, Phinney DG, Lacey MR, Barrilleaux BL, Meyertholen KE, O'Connor KC (2010) In vitro high-capacity assay to quantify the clonal heterogeneity in trilineage potential of mesenchymal stem cells reveals a complex hierarchy of lineage commitment. Stem Cells 28:788–798
115. Russell KC, Lacey MR, Gilliam JK, Tucker HA, Phinney DG, O'Connor KC (2011) Clonal analysis of the proliferation potential of human bone marrow mesenchymal stem cells as a function of potency. Biotechnol Bioeng 108:2716–2726
116. Sacchetti B, Funari A, Michienzi S, Di Cesare S, Piersanti S, Saggio I, Tagliafico E, Ferrari S, Robey PG, Riminucci M, Bianco P (2007) Self-renewing osteoprogenitors in bone marrow sinusoids can organize a hematopoietic microenvironment. Cell 131:324–336
117. Sacchetti B, Funari A, Remoli C, Giannicola G, Kogler G, Liedtke S, Cossu G, Serafini M, Sampaolesi M, Tagliafico E, Tenedini E, Saggio I, Robey PG, Riminucci M, Bianco P (2016) No identical "mesenchymal stem cells" at different times and sites: human committed progenitors of distinct origin and differentiation potential are incorporated as adventitial cells in microvessels. Stem Cell Rep 6:897–913
118. Saeed H, Qiu W, Li C, Flyvbjerg A, Abdallah BM, Kassem M (2015) Telomerase activity promotes osteoblast differentiation by modulating IGF-signaling pathway. Biogerontology 16:733–745
119. Salhotra A, Shah HN, Levi B, Longaker MT (2020) Mechanisms of bone development and repair. Nat Rev Mol Cell Biol 21:696–711
120. Sanghani-Kerai A, McCreary D, Lancashire H, Osagie L, Coathup M, Blunn G (2018) Stem cell interventions for bone healing: fractures and osteoporosis. Curr Stem Cell Res Ther 13:369–377
121. Sathiyanathan P, Samsonraj RM, Tan CLL, Ling L, Lezhava A, Nurcombe V, Stanton LW, Cool SM (2020) A genomic biomarker that identifies human bone marrow-derived mesenchymal stem cells with high scalability. Stem Cells 38:1124–1136
122. Sauer T, Facchinetti G, Kohl M, Kowal JM, Rozanova S, Horn J, Schmal H, Kwee I, Schulz AP, Hartwig S, Kassem M, Habermann JK, Gemoll T (2022) Protein expression of AEBP1, MCM4, and FABP4 differentiate osteogenic, adipogenic, and mesenchymal stromal stem cells. Int J Mol Sci 23:2568
123. Schäfer R, Schwab M, Siegel G et al (2020) Modulating endothelial adhesion and migration impacts stem cell therapies efficacy. EBioMedicine 60:102987
124. Schwab KE, Gargett CE (2007) Co-expression of two perivascular cell markers isolates mesenchymal stem-like cells from human endometrium. Hum Reprod 22:2903–2911
125. Shan SK, Lin X, Li F, Xu F, Zhong JY, Guo B, Wang Y, Zheng MH, Wu F, Yuan LQ (2019) Exosomes and bone disease. Curr Pharm Des 25:4536–4549
126. Sharma A, Goring A, Staines KA, Emery RJH, Pitsillides AA, Oreffo ROC, Mahajan S, Clarkin CE (2020) Raman spectroscopy links differentiating osteoblast matrix signatures to pro-angiogenic potential. Matrix Biol Plus 5:100018
127. Shi S, Gronthos S, Chen S, Reddi A, Counter CM, Robey PG, Wang CY (2002) Bone formation by human postnatal bone marrow stromal stem cells is enhanced by telomerase expression. Nat Biotechnol 20:587–591
128. Shi W, Xu C, Gong Y, Wang J, Ren Q, Yan Z, Mei L, Tang C, Ji X, Hu X, Qv M, Hussain M, Zeng LH, Wu X (2021) RhoA/Rock activation represents a new mechanism for inactivating Wnt/β-catenin signaling in the aging-associated bone loss. Cell Regen 10:8
129. Shih IM (1999) The role of CD146 (Mel-CAM) in biology and pathology. J Pathol 189:4–11
130. Siddappa R, Licht R, van Blitterswijk C, de Boer J (2007) Donor variation and loss of multipotency during in vitro expansion of human mesenchymal stem cells for bone tissue engineering. J Orthop Res 25:1029–1041
131. Sima LE (2017) Extracellular signals for guiding mesenchymal stem cells osteogenic fate. Curr Stem Cell Res Ther 12:139–144
132. Simonsen JL, Rosada C, Serakinci N, Justesen J, Stenderup K, Rattan SI, Jensen TG, Kassem M (2002) Telomerase expression extends the proliferative life-span and maintains the osteogenic potential of human bone marrow stromal cells. Nat Biotechnol 20:592–596
133. Sipp D, Robey PG, Turner L (2018) Clear up this stem-cell mess. Nature 561:455–457
134. Sivaraj KK, Jeong H-W, Dharmalingam B, Zeuschner D, Adams S, Potente M, Adams RH (2021) Regional specialization and fate specification of bone stromal cells in skeletal development. Cell Rep 36:109352
135. Smets J, Shevroja E, Hügle T, Leslie WD, Hans D (2021) Machine learning solutions for osteoporosis-a review. J Bone Miner Res 36:833–851
136. Stein GS, Lian JB, Gerstenfeld LG, Shalhoub V, Aronow M, Owen T, Markose E (1989) The onset and progression of osteoblast differentiation is functionally related to cellular proliferation. Connect Tissue Res 20:3–13
137. Stein GS, Lian JB, Owen TA (1990) Bone cell differentiation: a functionally coupled relationship

between expression of cell-growth- and tissue-specific genes. Curr Opin Cell Biol 2:1018–1027
138. Stein GS, Lian JB, Stein JL, Van Wijnen AJ, Montecino M (1996) Transcriptional control of osteoblast growth and differentiation. Physiol Rev 76:593–629
139. Stenderup K, Justesen J, Clausen C, Kassem M (2003) Aging is associated with decreased maximal life span and accelerated senescence of bone marrow stromal cells. Bone 33:919–926
140. Stenderup K, Rosada C, Justesen J, Al-Soubky T, Dagnaes-Hansen F, Kassem M (2004) Aged human bone marrow stromal cells maintaining bone forming capacity in vivo evaluated using an improved method of visualization. Biogerontology 5:107–118
141. Strzelecka-Kiliszek A, Mebarek S, Roszkowska M, Buchet R, Magne D, Pikula S (2017) Functions of Rho family of small GTPases and Rho-associated coiled-coil kinases in bone cells during differentiation and mineralization. Biochim Biophys Acta Gen Subj 1861:1009–1023
142. Taipaleenmäki H (2018) Regulation of bone metabolism by microRNAs. Curr Osteoporos Rep 16:1–12
143. Taipaleenmäki H, Bjerre Hokland L, Chen L, Kauppinen S, Kassem M (2012) Mechanisms in endocrinology: micro-RNAs: targets for enhancing osteoblast differentiation and bone formation. Eur J Endocrinol 166:359–371
144. Takeuchi Y, Kodama Y, Matsumoto T (1994) Bone matrix decorin binds transforming growth factor-beta and enhances its bioactivity. J Biol Chem 269:32634–32638
145. Tanavde V, Vaz C, Rao MS, Vemuri MC, Pochampally RR (2015) Research using mesenchymal stem/stromal cells: quality metric towards developing a reference material. Cytotherapy 17:1169–1177
146. Tang Y, Rowe RG, Botvinick EL, Kurup A, Putnam AJ, Seiki M, Weaver VM, Keller ET, Goldstein S, Dai J, Begun D, Saunders T, Weiss SJ (2013) MT1-MMP-dependent control of skeletal stem cell commitment via a β1-integrin/YAP/TAZ signaling axis. Dev Cell 25:402–416
147. Tavassoli M, Crosby WH (1968) Transplantation of marrow to extramedullary sites. Science 161:54–56
148. Terunuma A, Ashiba K, Takane T, Sakaguchi Y, Terunuma H (2019) Comparative transcriptomic analysis of human mesenchymal stem cells derived from dental pulp and adipose tissues. J Stem Cells Regen Med 15:8–11
149. Thomas S, Jaganathan BG (2022) Signaling network regulating osteogenesis in mesenchymal stem cells. J Cell Commun Signal 16:47–61
150. Tomasoni S, Longaretti L, Rota C, Morigi M, Conti S, Gotti E, Capelli C, Introna M, Remuzzi G, Benigni A (2013) Transfer of growth factor receptor mRNA via exosomes unravels the regenerative effect of mesenchymal stem cells. Stem Cells Dev 22:772–780
151. Tormin A, Li O, Brune JC, Walsh S, Schütz B, Ehinger M, Ditzel N, Kassem M, Scheding S (2011) CD146 expression on primary nonhematopoietic bone marrow stem cells is correlated with in situ localization. Blood 117:5067–5077
152. Twine NA, Chen L, Pang CN, Wilkins MR, Kassem M (2014) Identification of differentiation-stage specific markers that define the ex vivo osteoblastic phenotype. Bone 67:23–32
153. Twine NA, Harkness L, Adjaye J, Aldahmash A, Wilkins MR, Kassem M (2018) Molecular phenotyping of telomerized human bone marrow skeletal stem cells reveals a genetic program of enhanced proliferation and maintenance of differentiation responses. JBMR Plus 2:257–267
154. Viswanathan S, Keating A, Deans R, Hematti P, Prockop D, Stroncek DF, Stacey G, Weiss DJ, Mason C, Rao MS (2014) Soliciting strategies for developing cell-based reference materials to advance mesenchymal stromal cell research and clinical translation. Stem Cells Dev 23:1157–1167
155. Waddington RJ, Roberts HC, Sugars RV, Schönherr E (2003) Differential roles for small leucine-rich proteoglycans in bone formation. Eur Cell Mater 6:12–21; discussion 21
156. Wang Z, Li X, Yang J, Gong Y, Zhang H, Qiu X, Liu Y, Zhou C, Chen Y, Greenbaum J, Cheng L, Hu Y, Xie J, Yang X, Li Y, Schiller MR, Chen Y, Tan L, Tang SY, Shen H, Xiao HM, Deng HW (2021) Single-cell RNA sequencing deconvolutes the in vivo heterogeneity of human bone marrow-derived mesenchymal stem cells. Int J Biol Sci 17:4192–4206
157. Weivoda MM, Chew CK, Monroe DG, Farr JN, Atkinson EJ, Geske JR, Eckhardt B, Thicke B, Ruan M, Tweed AJ, McCready LK, Rizza RA, Matveyenko A, Kassem M, Andersen TL, Vella A, Drake MT, Clarke BL, Oursler MJ, Khosla S (2020) Identification of osteoclast-osteoblast coupling factors in humans reveals links between bone and energy metabolism. Nat Commun 11:87
158. Wewer UM, Ibaraki K, Schjørring P, Durkin ME, Young MF, Albrechtsen R (1994) A potential role for tetranectin in mineralization during osteogenesis. J Cell Biol 127:1767–1775
159. Whitfield MJ, Lee WC, Van Vliet KJ (2013) Onset of heterogeneity in culture-expanded bone marrow stromal cells. Stem Cell Res 11:1365–1377
160. Witwer KW, Van Balkom BWM, Bruno S, Choo A, Dominici M, Gimona M, Hill AF, De Kleijn D, Koh M, Lai RC, Mitsialis SA, Ortiz LA, Rohde E, Asada T, Toh WS, Weiss DJ, Zheng L, Giebel B, Lim SK (2019) Defining mesenchymal stromal cell (MSC)-derived small extracellular vesicles for therapeutic applications. J Extracell Ves 8:1609206
161. Xu L, Liu Y, Sun Y, Wang B, Xiong Y, Lin W, Wei Q, Wang H, He W, Wang B, Li G (2017) Tissue source determines the differentiation potentials of mesenchymal stem cells: a comparative study of human mesenchymal stem cells from bone marrow and adipose tissue. Stem Cell Res Ther 8:275
162. Yahao G, Xinjia W (2021) The role and mechanism of exosomes from Umbilical cord mesenchymal stem

cells in inducing osteogenesis and preventing osteoporosis. Cell Transplant 30:9636897211057465
163. Yang L, Li Q, Zhang J, Li P, An P, Wang C, Hu P, Zou X, Dou X, Zhu L (2021) Wnt7a promotes the osteogenic differentiation of human mesenchymal stem cells. Int J Mol Med 47:94
164. Zhang H, Ahmad M, Gronowicz G (2003) Effects of transforming growth factor-beta 1 (TGF-beta1) on in vitro mineralization of human osteoblasts on implant materials. Biomaterials 24:2013–2020
165. Zhang Z, Garron TM, Li XJ, Liu Y, Zhang X, Li YY, Xu WS (2009) Recombinant human decorin inhibits TGF-beta1-induced contraction of collagen lattice by hypertrophic scar fibroblasts. Burns 35:527–537

5 Potency Assay Considerations for Cartilage Repair, Osteoarthritis and Use of Extracellular Vesicles

Lucienne A. Vonk

5.1 Articular Cartilage

Articular cartilage is a hyaline cartilage that covers the end of bones in synovial joints. It provides a smooth firm surface for the movement of articulating bones withstanding compressive and shear forces and helps distribute these forces onto the subchondral bone. Hyaline cartilage contains only a small number of chondrocytes (<10%), the extracellular matrix being composed of mainly of type II collagen and glycosaminoglycan containing proteoglycans. It has no blood supply and is not innervated by nerves or lymphatic vessels [64]. Cartilage has a very poor ability to repair itself and damage tends to progress into osteoarthritis (OA) if left untreated. Therefore, articular cartilage damage usually requires surgical treatment [54].

MSC were adopted for clinical cartilage repair almost 15 years ago [33]. They are mainly used for the treatment of (medium- to large-sized, >2cm^2) non-arthritic cartilage defects, focal areas where the cartilage is damaged or absent, and osteoarthritis (OA) (Fig. 5.1). In adults, the main cause of a cartilage defect is trauma, while OA is more precepted as an aging disease associated with wear and tear. In OA the cartilage gradually and progressively degenerates, which is accompanied by subchondral bone remodeling, bone marrow lesions, meniscus degeneration, synovitis, and osteophyte formation (Fig. 5.1). While there might be an initial inflammatory response to the damage in a focal cartilage defect, the inflammatory component is much more pronounced in OA [35].

5.2 Cell-Based Treatment of Cartilage Defects

Autologous chondrocyte implantation (ACI) has been used since 1987 for the treatment of cartilage defects (>2 cm^2) [8]. ACI is a two-step procedure. In a first surgery, small biopsies of healthy cartilage from a non-weight bearing site of the cartilage are taken. Subsequently chondrocytes are isolated from the biopsies and culture expanded. In a second surgery, the culture expanded autologous chondrocytes are implanted in the cartilage defect. By implanting chondrocytes directly into the defect, the defect will be filled with new hyaline cartilage tissue. Generally, ACI provides good to satisfactory results and it is a well proven treatment with level 1 evidence [45].

Of three ACI products receiving EU market authorisation, one is currently available for use in Europe. ChondroCelect (withdrawn from use in the EU in 2016, at the request of the marketing authorisation holder, TiGenix NV, for commer-

L. A. Vonk (✉)
Department of Orthopaedics, University Medical Center Utrecht, Utrecht, The Netherlands
e-mail: L.a.vonk-3@umcutrecht.nl

J. S. Burns (ed.), *Potency Assays for Advanced Stem Cell Therapy Medicinal Products*, Advances in Experimental Medicine and Biology 1420, https://doi.org/10.1007/978-3-031-30040-0_5

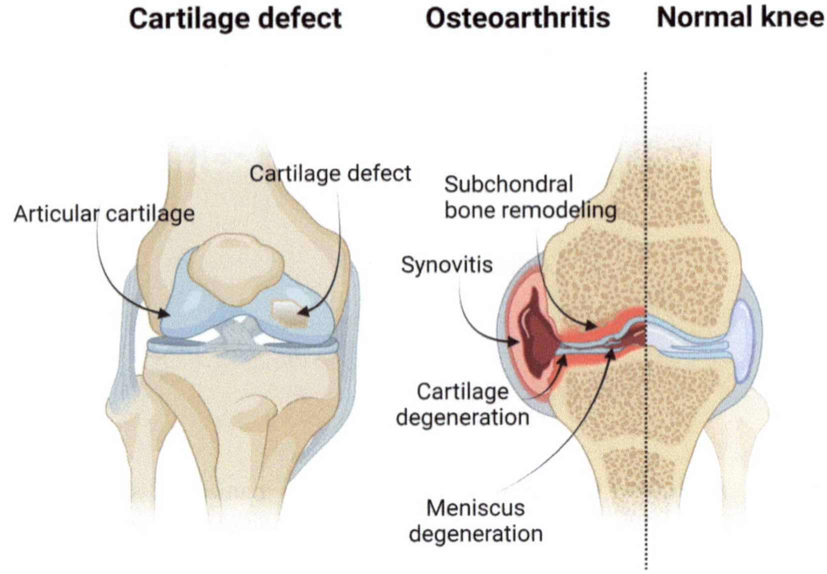

Fig. 5.1 A cartilage defect is a focal area where the articular cartilage is damaged. In osteoarthritis, there is progressive degeneration of the articular cartilage and meniscus, subchondral bone remodeling and synovitis

cial reasons) selected autologous chondrocytes based on a set of positive and negative molecular markers that predicted good hyaline cartilage production by the cells [85]. Initially, the expanded chondrocytes were implanted under an autologous periosteal patch. However, the periost often caused hypertrophy and was replaced by collagen membranes. The approach of matrix-induced autologous chondrocyte implantation (MACI®), Sanofi / Genzyme (withdrawn from use in the EU, although currently approved for use in the US as a product from Vericel Corporation), involved the seeding of characterised viable expanded autologous chondrocytes, that expressed chondrocyte-specific marker genes, onto a porcine derived type I / III collagen membrane [9]. Spherox (CO.DON AG, authorised for use in the EU) are spheroids (spherical aggregates) of expanded autologous chondrocytes and their self-synthesised extracellular matrix [40]. These spheroids are self-adhesive to the subchondral bone when applied to a cartilage defect. Thus, for the cell culture and implantation no animal derived material is required, and it is a fully autologous product.

In addition to autologous chondrocytes, several clinical studies and case reports have been published where (part of) the chondrocytes were replaced by autologous or allogeneic MSCs from various tissue sources [2, 48, 56, 70, 91, 97, 98, 108]. The MSCs were implanted and retained locally in a variety of scaffolds and hydrogels. Especially when using allogeneic MSCs, there can be clear benefits from the availability of an off-the-shelf product; a true one-step treatment (only implantation, no prior harvesting of tissue) and relatively economic, as allogeneic MSCs can be expanded up to millions of cells, cryopreserved and used for the treatment of multiple patients [89]. So far, only positive results have been published on MSC-based cartilage defect repair with clinical improvement and filling of the defects with new tissue [69].

5.2.1 Potency Assays Used for Autologous Chondrocyte Implantation

The idea of ACI is that by filling the defect with chondrocytes, the chondrocytes will produce new hyaline cartilage tissue that will repair the defect. However, the chondrocytes need to be multiplied ex vivo to reach a clinical dose [8]. When put into expansion culture, chondrocytes dedifferentiate [19], associated with a morphological change from round to more elongated spindle shaped cells. Concomitantly, expression levels of proteoglycans and type II collagen decrease while expression of type I collagen increases. To be able to produce hyaline cartilage, the expanded chondrocytes need to 're-differentiate' towards hyaline cartilage producing chondrocytes. Thus, for ACI, a potency assay should be able to predict the cartilage regeneration capacity of the chondrocytes after implantation at a time-point before implantation. Moreover, as also stated by the Committee for Advances Therapies (CAT) of the European Medicines Agency, due to time constraints an assay based on a surrogate marker could be used for batch release [78]. However, there should be a correlation between a surrogate marker and a functional assay.

In 2001, a set of positive and negative molecular markers that could predict the outcome of an in vivo Ectopic Cartilage Forming Assay were identified [22]. This approach was used in the development of a potency assay for ChondroCelect by comparing several chondrocyte batches with varying cartilage-forming capacities for in vivo ectopic to orthotopic cartilage formation, and a gene expression array at the molecular level [7]. Since correlations between molecular markers and the in vivo assays were observed, the gene expression of specific molecular markers could be used as surrogate potency markers (Fig. 5.2a).

With respect to MACI® it was reported that expression of the aggrecan gene, encoding one of the main proteoglycans in articular cartilage, could be used as a potency marker [67]. It was shown that the MACI® cultured chondrocytes expressed relatively higher levels of the aggrecan gene than dermal fibroblasts and when cells were cultured in 3D (as cell pellets or in alginate), the chondrocytes produced type II collagen. It was not disclosed whether these findings were further developed into the VIP (viability, identity and potency) assay of MACI®.

A more recent study found a direct correlation between the expression of S100A1 and S100B by chondrocytes in monolayer and their subsequent capacity to produce neocartilage when cultured in vitro in a 3D regeneration culture [23].

Furthermore, researchers from the biopharmacy company CO.DON AG reported a human ex vivo functional potency assay whereby spheroids of culture expanded chondrocytes were implanted in a chondral defect created in a chip of human osteochondral tissue [3]. Notably, protein expression and potentially gene expression of aggrecan could be used as surrogate potency markers for this functional assay (Fig. 5.2b).

5.2.2 Mechanism of Action (MoA) of MSC-Based Cartilage Defect Repair

To develop and establish potency assays for ATMP products, the mode and mechanism of action of the products should be known. Potency is the quantitative measure of biological activity based on the critical attribute of the product, linked to the relevant biological properties. Moreover, a potency assay should demonstrate the biological activity based on the intended biological effect and ideally be related to the clinical response [102]. However, the mechanism of action of MSCs for cartilage defect repair and OA is not fully understood. Moreover, not even the cell fate of transplanted MSCs is fully elucidated.

5.2.3 MoA: Differentiation Versus Paracrine Signalling

Initially it was believed that MSCs would differentiate into chondrocytes and produce and engraft new cartilage tissue. Much effort was put in finding the MSC source with most effective chondrogenic differentiation and in creating the optimal circumstances to differentiate MSCs into the chondrogenic lineage. This kind of research focused for instance, on comparing the differentiation capacity of different MSC clones, the use of various growth factors and culture conditions, and on cocultures of chondrocytes and MSCs, where it was believed the chondrocytes would instruct the MSCs to differentiate. However, in 2006, it was proposed that MSCs do not differentiate, but rather produce molecules that have an immunomodulatory and a pro-regenerative effect [11]. In cartilage regeneration this feature is

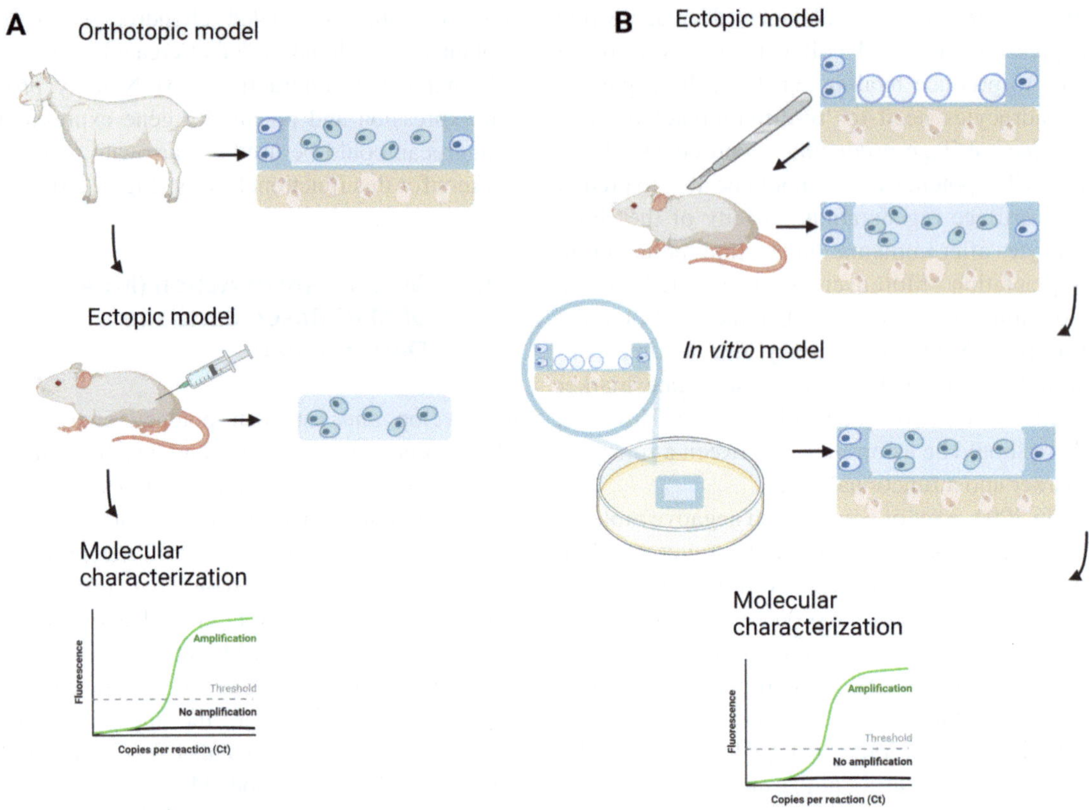

Fig. 5.2 Surrogate models testing cartilage regenerative capacity. (**a**) For ChondroCelect, several batches of chondrocytes, ranging from phenotypically stable chondrocytes to dedifferentiated chondrocytes, were tested for their cartilage regenerative capacity in goats in an orthotopic model. Subsequently, similar chondrocyte populations were investigated in an ectopic cartilage formation assay where they were injected in the adductor muscle in immunocompromised mice. Finally, a correlation was found between the in vivo cartilage regenerative capacity and a set of molecular markers. (**b**) For Spherox, the cartilage regenerative capacity was tested in chips of human osteochondral tissue in which defects were created. The spheroids were implanted in the chips and the constructs were subcutaneously implanted in immunocompromised mice. Subsequently, it was shown this model is suitable for in vitro use, where the amount of newly formed tissue is a measure of the regenerative capacity and this correlated with aggrecan protein expression

called a chondroinduction [105]. Although research on the (chondrogenic) differentiation of MSCs is still ongoing, emphasis on intercellular mediation caused a shift towards implementing chondroinductive effects of MSCs [95].

5.2.4 Cell Fate of MSCs Used to Treat Cartilage Defects

Two main application routes of MSCs for cartilage defect repair have been investigated: either via intra-articular injection into the joint space, supposing they would home to the defect site, or via direct application into the defect in a cell carrier (Fig. 5.3). Also cocultures of MSCs and chondrocytes have been used in a cell carrier [95].

Determining the cell fate of MSCs after intra-articular injection has been mostly studied in vivo, sometimes in combination with general biodistribution analyses [27, 50, 53, 71, 107]. There is no conclusive answer to the question whether MSCs home to the site of a cartilage defect and if so, how long they stay present. Using immunocompetent transgenic rats that express a stable or heat-instable form of the human placental alkaline phosphatase, it

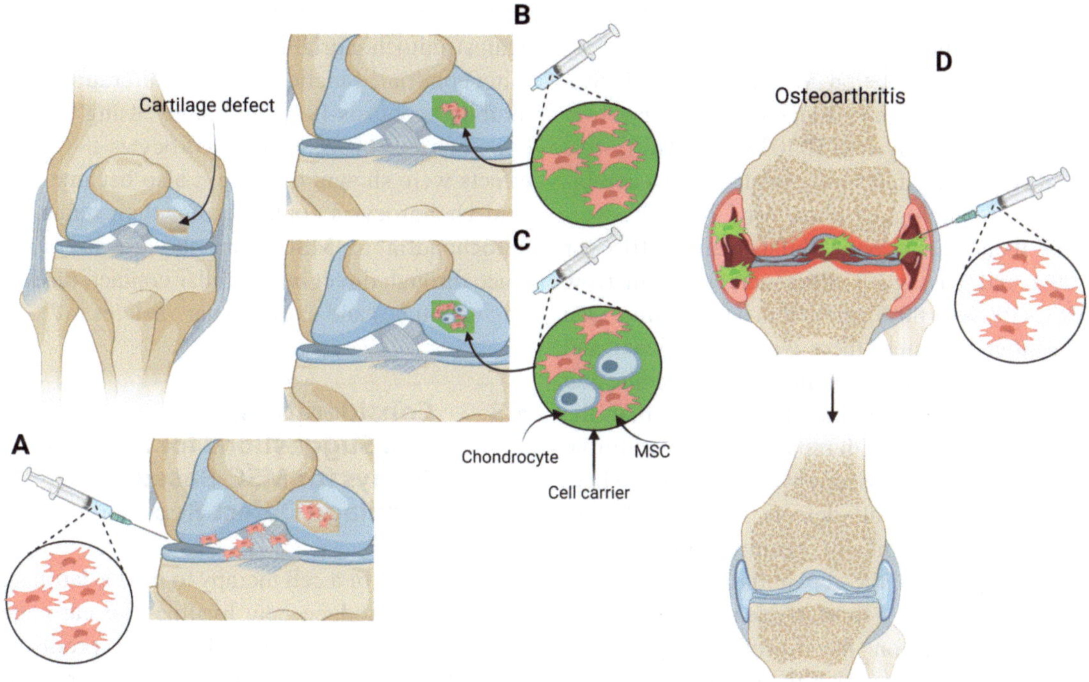

Fig. 5.3 Mesenchymal stromal cells (MSC) have been administered in various ways for cartilage defect repair. They have been administered in a suspension by intra-articular injection (**a**). For this method there is no consensus whether the MSCs home to the defect. In addition, they have been applied in a cell carrier, such as a hydrogel, directly onto the defect (**b**). Also, cocultures of MSCs and chondrocytes have been applied directly to the defect in a hydrogel (**c**). For osteoarthritis, the preferred method is intra-articular injection of a suspension with MSCs, so the MSCs can reach all affected tissues (**d**)

was shown that a few intra-articular injected MSCs migrate to a focal cartilage defect [71]. However, the majority of the injected MSCs were traced back elsewhere in the joint where they formed cell aggregates. Homing of the MSCs to a cartilage defect was supported by three additional studies using rabbits [107], pigs [50], and mice when MSCs from the superhealer MRL/MpJ mice were injected into C57BL6 mice [53]. However, when MSCs from C57BL6 mice were injected in C57BL6 mice, they did not migrate to the defect [53]. Others also failed to find any rat synovium derived MSCs had migrated to a partial thickness cartilage defect in rats [27]. All studies did report a positive effect of intra-articular injection of MSCs on cartilage defect repair, regardless of homing to the defect site. In addition, most of the studies agreed that the number of MSCs found in the knee joint decreased over time. One study in rats indicated MSC presence after one month, but this was no longer the case after two and six months [71], whilst others indicated they could not be found in the joint after only one week [27]. Alternatively, in rabbits, the number of MSCs decreased over 14 days [107]. Only one study reported autologous labelled MSCs in the repair tissue of partial thickness cartilage defects created in pigs after three months [50].

In clinical use, MSCs have mostly been applied in a cell carrier directly onto the cartilage defect, but only a very limited number of studies have tried to determine the cell fate of MSCs implanted in a cell carrier. In an osteochondral defect rabbit model, it was shown that the number of human umbilical cord blood MSCs in a hyaluronic acid hydrogel decreased from two to eight weeks and were no longer detectable after 16 weeks [63]. However, it must be noted that this was a xenograft transplantation in immunocompetent rabbits.

Clinical evidence on the cell fate of MSCs implanted in human cartilage defects came from the IMPACT study (NCT02037204). Thirty-five focal chondral defects were treated with a combination of autologous chondrons (chondrocytes with their native pericellular matrix, 10% or 20%) and passage 3 allogeneic bone marrow MSCs (90% or 80%) applied in a fibrin glue. One year after treatment there was a significant and meaningful improvement in clinical outcome and magnetic resonance imaging showed that the defects were filled with repair tissue. In addition, a second look arthroscopy was performed where it was confirmed all defects were filled with mostly macroscopically healthy cartilage and small biopsies of the repair tissue were taken. Besides histology, short tandem repeat (STR) analyses based on the EuroChimerism STR marker panel showed that the biopsies only contained autologous genomic DNA. This supported the idea that MSCs do not differentiate to constitute the newly repaired tissue, but they acted as cellular moderators stimulating the autologous cells to proliferate and produce new tissue, thereby restoring the cartilage defect [97, 98]. Also in vitro and in vivo, most studies found a progressive loss of MSCs accompanied by an increase of the number of chondrocytes and hyaline cartilage formation in cocultures [1, 62, 96, 105, 106]. Only one study using cocultures of immortalised MSC and chondrocyte cell lines found an increase in the number of MSCs that also expressed cartilage markers [14].

However, the exact biological process by which the MSCs disappear, is not clear yet; apoptosis and / or autophagy seem to play a role in this [62, 105]. In addition, there is also still little understanding on the signals and mechanism underlying the chondroinductive effects in cocultures. It has been suggested that various growth factors, such as fibroblast growth factor and bone morphogenetic proteins are responsible for the proliferative effect on chondrocytes [62, 104]. In addition, the transfer of mitochondria is a newly proposed mechanism. With rat cells it has been shown that mitochondrial transfer from bone marrow MSCs to OA chondrocytes protects against mitochondrial dysfunction and improves mitochondrial function, cell proliferation and inhibits apoptosis [92]. In a study using human bone marrow MSCs and chondrocytes the same effects were shown, but here it also became evident chondrocytes could transfer their damaged mitochondria to MSCs and the transfer took place through direct intercellular contact, tunnelling nanotubes and extracellular vesicles [47].

5.3 Considerations and Suggestions for Potency Assays for MSC-Based Cartilage Defect Repair

The main goal of a cell therapy to treat a cartilage defect is to produce new hyaline-like cartilage tissue at the defect site. To show proof-of-principle and effectivity of treatments for cartilage defect repair, large animal models including pigs, sheep, goats and horses have been used. However, there are multiple reasons that these animal models remain unsuitable as potency assays for human MSCs. Not only are they expensive, take a long time from treatment to result (3–6 months) and are demanding in terms of maintenance plus care, but also xenogeneic MSC transplantation might induce an immunological reaction that could influence the therapeutical effects [55, 66]. Thus, cartilage defect models in larger animals have been useful to show initial efficacy of a treatment, but are not appropriate as a (batch) potency assay for human MSCs for cartilage repair. For smaller animal species, such as mouse and rat, immunocompromised or humanised animals are available. However, orthotopic cartilage defect models are too complicated because of their size; it is almost impossible to create a focal defect on the cartilage surface without undesirably damaging any other tissues or the subchondral bone.

One application for immunocompromised small animals has involved ectopic cartilage formation assays. For instance, for the cell product ChondroCelect, chondrocytes in suspension were injected intramuscularly into the adductor muscles of the thighs of female immunodeficient mice [7, 22]. Alternatively, for the combination of chondrons and MSCs in the IMPACT study, cocultures of these cells in fibrin glue constructs were subcutaneously implanted in nude mice [4]. However, it is unlikely these assays of themselves would suffice to assess the cartilage regenerative effect of MSCs. Neither location, the adductor muscle nor the dorsal subcutaneous pockets, manages to mimic a (knee) joint environment. So far, no cells that were not (pre-)committed to the chondrogenic lineage were able to form tissue containing cartilage components in these ectopic assays [4, 22]. Thus, regardless of the exact MSC mode of action, these assays would not be feasible as potency assays for MSCs.

For MSC and chondrocyte cocultures, 3D growth as cell pellets or cells in fibrin glue using culture medium without supraphysiological concentrations of chondrogenic growth factors, was a functional assay that provided insight into the production of new cartilage tissue [96]. At least one of the components of native hyaline cartilage, such as type II collagen, proteoglycans or more specifically aggrecan seemed to correlate in a quantitative manner. However, this has yet to be validated. In addition, the coculture assay more pragmatically fits a one-step procedure when autologous chondrocytes are combined with off-the-shelf allogeneic MSCs to implant the cell product in the time-frame of a single surgery [89]. To determine the potency of various batches of allogeneic MSCs for chondroinduction, it would be desirable to have the same chondrocytes in the cocultures to avoid donor variation discrepancies influencing measurement of chondrocyte capacity. A practical approach to improve standardisation would be using a chondrocyte cell line. Several immortalised chondrocyte cell lines have been developed, but they often show suboptimal cartilage production and/or responsiveness to growth factors, which are typical outcomes in such an assay [18]. Another possibility would be to establish a standardised induced Pluripotent Stem Cell (iPSC)-derived chondrocyte strain that shows promising hyaline cartilage production [6].

When considering just MSC alone, their implantation into a cartilage defect created in a piece of osteochondral tissue provided a good way to measure production of new cartilage [42]. This approach would accommodate both potential mechanisms for cartilage production, either by direct MSC differentiation or through MSC mediated chondroinduction. A similar assay was also used for the German ACI product chondrosphere [72]. To investigate formation of new cartilage, including integration into the surrounding native cartilage and subchondral bone in vivo, spheroids of ex vivo expanded chondrocytes were applied to a cartilage defect in an osteochondral chip and this whole unit was subcutaneously implanted into immunocompromised SCID mice [72]. After 24 weeks, new hyaline cartilage tissue had filled the defect and this tissue was well integrated in the surrounding tissues of the osteochondral chip. To assess the potency of the chondrocyte spheroids, the investigators used the cartilage defect in an osteochondral chip that was cultured ex vivo for 12 weeks [3]. Here the amount of formed repair tissue was used as an outcome measure. In addition, a positive correlation ($r = 0.55$; $p < 0.025$) between aggrecan protein expression in spheroids before implantation and newly formed tissue was observed.

A similar approach with the osteochondral chip assay has also been used currently as a functional assay for new cartilage production following MSCs implantation [83, 90]. This assay could be a good starting point to further develop a potency assay, identify additional quantitative measures, and validate them. Subsequently it can also be used to identify surrogate biomarkers or biomarker sets. Ultimately, additional in vivo tracing experiments for MSC products, identifying their cell fate, will provide more confidence as to whether a surrogate potency assay could be sought in the differentiation capacity of MSC, their trophic signalling, or both.

5.3.1 Potency Assays for Differentiation

Although most evidence points towards trophic signalling as the mechanism of action of MSCs [95, 97], (partial) differentiation cannot be fully excluded at this point. Moreover, for autologous MSCs no attempts have been made so far to trace their fate after clinical use for cartilage repair.

The ability to differentiate into the chondrogenic lineage is one of the minimal criteria MSC must adhere to [24], however, it is being increasingly appreciated that there is a lot of variability in the chondrogenic differentiation capacity of MSCs. This may reflect donor variability and the methods of cell culture plastic-expanded MSCs.

Generally, chondrogenic differentiation medium contains ascorbic acid to facilitate collagen synthesis and supraphysiological concentrations of one of the transforming growth factor beta (TFG-β) isoforms and dexamethasone to steer the differentiation [44, 109]. For the differentiation, three-dimensional pellets rather than monolayer cultures are advised and often provide more chondrogenic conditions for poorly-differentiating MSC. After performing the differentiation for 21 or 28 days, the principal components of hyaline cartilage, proteoglycans and type II collagen, can be semi-quantitatively determined as a measure of the chondrogenic differentiation capacity. Since it takes time before these hyaline cartilage extracellular matrix components are deposited, assays that predict the chondrogenic differentiation capacity were developed. A notable gene expression reporter assay involved co-transfection of a plasmid constructed with the type II collagen promotor upstream of Metridia luciferase and a control plasmid with Renilla luciferase. Both in monolayer and in pellet cultures, the chondrogenic induction could be determined optically after 3 days and longer-term pellet cultures demonstrated a correlation with good and poor chondrogenic potency [58]. In addition, it has been proposed that monitoring the expression levels of the TFG-β receptors, TGFBRI and TGFBRII, could predict the differentiation potency. However, it remains unclear to what extent this correlates with in vivo chondrogenic differentiation. In essence this assay measures the ability of a cell to respond to TFG-β. However, especially for adipose derived MSC, even cells from low-chondrogenic potency donors could be driven to differentiate using a combination of TFG-β and bone morphogenetic protein (BMP)-6 [39]. Moreover, besides the use of growth factors, multiaxial loading can be used to steer chondrogenic differentiation. Although MSCs subjected to multiaxial load produce endogenous TFG-β and there are clear similarities with TFG-β induced differentiation, there are also distinct differences, e.g. in nitric oxide production [30]. Therefore, results on growth factor-induced chondrogenic differentiation need to be carefully interpreted; correlation to in vivo differentiation requires validation.

5.4 Treatment of Osteoarthritis

Currently, we lack an effective disease-modifying therapy for OA and existing treatments are largely unsatisfactory. Most therapies are aimed at symptom relief, but fail to restore the joint tissues. ACI has been applied in OA, but with a high failure rate. Chondrocytes seem ineffective against the ongoing inflammation and progressive cartilage degradation and currently patients with end-stage OA receive a total knee replacement. This treatment is generally effective in reducing pain and restoring function. It works well for approximately 15–20 years in 90% of the cases [10]. However, after this time a revision surgery is required with less success [16, 37]. Therefore, treatment options to postpone or even avoid the need for a total joint replacement are required.

Recently, so-called orthobiologics with 'minimal manipulation', such as autologous platelet rich plasma, bone marrow aspirate concentrates and the stromal vascular fraction from adipose tissue, have become popular for the treatment of OA [93]. These therapeutic agents can be prepared in the operation theatre and have been shown to be safe with some short-term beneficial effects. However, high-quality efficacy studies and proper recording of treatment failures and adverse reactions are still lacking.

5.4.1 MSC-Based Treatment of Osteoarthritis

Studies with culture-expanded MSC for the treatment of OA increasingly emerged since it became known MSC could exert anti-inflammatory and pro-regenerative effects [35, 103], as both inflammation and cartilage degradation need to be addressed in OA. In the early phases of OA, the innate immune cells play the most important role. Natural killer cells in the synovial tissue produce granzymes and perforins that induce apoptosis of chondrocytes in the articular cartilage and likely also cells in the meniscus. In the chronic inflammatory phase, many cells of the adaptive immune system are involved and those are mainly infiltrating the inflamed synoval tissue. Activated M1 macrophages play an especially important role, producing pro-inflammatory cytokines that stimulate the production of extracellular matrix degrading enzymes such as collagenase-3 (MMP13) and aggrecanases (ADAMTS4 and 5).

For the treatment of OA both autologous [12, 20, 25, 26, 43, 49, 59, 65, 75] and allogeneic [36, 87] MSCs have been applied. All these studies reported safety, feasibility, and improvement in clinical outcomes after intra-articular MSC injection. As with the use of MSCs for cartilage defect repair (Sect. 5.2.2), no clear mechanism of action has been defined.

5.4.2 Tracking MSC After Intra-Articular Injection in Osteoarthritic Joint

As OA is a disease that affects the whole joint, an effective treatment should target all joint tissues and restore normal joint homeostasis. Therefore, for the treatment of OA, MSCs are usually delivered via an intra-articular injection into the synovial fluid from where they can reach all joint tissues [76].

The biodistribution of intra-articular injected human MSCs has been explored in joints of SCID mice. In the first month, 15% of the introduced MSCs could be found in the joints and this decreased to 1.5% after 6 months. In addition, MSCs were found in stem cell niches such as the bone marrow, adipose and muscle tissue [81]. In a subsequent study, the investigators injected human MSCs in immunocompetent mouse models for arthritis and OA, and their controls. The MSCs had a positive effect on both the arthritis and OA scores in the animal models, but no difference was found in the number of MSCs that could be traced in the joints between the experimental and control animals [80]. Similarly, another study reported no difference in the time the MSC resided in the joint, when comparing nude mice with and without induced OA [73]. In contrast, in immunocompetent rats the MSCs were diminished after 28 days in normal joints, while MSCs were detectable in OA joints for over 70 days [51].

5.5 Considerations and Suggestions for Potency Assays for MSC-Based Treatment of Osteoarthritis

In ways similar to MSC-based treatment of cartilage defects, the safety and efficacy of autologous and allogeneic MSC-based treatments for OA can also be shown in large animal models. However, xenogeneic transplantations of human MSCs might induce an immune response, that could influence results [55, 66]. In addition, use of immunocompromised or immunodeficient small animals might not be biologically suitable to investigate the potency of human MSCs to treat OA. Even when small animal size issues are overcome in well-established small animal models of surgically- and chemically-induced OA with feasible intra-articular injections, the inflammatory component may be mimicked poorly in immunocompromised or immunodeficient animals. Either way, there have been positive reports on the treatment of arthritic diseases, including OA, in small immunocompetent animals such as mice and rats using human MSCs [80, 81]. Although there are currently no (published) validated potency assays for the use of MSCs for OA available, there are some main mechanisms by which MSCs can alleviate OA with relevant functional assays available. Those functional assays may lay the foundation for potency assays given validation and appropriate establishment of a reference standard for calibration.

5.5.1 Effects on Macrophage Polarisation

Macrophages in the synovial tissue play an important role in the symptoms and progression of OA [35]. Especially the M1/M2 subtype ratio is associated with OA severity. M1 macrophages produce pro-inflammatory cytokines and attract more immune cells. Chondrocytes respond to this by secreting more pro-inflammatory cytokines and enzymes that can degrade cartilage tissue such as matrix metalloproteinases (MMPs). In addition, it has been shown that reprogramming macrophages from M1 to the more anti-inflammatory M2 subtype relieve pain and protects against cartilage degradation, synovitis and osteophyte formation [110]. MSCs can stimulate the polarisation of macrophages to the M2 phenotype, partly by the secretion of anti-inflammatory cytokines, prostaglandin E2 (PGE2) and TFG-β (Fig. 5.4). This can be assessed by coculturing macrophages or CD14+ monocytes differentiated towards macrophages with MSCs or MSC-conditioned medium. Subsequently expression of the M1 phenotype markers CD40 and CD86 and the M2 phenotype marker CD206 can be determined by flow cytometry. This can be supported by measurement of the concentrations of pro-inflammatory cytokines in the cell culture supernatant by (multiplex) Enzyme-linked immunoadsorbant assay (ELISA) [29, 74, 94].

5.5.2 Effects on NK Cells

NK (Natural Killer) cells are one of the main immune cells infiltrating the synovial tissue in OA [41]. Compared to NK cells found in blood, NK cells in the synovium have relatively low cytotoxic activity, but they express high levels of the pro-inflammatory protease granzyme A, that may induce or maintain the inflammatory conditions in OA [35, 103]. MSCs can secrete indoleamine 2,3-dioxygenase (IDO), PGE2 and TFGβ that in turn can suppress NK cell function by decreasing their proliferation, cytotoxic activity and secretion of pro-inflammatory cytokines [15, 38, 57] (Fig. 5.4).

Assays that can be used to determine the effects of MSCs on NK cell function relevant for OA measure NK cell proliferation, NK cell cytotoxic activity, NK cell receptor activation, cytotoxic molecule expression and pro-inflammatory cytokine production [84]. In NK proliferation assays, cells are stained with membrane labels (such as PKH-67 or CFSE) before the NK cells are stimulated with, e.g. interleukin (IL)-15 in the absence or presence of different MSC ratios. During division of the NK cells, the stable label divides equally when the cells divide and therefore dilutes. After approximately 5 days (about 4 cell divisions), the percentage of cells with low label intensity (such as PKH-67^{low} or CFSE^{low}) can be determined by flow cytometry and used as quantitative outcome. For measuring the NK cell cytotoxic activity in vitro, the radioactive ^{51}chromium-release assay has represented the golden standard, whereby target cells loaded with ^{51}Cr are cultured with NK cells stimulated with, e.g. IL-15 in the presence and absence of different MSC ratios. Direct NK cell-mediated lysis is subsequently determined by the amount of ^{51}Cr released into the cell culture supernatant. The radioactive loading of the target cells can be replaced by fluorophore-labelling, bearing in mind that inconsistent dye uptake and dye leakage can give intra- and inter-assay variability.

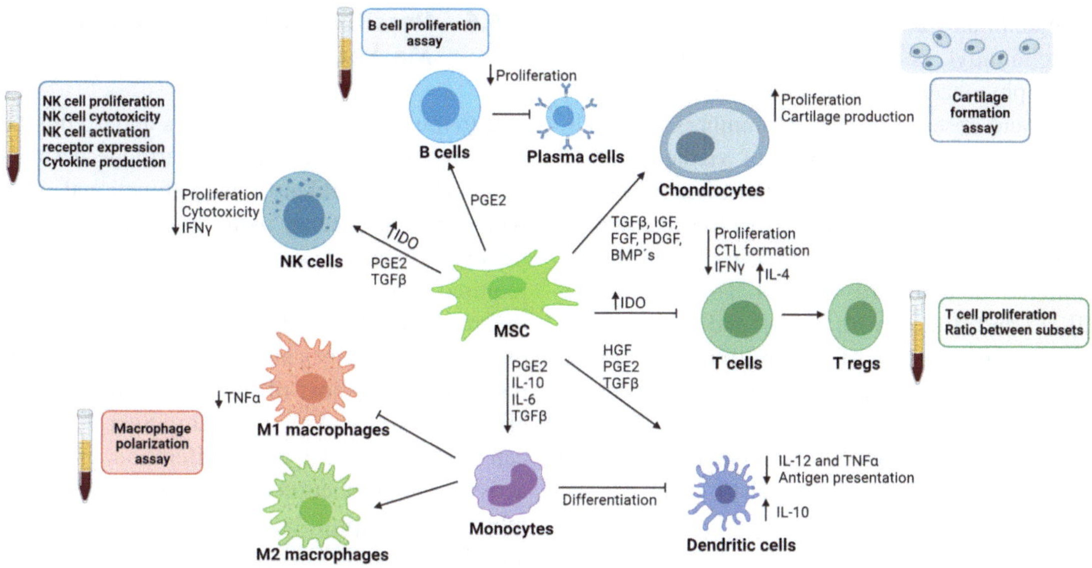

Fig. 5.4 For the treatment of osteoarthritis, mesenchymal stromal cells (MSCs) can have an anti-inflammatory effect. Amongst others, due to the secretion of prostaglandin E2 (PGE2), interleukin (IL)-10 and IL-2, and transforming growth factors beta (TGFβ), monocytes are stimulated to differentiate more towards the anti-inflammatory M2 macrophages and not the pro-inflammatory M1 macrophage. The effect of MSC preparations on monocytes and macrophages can be tested in vitro with a macrophage polarisation assay. Via an increase in indoleamine 2,3-dioxygenase (IDO), PGE2 and TGFβ, MSCs can inhibit the proliferation, cytotoxicity and interferon gamma (IFNγ) secretion of natural killer (NK) cells. This can be tested in vitro via NK cell proliferation, cytotoxicity and activation receptor expression assays and by cytokine release into the medium. Via PGE2 the proliferation of B cells and their differentiation into plasma cells can be inhibited, which can be tested via a B cell proliferation assay. The increased IDO can also inhibit T cell proliferation, cytotoxic T lymphocyte (CTL) formation and IFNγ secretion by T cells, while increasing IL-4 expression and the formation of T regulatory (T reg) cells. This can be tested with a T cell proliferation assay and in there the ratio of T cell subsets can be determined. Finally, MSCs can stimulate chondrocytes to proliferate and produce new cartilage via the secretion of TGFβ, insulin growth factor (IGF), fibroblast growth factor (FGF), platelet derived growth factor (PDF) and bone morphogenetic proteins (BMPs). This can be tested in vitro with cartilage formation assays

Nonetheless, there are several commercially available target cell lines with stable endogenous expression of fluorescent proteins that can be used. NK cell activation receptor and cytotoxic protein expression can be determined by activating the NK cells again in the presence and absence of different ratios MSCs and after approximately 4 days of culture, the percentage of NK cells expressing a specific receptor or cytotoxic molecule, such as Granzymes, can be determined with flow cytometry. To determine the secretion of pro-inflammatory cytokines, NK cells can be activated in the presence and absence of different ratios MSCs and after a few days the concentrations of pro-inflammatory cytokines in the cell culture supernatant can be determined with a (multiplex) ELISA.

5.5.3 Effects on T Cells

Synovial tissue is also infiltrated by T cells in OA [38]. Especially type 1 (Th1) and type 17 (Th17) helper cells and cytotoxic T cells are increased in the synovium. MSCs can inhibit T cell proliferation and influence the ratio between subtypes of Th cells [35, 103] (Fig. 5.4). The pro-inflammatory cytokine expression by T cells is a concern and especially the interferon gamma secretion by Th1 cells, as it can create a positive feedback loop activating M1 macrophages.

Suitable assays to determine the effects of MSCs on T cells would be T cell proliferation and T cell differentiation assays. For T cell proliferation, either peripheral blood mononuclear cells (PBMCs) or isolated T cells are stained with a membrane label such as Carboxyfluorescein succinimidyl ester (CFSE) and activated (e.g. with phytohaemagglutinin (PHA)) in the presence and absence of various numbers of MSCs. After 4–7 days, the number of viable CD3+ T cells with low label intensity (CFSE-diminished) can be determined. To investigate the effects on T cell differentiation, isolated CD4+ T cells can be differentiated towards Th1 or Th17 cells with IL-12 and IL-2 or with TFG-β and IL-2, respectively. After 3 days of culture, the cells can be stimulated with a leukocyte activation cocktail and after a few hours the percentage CD4 and interferon gamma (IFNγ) positive Th1 cells and CD4 and IL-17A positive Th17 cells can be determined [84].

5.5.4 Effects on B Cells

Although OA is not an auto-immune disease, B cells have been found in the synovium of OA patients in combination with auto-antibodies against components of cartilage [35, 103]. In addition, B cells can activate humoral immunity, leading to a disbalanced joint homeostasis. MSCs can inhibit B cell proliferation [82] (Fig. 5.4). This effect can be determined by stimulating membrane stained PBMCs with CpG in the absence and presence of different numbers of MSCs. After approximately one week, B cells can be collected and CD19 can be used to determine the percentage B cells with diluted membrane staining and a combination of CD19 and CD27 to specify plasma cells [82].

5.5.5 Effects on Cartilage Formation

A few studies have suggested that there is new cartilage formation after intra-articular injection of MSCs in OA joints [21]. As for cartilage defect repair, this is probably due to trophic signalling by MSCs, but differentiation of MSCs cannot be excluded at this point (Fig. 5.4). Functional assays that have been described in Sects. 5.3 and 5.3.1 can be used to assess new cartilage formation and can be transferred to OA as well.

5.5.6 Possible Surrogate Potency Markers

Due to multifactorial mechanisms underlying the inflammation and progressive cartilage destruction in OA, it is unlikely that one functional assay will suffice as potency assay for using MSCs to treat OA. In that case, multiple assays would be required to determine the potency of MSCs for the treatment of OA. However, it can be that a set of secreted cytokines and signalling molecules is responsible for most of the immunomodulatory effects by MSCs. As explained in Sect. 5.5, IDO can inhibit NK and T cell proliferation, cytotoxic activity and production of IFNγ, while PGE2 can inhibit B cell proliferation and differentiation, and PGE2 and TFG-β can inhibit the formation of antigen presenting dendritic cells and stimulate the differentiation of monocytes towards M2 macrophages. In addition, TFG-β can signal some resident chondrocytes to proliferate and produce neocartilage tissue (Fig. 5.4).

So far, only one study investigated in vitro anti-inflammatory effects of MSCs to clinical outcome after injecting them in OA knee joints [12]. The investigators licensed a portion of culture expanded bone marrow MSCs, that were used in an autologous treatment, with IFNγ and TNFα and studied their cytokine expressions in vitro. It was found that an increased TSG-6 protein expression and increased gene expression of PGE2, PDL1, IDO, IL-10, HGF and TFG-β were a significant predictor of better patient reported outcome measures [12]. Such a set-up might provide a basis to determine surrogate potency markers that correlate to one or more functional assays.

5.6 Extracellular Vesicles

The evidence that the therapeutic efficacy of MSCs relies on paracrine signalling rather than engraftment and differentiation is growing. Part of the paracrine signalling is attributed to the secretion of extracellular vesicles (EVs) [88]. EVs are membrane enclosed structures, without functional nucleus, that are released by cells. The

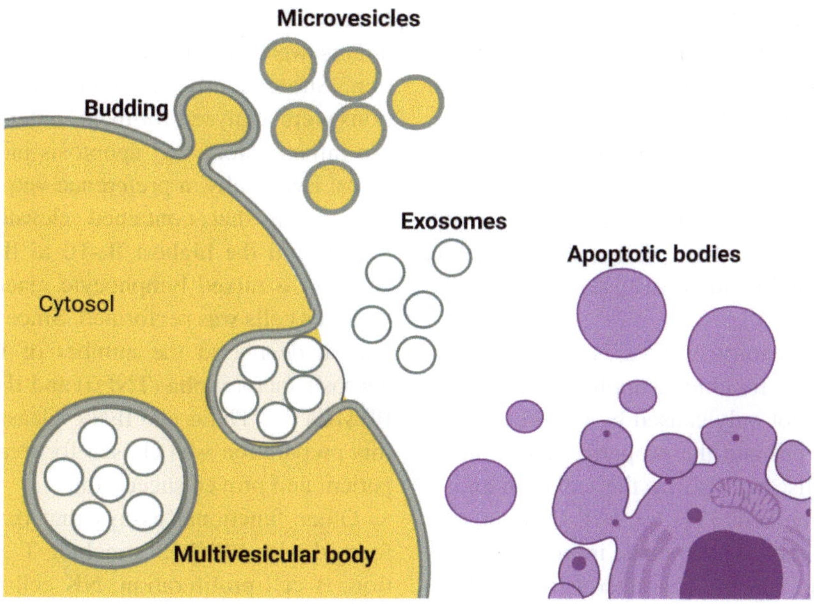

Fig. 5.5 Part of the paracrine effects of mesenchymal stromal cells are due to the production of extracellular vesicles. Extracellular vesicles are a collective term for membrane enclosed particles that are synthesised by cells in various ways. Exosomes are released after the fusion of multivesicular bodies with the plasma membrane, while microvesicles bud directly of the plasma membrane. Apoptotic bodies are formed due to blebbing of the plasma membrane in a response to apoptosis. Most apoptotic bodies are larger than exosomes and microvesicles, but they have also been observed in the same size range

term extracellular vesicle is a collective name for several EV subtypes and is endorsed by the International Society for Extracellular Vesicles (ISEV), as specific markers for the subtypes are still missing [101] (Fig. 5.5). A specifically sized subset of extracellular vesicles termed exosomes (30–150 nm) are formed by the invagination of the membrane of early endosomes that mature into multivesicular bodies (MVBs). When multivesicular bodies are not sent to the lysosome, they fuse with the cell membrane whereby the content, including the exosomes, are released into the extracellular space [86].

In contrast, the EV subtype termed microvesicles (MVs, 100–1000 nm) are directly budded of the plasma membrane. Therefore, they contain mainly cytosolic and plasma membrane associated proteins, including cytoskeletal and heat shock proteins, integrins and post translationally modified proteins (glycosylated and phosphorylated). Proteins associated with other cell organelles are hardly abundant [86].

Apoptotic bodies (50–5000 nm) are released by cells undergoing apoptosis. The majority of apoptotic bodies are larger in size compared to exosomes and MVs (1–5 μm), but smaller apoptotic bodies have been described. Apoptotic bodies are released by separation of the plasma membrane from the cytoskeleton after apoptotic cells contract. Due to their biogenesis, apoptotic bodies can contain intact cell organelles and their contents are quite similar to that of cell lysates [86].

For years it was believed that EVs were part of a dumping mechanism by which a cell would get rid of unwanted material. However, more recently it became clear that cells use EVs for intercellular communication between local and distant cells. EVs can contain a variety of molecules such as nucleic acids (DNA fragments, various types of RNA), cytokines, lipids, enzymes and other proteins. Particularly, the small EVs (50–200 nm) have been shown to be therapeutically effective in various studies [99].

The secretion of EVs is not exclusive by MSCs, but since EVs generally reflect the properties of their donor cells, there is an interest for MSC-derived EVs for regenerative medicine. The number and content of secreted EVs does not only depend on the type of the donor cell, but also on the state and microenvironment. Therefore, the content of EVs and their signaling message are highly adaptive [31, 32, 68, 82, 99, 100].

The heterogeneity of EVs is a major challenge, especially for clinical application. The heterogeneity is not only caused by the presence of various subtypes and the properties and state of the donor cells, but also by the isolation and/or purification methods [31, 60, 68]. Dealing with this heterogeneity is even made more difficult by the absence of specific characterisation methods [79]. Therefore, it is highly recommended to first develop and establish an EV production and isolation procedure in combination with release criteria before treating patients [28]. Even then, despite using the same MSC donors and standardised procedures for production, there can be functional heterogeneity among independent preparations [52]. This underlines the need to test each MSC-EV preparation for potency before clinical use.

5.6.1 Functional Assays for EVs

Many pathologies, including OA, are complex and likely to respond to a multifaceted mode of treatments. Thus, as in the case of MSCs, several functional assays might be required to determine functionality and potency of MSC-EVs for a certain pathology. Although the knowledge of MSC-EVs is not sufficient to establish definitive potency tests for clinical treatment yet, functionality of independent MSC-EV preparations can be determined with functional assays that are also used for MSCs.

When the first clinical treatment with MSC-EVs was performed, the investigators used multiple functional assays to select their best MSC-EV preparation [46]. To treat a therapy-refractory graft-versus-host disease patient, bone marrow MSCs from four different unrelated donors were cultured and used for EV preparation. Subsequently, the four EV-enriched fractions were analysed for their content of anti- and pro-inflammatory and apoptosis-inducing molecules. Eventually, a preference was given to one preparation that contained elevated levels of TFG-β and the highest IL-10 to IFNγ ratio. In addition, a mixed lymphocyte reaction with the patient's cells was performed. Since the EV preparation decreased the number of IL-1β, tumor necrosis factor alpha (TNFα) and IFNγ releasing PBMCs and TNFα and IFNγ releasing NK cells, this preparation was chosen for treatment of the patient and proven successful.

Other functional assays that have been performed with MSC-EVs include T cell proliferation, B cell proliferation, NK cell proliferation, macrophage polarisation, cytokine release patterns and angiogenesis (Table 5.1). The most reported assay with EVs are the various T cell proliferation assays. Only one study used purified T cells for this, but also investigated the effects of MSC-EVs on T cell proliferation of PHA stimulated PBMCs [82]. All other studies investigated the proliferation of T cells from PBMCs and several activators were used such as PHA, concanavalin A (ConA), mixed lymphocyte reaction (MLR) or CD3/CD28 activation. Several studies reported a decreased T cell proliferation when EVs were added to the assay and some even confirmed a dose dependent effect [5, 17, 60, 77]. However, other studies did not find an effect on T cell proliferation when EVs were added [13, 34, 82]. One of these studies, performed by Di Trapani et al., compared EVs derived from primed (the donor MSCs were pre-treated with pro-inflammatory stimuli, to invoke an anti-inflammatory response) and non-primed MSCs [82]. Although no effect on T cell proliferation was observed, there was an effect on B cell proliferation, where the suppressive effects of EVs from primed MSCs were more pronounced.

Overall, all functional assays have been used successfully with MSC-EVs, but not all MSC-EV preparations were successful in decreasing T cell proliferation. As also recently observed by Madel et al., although standardised methods and same

Table 5.1 Functional assays used with mesenchymal stromal cell-derived extracellular vesicles

Details of donor cells	Number of cell donors or preparations	Assays	References
BM, passage 3	4/1	Content of anti- and proinflammatory cytokines and apoptosis inducing molecules (4) Number of PBMCs releasing IL-1β, TNFα and IFNγ upon stimulation (1) Number of NK cells releasing TNFα and IFNγ upon stimulation (1)	Kordelas et al. [46]
BM, passage 2–3	12	T cell proliferation by PHA In vitro B cell proliferation and differentiation with CpG	Conforti et al. [17]
BM UC	3	T cell proliferation stimulated by PHA T cell proliferation induced by alloantigen-driven MLR	Pachler et al. [60]
WJ, passage 4		Macrophage polarisation assay	Willis et al. [94]
BM, passage 3		Angiogenesis - tube formation assay T cell proliferation by CD3	Teng et al. [77]
BM, passage 2–7	14	Lymphocyte proliferation by PBMC stimulation with PHA T cell proliferation assay by stimulation with CD3 and CD28 antibodies B cell proliferation stimulated by CpG NK cell proliferation stimulated by IL-2	Di Trapani et al. [82]
BM, passage 4 or 5	7 preparations from 1 donor	Macrophage polarisation assay	Pacienza et al. [61]
AT, passage 2 or 3	18	In vivo angiogenic assay Macrophage polarisation assay	Lo Sicco et al. [74]
AT, BM	3	Lymphocyte proliferation by PBMC stimulation with CD3/CD28	Gouveia de Andrade et al. [34]
BM, max passage 8		Apoptosis of ConA stimulated PBMCs and T cells Proliferation of ConA treated PBMCs	Chen et al. [13]
		T cell differentiation	
		Cytokine quantification by PBMCs (IL-1β, TNFα and TGFβ)	
		IDO activity	
AT	2 2	T cell activation by stimulating PBMCs with antiCD2/antiCD3/anti CD28	Blazquez et al. [5]
		T cell proliferation in stimulated PBMCs	
		T cell subset distribution of stimulated PBMCs	
		Intracellular IFNγ expression after PBMCs stimulation	

BM bone marrow, *UC* umbilical cord, *WJ* wharton's jelly, *AT* adipose tissue, *PBMC* peripheral blood mononuclear cells, *IL* interleukin, *TNFα* tumor necrosis factor alpha, *IFNγ* interferon gamma, *NK* natural killer, *PHA* phytohaemagglutinin, *MLR* mixed lymphocyte reaction, *ConA* Concanavalin-A, *IDO* indoleamine 2,3-dioxygenase, *TGFβ* transforming growth factor beta

MSC donors were used, not all MSC-EV preparation showed functionality [52]. This underlines again the need to use at least functional assays for each MSC-EV preparation used for clinical application and the development of potency assays will aid the standardisation.

For the treatment of OA, safety and efficacy of MSC-EVs have been shown in small animal models [86]. In addition, functionality of MSC-EVs on immunomodulation, proliferation and cartilage tissue production of chondrocytes have been shown in vitro [88]. For immunomodulation it was shown that MSC-EVs inhibited the TNFα-induced nuclear translocation of p65 subunit of nuclear factor kappa-light-chain-enhancer of activated B cells (NF-κB). In addition, by using an EdU incorporation-assay, it was observed that the proliferation of chondrocytes increased when MSC-EVs were added to in vitro cultures, both in the presence and absence of TNFα. Furthermore, the addition of MSC-EVs to the culture medium of chondrocytes in 3D-fibrin constructs increased the deposition of the cartilage components proteoglycans and type II collagen.

5.7 Summary

MSCs and MSC-EVs are emergent promising therapeutics for the treatment of focal cartilage defects and OA. Whereas the treatment of focal cartilage defects aims to have production of new cartilage tissue, for OA it relies more on the anti-inflammatory properties. Especially for MSCs, numerous in vivo studies and early phase clinical studies have been performed and/or are ongoing. While the focus of those studies lies on proof-of-concept, safety and efficacy, the modes of action and even the cell fate of implanted MSCs are still not fully elucidated. The modes of action are complex and likely multifaceted, modulating several pathological processes. As such, an array of multiple potency assays might be required.

For cartilage defect repair, potency tests that were developed for autologous chondrocyte implantation can facilitate the development of functional assays for MSCs and MSC-EVs. Furthermore, there are several functional assays aiming at immunomodulation that may be suited for OA as well. However, potency assays must measure the biological activity based on the intended biological effect and preferably reflect the clinical mode of action. Although, based on the current knowledge, these functional assays seem suitable to measure the biologic activity of MSCs and MSC-EVs for cartilage defect repair and OA, this has yet to be scientifically confirmed.

References

1. Acharya C, Adesida A, Zajac P, Mumme M, Riesle J, Martin I, Barbero A (2012) Enhanced chondrocyte proliferation and mesenchymal stromal cells chondrogenesis in coculture pellets mediate improved cartilage formation. J Cell Physiol 227:88–97. https://doi.org/10.1002/jcp.22706
2. Adachi N, Ochi M, Deie M, Ito Y (2005) Transplant of mesenchymal stem cells and hydroxyapatite ceramics to treat severe osteochondral damage after septic arthritis of the knee. J Rheumatol 32:1615–1618
3. Bartz C, Meixner M, Giesemann P, Roël G, Bulwin GC, Smink JJ (2016) An ex vivo human cartilage repair model to evaluate the potency of a cartilage cell transplant. J Transl Med 14:1–15. https://doi.org/10.1186/s12967-016-1065-8
4. Bekkers JEJ, Tsuchida AI, Van Rijen MHP, Vonk LA, Dhert WJA, Creemers LB, Saris DBF (2013) Single-stage cell-based cartilage regeneration using a combination of chondrons and mesenchymal stromal cells: comparison with microfracture. Am J Sports Med 41:2158–2166. https://doi.org/10.1177/0363546513494181
5. Blázquez R, Sánchez-Margallo FM, Álvarez V, Usón A, Marinaro F, Casado JG (2018) Fibrin glue mesh fixation combined with mesenchymal stem cells or exosomes modulates the inflammatory reaction in a murine model of incisional hernia. Acta Biomater 71:318–329. https://doi.org/10.1016/j.actbio.2018.02.014
6. Boreström C, Simonsson S, Enochson L, Bigdeli N, Brantsing C, Ellerström C, Hyllner J, Lindahl A (2014) Footprint-free human induced pluripotent stem cells from articular cartilage with redifferentiation capacity: a first step toward a clinical-grade cell source. Stem Cells Transl Med 3:433–447. https://doi.org/10.5966/sctm.2013-0138
7. Bravery CA, Carmen J, Fong T, Oprea W, Hoogendoorn KH, Woda J, Burger SR, Rowley JA, Bonyhadi ML, Van'T Hof W (2013) Potency assay development for cellular therapy products: an ISCT* review of the requirements and experiences in the industry. Cytotherapy 15:9–19. https://doi.org/10.1016/j.jcyt.2012.10.008
8. Brittberg M, Lindahl A, Nilsson A, Ohlsson C, Isaksson O, Peterson L (1994) Treatment of deep cartilage defects in the knee with autologous chondrocyte transplantation. N Engl J Med 331:889–895
9. Brittberg M, Recker D, Ilgenfritz J, Saris DBF (2018) Matrix-applied characterized autologous cultured chondrocytes versus microfracture: five-year follow-up of a prospective randomized trial. Am J Sports Med 46:1343–1351. https://doi.org/10.1177/0363546518756976
10. Callahan CM, Drake BG, Heck DA, Dittus RS (1994) Patient outcomes following tricompartmental total knee replacement: a meta-analysis. JAMA

271:1349–1357. https://doi.org/10.1001/jama.1994.03510410061034
11. Caplan AI, Dennis JE (2006) Mesenchymal stem cells as trophic mediators. J Cell Biochem 98:1076–1084. https://doi.org/10.1002/jcb.20886
12. Chahal J, Gómez-Aristizábal A, Shestopaloff K, Bhatt S, Chaboureau A, Fazio A, Chisholm J, Weston A, Chiovitti J, Keating A, Kapoor M, Ogilvie-Harris DJ, Syed KA, Gandhi R, Mahomed NN, Marshall KW, Sussman MS, Naraghi AM, Viswanathan S (2019) Bone marrow mesenchymal stromal cells in patients with osteoarthritis results in overall improvement in pain and symptoms and reduces synovial inflammation. Stem Cells Transl Med 8:746–757. https://doi.org/10.1002/sctm.18-0183
13. Chen W, Huang Y, Han J, Yu L, Li Y, Lu Z, Li H, Liu Z, Shi C, Duan F, Xiao Y (2016) Immunomodulatory effects of mesenchymal stromal cells-derived exosome. Immunol Res 64:831–840. https://doi.org/10.1007/s12026-016-8798-6
14. Chen WH, Lai MT, Wu ATH, Wu CC, Gelovani JG, Lin CT, Hung SC, Chiu WT, Deng WP (2009) In vitro stage-specific chondrogenesis of mesenchymal stem cells committed to chondrocytes. Arthritis Rheum 60:450–459. https://doi.org/10.1002/art.24265
15. Choi H, Lee RH, Bazhanov N, Oh JY, Prockop DJ (2011) Anti-inflammatory protein TSG-6 secreted by activated MSCs attenuates zymosan-induced mouse peritonitis by decreasing TLR2/NF-κB signaling in resident macrophages. Blood 118:330–338. https://doi.org/10.1182/blood-2010-12-327353
16. Civinini R, Carulli C, Matassi F, Cozzi Lepri A, Sirleo L, Innocenti M (2017) The survival of total knee arthroplasty: current data from registries on tribology: review article. HSS J 13:28–31. https://doi.org/10.1007/s11420-016-9513-9
17. Conforti A, Scarsella M, Starc N, Giorda E, Biagini S, Proia A, Carsetti R, Locatelli F, Bernardo ME (2014) Microvescicles derived from mesenchymal stromal cells are not as effective as their cellular counterpart in the ability to modulate immune responses in vitro. Stem Cells Dev 23:2591–2599. https://doi.org/10.1089/scd.2014.0091
18. Dale TP, De Castro A, Kuiper NJ, Parkinson EK, Forsyth NR (2015) Immortalisation with hTERT impacts on sulphated glycosaminoglycan secretion and immunophenotype in a variable and cell specific manner. PLoS One 10. https://doi.org/10.1371/journal.pone.0133745
19. Darling EM, Athanasiou KA (2005) Rapid phenotypic changes in passaged articular chondrocyte subpopulations. J Orthop Res 23:425–432. https://doi.org/10.1016/j.orthres.2004.08.008
20. Davatchi F, Sadeghi Abdollahi B, Mohyeddin M, Nikbin B (2016) Mesenchymal stem cell therapy for knee osteoarthritis: 5 years follow-up of three patients. Int J Rheum Dis 19:219–225. https://doi.org/10.1111/1756-185X.12670
21. Delco ML, Goodale M, Talts JF, Pownder SL, Koff MF, Miller AD, Nixon B, Bonassar LJ, Lundgren-Åkerlund E, Fortier LA (2020) Integrin α10β1-selected mesenchymal stem cells mitigate the progression of osteoarthritis in an equine Talar Impact model. Am J Sports Med 036354651989908. https://doi.org/10.1177/0363546519899087
22. Dell'Accio F, De Bari C, Luyten FP (2001) Molecular markers predictive of the capacity of expanded human articular chondrocytes to form stable cartilage in vivo. Arthritis Rheum 44:1608–1619. https://doi.org/10.1002/1529
23. Diaz-Romero J, Kürsener S, Kohl S, Nesic D (2017) S100B + A1 CELISA: a novel potency assay and screening tool for redifferentiation stimuli of human articular chondrocytes. J Cell Physiol 232:1559–1570. https://doi.org/10.1002/jcp.25682
24. Dominici M, Le Blanc K, Mueller I, Slaper-Cortenbach I, Marini F, Krause D, Deans R, Keating A, Prockop D, Horwitz E (2006) Minimal criteria for defining multipotent mesenchymal stromal cells. The International Society for Cellular Therapy position statement. Cytotherapy 8:315–317. https://doi.org/10.1080/14653240600855905
25. Emadedin M, Aghdami N, Taghiyar L, Fazeli R, Moghadasali R, Jahangir S, Farjad R, Eslaminejad MB (2012) Intra-articular injection of autologous mesenchymal stem cells in six patients with knee osteoarthritis. Arch Iran Med 15:422–428. 012157/AIM.0010
26. Emadedin M, Liastani MG, Fazeli R, Mohseni F, Moghadasali R, Mardpour S, Hosseini SE, Niknejadi M, Moeininia F, Fanni AA, Eslaminejhad RB, Dizaji AV, Labibzadeh N, Bafghi AM, Baharvand H, Aghdami N (2015) Long-term follow-up of intra-articular injection of autologous mesenchymal stem cells in patients with knee, ankle, or hip osteoarthritis. Arch Iran Med 18:336–344. 015186/AIM.003
27. Enomoto T, Akagi R, Ogawa Y, Yamaguchi S, Hoshi H, Sasaki T, Sato Y, Nakagawa R, Kimura S, Ohtori S, Sasho T (2020) Timing of intra-articular injection of synovial mesenchymal stem cells affects cartilage restoration in a partial thickness cartilage defect model in rats. Cartilage 11:122–129. https://doi.org/10.1177/1947603518786542
28. Forsberg MH, Kink JA, Hematti P, Capitini CM (2020) Mesenchymal stromal cells and exosomes: progress and challenges. Front Cell Dev Biol 8:1–11. https://doi.org/10.3389/fcell.2020.00665
29. Galipeau J (2021) Macrophages at the nexus of mesenchymal stromal cell potency: the emerging role of chemokine cooperativity. Stem Cells 1–10. https://doi.org/10.1002/stem.3380
30. Gardner OFW, Fahy N, Alini M, Stoddart MJ (2016) Differences in human mesenchymal stem cell secretomes during chondrogenic induction. Eur Cells Mater 31:221–235. https://doi.org/10.22203/eCM.v031a15
31. Gimona M, Brizzi MF, Choo ABH, Dominici M, Davidson SM, Grillari J, Hermann DM, Hill AF, de Kleijn D, Lai RC, Lai CP, Lim R, Monguió-

Tortajada M, Muraca M, Ochiya T, Ortiz LA, Toh WS, Yi YW, Witwer KW, Giebel B, Lim SK (2021) Critical considerations for the development of potency tests for therapeutic applications of mesenchymal stromal cell-derived small extracellular vesicles. Cytotherapy 000. https://doi.org/10.1016/j.jcyt.2021.01.001
32. Gimona M, Pachler K, Laner-Plamberger S, Schallmoser K, Rohde E (2017) Manufacturing of human extracellular vesicle-based therapeutics for clinical use. Int J Mol Sci 18. https://doi.org/10.3390/ijms18061190
33. Goldberg A, Mitchell K, Soans J, Kim L, Zaidi R (2017) The use of mesenchymal stem cells for cartilage repair and regeneration : a systematic review. J Orthop Surg Res 12:39. https://doi.org/10.1186/s13018-017-0534-y
34. Gouveia De Andrade AV, Bertolino G, Riewaldt J, Bieback K, Karbanová J, Odendahl M, Bornhäuser M, Schmitz M, Corbeil D, Tonn T (2015) Extracellular vesicles secreted by bone marrow- and adipose tissue-derived mesenchymal stromal cells fail to suppress lymphocyte proliferation. Stem Cells Dev 24:1374–1376. https://doi.org/10.1089/scd.2014.0563
35. Griffin TM, Scanzello CR (2019) Innate inflammation and synovial macrophages in osteoarthritis pathophysiology. Clin Exp Rheumatol 37:57–63
36. Gupta PK, Chullikana A, Rengasamy M, Shetty N, Pandey V, Agarwal V, Wagh SY, Vellotare PK, Damodaran D, Viswanathan P, Thej C, Balasubramanian S, Sen MA (2016) Efficacy and safety of adult human bone marrow-derived, cultured, pooled, allogeneic mesenchymal stromal cells (Stempeucel®): preclinical and clinical trial in osteoarthritis of the knee joint. Arthritis Res Ther 18:1–18. https://doi.org/10.1186/s13075-016-1195-7
37. Hamilton DF, Howie CR, Burnett R, Simpson AHRW, Patton JT (2015) Dealing with the predicted increase in demand for revision total knee arthroplasty: challenges, risks and opportunities. Bone Jt J 97-B:723–728. https://doi.org/10.1302/0301-620X.97B6.35185
38. Harrell CR, Markovic BS, Fellabaum C, Arsenijevic A, Volarevic V (2019) Mesenchymal stem cell-based therapy of osteoarthritis: current knowledge and future perspectives. Biomed Pharmacother 109:2318–2326
39. Hennig T, Lorenz H, Thiel A, Goetzke K, Dickhut A, Geiger F, Richter W (2007) Reduced chondrogenic potential of adipose tissue derived stromal cells correlates with an altered TGFb receptor and BMP profile and is overcome by BMP-6. J Cell Physiol 211:682–691. https://doi.org/10.1002/JCP
40. Hoburg A, Niemeyer P, Laute V, Zinser W, Becher C, Kolombe T, Fay J, Pietsch S, Kuźma T, Widuchowski W, Fickert S (2020) Matrix-associated autologous chondrocyte implantation with spheroid technology is superior to arthroscopic microfracture at 36 months regarding activities of daily living and sporting activities after treatment. Cartilage Adv. https://doi.org/10.1177/1947603519897290
41. Jaime P, García-guerrero N, Estella R, Pardo J, García-Alvarez F, Martinez-Lostao L (2017) CD56+/CD16- Natural Killer cells expressing the inflammatory protease granzyme A are enriched in synovial fluid from patients with osteoarthritis. Osteoarthr Cartil 25:1708–1718. https://doi.org/10.1016/j.joca.2017.06.007
42. Jeon JE, Vaquette C, Theodoropoulos C, Klein TJ, Hutmacher DW (2014) Multiphasic construct studied in an ectopic osteochondral defect model. J R Soc Interface 11. https://doi.org/10.1098/rsif.2014.0184
43. Jo CH, Chai JW, Jeong EC, Oh S, Shin JS, Shim H, Yoon KS (2017) Intra-articular injection of mesenchymal stem cells for the treatment of osteoarthritis of the knee: a 2-year follow-up study. Am J Sports Med 45:2774–2783. https://doi.org/10.1177/0363546517716641
44. Johnstone B, Hering TM, Caplan AI, Goldberg VM, Yoo JU (1998) In vitro Chondrogenesis of bone marrow-derived mesenchymal progenitor cells. Exp Cell Res 238:265–272
45. Jones KJ, Kelley BV, Arshi A, McAllister DR, Fabricant PD (2019) Comparative effectiveness of cartilage repair with respect to the minimal clinically important difference. Am J Sports Med 47:3284–3293. https://doi.org/10.1177/0363546518824552
46. Kordelas L, Rebmann V, Ludwig AK, Radtke S, Ruesing J, Doeppner TR, Epple M, Horn PA, Beelen DW, Giebel B (2014) MSC-derived exosomes: a novel tool to treat therapy-refractory graft-versus-host disease. Leukemia 28:970–973. https://doi.org/10.1038/leu.2014.41
47. Korpershoek JV, Rikkers M, Wallis F, Dijkstra K, Saris DBF, Vonk LA (2022) Mitochondrial transport between chondrocytes and mesenchymal stromal cells. In: International Cartilage Regeneration and Joint Preservation Society, Berlin
48. Kuroda R, Ishida K, Matsumoto T, Akisue T, Fujioka H, Mizuno K, Ohgushi H, Wakitani S, Kurosaka M (2007) Treatment of a full-thickness articular cartilage defect in the femoral condyle of an athlete with autologous bone-marrow stromal cells. Osteoarthr Cartil 15:226–231. https://doi.org/10.1016/j.joca.2006.08.008
49. Lamo-Espinosa JM, Mora G, Blanco JF, Granero-Moltó F, Núñez-Córdoba JM, López-Elío S, Andreu E, Sánchez-Guijo F, Aquerreta JD, Bondía JM, Valentí-Azcárate A, Consuelo del Cañizo M, Villarón EM, Valentí-Nin JR, Prósper F (2018) Intra-articular injection of two different doses of autologous bone marrow mesenchymal stem cells versus hyaluronic acid in the treatment of knee osteoarthritis: long-term follow up of a multicenter randomized controlled clinical trial (phase I/II). J Transl Med 16:1–5. https://doi.org/10.1186/s12967-018-1591-7
50. Lee KBL, Hui JHP, Song IC, Ardany L, Lee EH (2007) Injectable mesenchymal stem cell therapy for large cartilage defects-A Porcine model. Stem

Cells 25:2964–2971. https://doi.org/10.1634/stemcells.2006-0311
51. Li M, Luo X, Lv X, Liu V, Zhao G, Zhang X, Cao W, Wang R, Wang W (2016) In vivo human adipose-derived mesenchymal stem cell tracking after intra-articular delivery in a rat osteoarthritis model. Stem Cell Res Ther 7:1–13. https://doi.org/10.1186/s13287-016-0420-2
52. Madel RJ, Börger V, Dittrich R, Bremer M, Tertel T, Thi Phuong NN, Baba HA, Kordelas L, Buer J, Horn PA, Westendorf AM, Brandau S, Kirschning CJ, Giebel B (2020) Independent human mesenchymal stromal cell-derived extracellular vesicle preparations differentially affect symptoms in an advanced murine Graft-versus-Host-Disease model. bioRxiv. https://doi.org/10.1101/2020.12.21.423658
53. Mak J, Jablonski CL, Leonard CA, Dunn JF, Raharjo E, Matyas JR, Biernaskie J, Krawetz RJ (2016) Intra-articular injection of synovial mesenchymal stem cells improves cartilage repair in a mouse injury model. Sci Rep 6:1–12. https://doi.org/10.1038/srep23076
54. Mandelbaum BR, Browne JE, Fu F, Micheli L, Mosely JB, Erggelet C, Minas T, Peterson L (1998) Articular cartilage lesions of the knee. Am J Sports Med 26:853–861. https://doi.org/10.1177/03635465980260062201
55. Marquina M, Collado JA, Pérez-Cruz M, Fernández-Pernas P, Fafián-Labora J, Blanco FJ, Mánez R, Arufe MC, Costa C (2017) Biodistribution and immunogenicity of allogeneic mesenchymal stem cells in a Rat Model of intraarticular chondrocyte xenotransplantation. Front Immunol 8:1465. https://doi.org/10.3389/fimmu.2017.01465
56. Nejadnik H, Hui JH, Feng Choong EP, Tai B-C, Lee EH (2010) Autologous bone marrow-derived mesenchymal stem cells versus autologous chondrocyte implantation: an observational cohort study. Am J Sports Med 38:1110–1116. https://doi.org/10.1177/0363546509359067
57. Németh K, Leelahavanichkul A, Yuen PST, Mayer B, Parmelee A, Doi K, Robey PG, Leelahavanichkul K, Koller BH, Brown JM, Hu X, Jelinek I, Star RA, Mezey É (2009) Bone marrow stromal cells attenuate sepsis via prostaglandin E 2-dependent reprogramming of host macrophages to increase their interleukin-10 production. Nat Med 15:42–49. https://doi.org/10.1038/nm.1905
58. Oberbauer E, Steffenhagen C, Feichtinger G, Hildner F, Hacobian A, Danzer M, Gabriel C, Redl H, Wolbank S (2016) A Luciferase-based quick potency assay to predict chondrogenic differentiation. Tissue Eng Part C Methods 22:487–495. https://doi.org/10.1089/ten.tec.2015.0435
59. Orozco L, Munar A, Soler R, Alberca M, Soler F, Huguet M, Sentís J, Sánchez A, García-Sancho J (2013) Treatment of knee osteoarthritis with autologous mesenchymal stem cells: a pilot study. Transplantation 95:1535–1541. https://doi.org/10.1097/TP.0b013e318291a2da
60. Pachler K, Ketterl N, Desgeorges A, Dunai ZA, Laner-Plamberger S, Streif D, Strunk D, Rohde E, Gimona M (2017) An in vitro potency assay for monitoring the immunomodulatory potential of stromal cell-derived extracellular vesicles. Int J Mol Sci 18:1–11. https://doi.org/10.3390/ijms18071413
61. Pacienza N, Lee RH, Bae EH, Kim D, Ki LQ, Prockop DJ, Yannarelli G (2019) In vitro macrophage assay predicts the in vivo anti-inflammatory potential of exosomes from human mesenchymal stromal cells. Mol Ther Methods Clin Dev 13:67–76. https://doi.org/10.1016/j.omtm.2018.12.003
62. Paggi CA, Dudakovic A, Fu Y, Garces CG, Hevesi M, Galeano Garces D, Dietz AB, van Wijnen A, Karperien M (2020) Autophagy is involved in mesenchymal stem cell death in Coculture with chondrocytes. Cartilage. https://doi.org/10.1177/1947603520941227
63. Park YB, Ha CW, Kim JA, Han WJ, Rhim JH, Lee HJ, Kim KJ, Park YG, Chung JY (2017) Single-stage cell-based cartilage repair in a rabbit model: cell tracking and in vivo chondrogenesis of human umbilical cord blood-derived mesenchymal stem cells and hyaluronic acid hydrogel composite. Osteoarthr Cartil 25:570–580. https://doi.org/10.1016/j.joca.2016.10.012
64. Pearle AD, Warren RF, Rodeo SA (2005) Basic science of articular cartilage and osteoarthritis. Clin Sports Med 24:1–12. https://doi.org/10.1016/j.csm.2004.08.007
65. Pers YM, Quentin J, Feirreira R, Espinoza F, Abdellaoui N, Erkilic N, Cren M, Dufourcq-Lopez E, Pullig O, Nöth U, Jorgensen C, Louis-Plence P (2018) Injection of adipose-derived stromal cells in the knee of patients with severe osteoarthritis has a systemic effect and promotes an anti-inflammatory phenotype of circulating immune cells. Theranostics 8:5519–5528. https://doi.org/10.7150/thno.27674
66. Pigott JH, Ishihara A, Wellman ML, Russell DS, Bertone AL (2013) Investigation of the immune response to autologous, allogeneic, and xenogeneic mesenchymal stem cells after intra-articular injection in horses. Vet Immunol Immunopathol 156:99–106. https://doi.org/10.1016/j.vetimm.2013.09.003
67. Rapko S, Parker A, Mortelliti C, Duguay SJ (2007) P192 Aggrecan gene expression as a potency marker for matrix-induced autologous chondroctye implantation (MACI). Osteoarthr Cartil 15:B136. https://doi.org/10.1016/s1063-4584(07)61547-7
68. Rohde E, Pachler K, Gimona M (2019) Manufacturing and characterization of extracellular vesicles from umbilical cord–derived mesenchymal stromal cells for clinical testing. Cytotherapy 21:581–592. https://doi.org/10.1016/j.jcyt.2018.12.006
69. Saris DBF, de Windt TS, Vonk LA, Krych AJ, Terzic A (2018) Regenerative musculoskeletal care: ensuring practice implementation. Clin Pharmacol Ther 103:50–53. https://doi.org/10.1002/cpt.883

70. Saris TFF, de Windt TS, Kester EC, Vonk LA, Custers RJH, Saris DBF (2021) Five-year outcome of 1-stage cell-based cartilage repair using recycled autologous Chondrons and allogenic mesenchymal stromal cells: a first-in-human clinical trial. Am J Sports Med 49:941–947. https://doi.org/10.1177/0363546520988069
71. Satué M, Schüler C, Ginner N, Erben RG (2019) Intra-articularly injected mesenchymal stem cells promote cartilage regeneration, but do not permanently engraft in distant organs. Sci Rep 9:1–10. https://doi.org/10.1038/s41598-019-46554-5
72. Schubert T, Anders S, Neumann E, Schölmerich J, Hofstädter F, Grifka J, Libera J, Schedel J (2009) Long-term effects of chondrospheres on cartilage lesions in an autologous chondrocyte implantation model as investigated in the SCID mouse model. Int J Mol Med 23:455–460. https://doi.org/10.3892/ijmm_00000151
73. Shim G, Lee S, Han J, Kim G, Jin H, Miao W, Yi TG, Cho YK, Song SU, Oh YK (2015) Pharmacokinetics and in vivo fate of intra-articularly transplanted human bone marrow-derived clonal mesenchymal stem cells. Stem Cells Dev 24:1124–1132. https://doi.org/10.1089/scd.2014.0240
74. lo Sicco C, Reverberi D, Balbi C, Ulivi V, Principi E, Pascucci L, Becherini P, Bosco MC, Varesio L, Franzin C, Pozzobon M, Cancedda R, Tasso R (2017) Mesenchymal stem cell-derived extracellular vesicles as mediators of anti-inflammatory effects: endorsement of macrophage polarization. Stem Cells Transl Med 6:1018–1028
75. Soler R, Orozco L, Munar A, Huguet M, López R, Vives J, Coll R, Codinach M, Garcia-Lopez J (2016) Final results of a phase I–II trial using ex vivo expanded autologous mesenchymal stromal cells for the treatment of osteoarthritis of the knee confirming safety and suggesting cartilage regeneration. Knee 23:647–654. https://doi.org/10.1016/j.knee.2015.08.013
76. Tan SHS, Kwan YT, Neo WJ, Chong JY, Kuek TYJ, See JZF, Wong KL, Toh WS, Hui JHP (2021) Intra-articular injections of mesenchymal stem cells without adjuvant therapies for knee osteoarthritis: a systematic review and meta-analysis. Am J Sports Med 49:3113–3124. https://doi.org/10.1177/0363546520981704
77. Teng X, Chen L, Chen W, Yang J, Yang Z, Shen Z (2015) Mesenchymal stem cell-derived exosomes improve the microenvironment of infarcted myocardium contributing to angiogenesis and anti-inflammation. Cell Physiol Biochem 37:2415–2424. https://doi.org/10.1159/000438594
78. The Committee for Advanced Therapies (CAT) (2010) Reflection paper on in-vitro cultured chondrocyte containing products for cartilage repair of the knee Reflection paper on in-vitro cultured chondrocyte containing products for cartilage repair of the knee. EMA/CAT/CPWP/568181/2009
79. Théry C, Witwer KW, Aikawa E, Alcaraz MJ, Anderson JD, Andriantsitohaina R, Antoniou A, Arab T, Archer F, Atkin-Smith GK, Ayre DC, Bach JM, Bachurski D, Baharvand H, Balaj L, Baldacchino S, Bauer NN, Baxter AA, Bebawy M, Beckham C, Bedina Zavec A, Benmoussa A, Berardi AC, Bergese P, Bielska E, Blenkiron C, Bobis-Wozowicz S, Boilard E, Boireau W, Bongiovanni A, Borràs FE, Bosch S, Boulanger CM, Breakefield X, Breglio AM, Brennan M, Brigstock DR, Brisson A, Broekman MLD, Bromberg JF, Bryl-Górecka P, Buch S, Buck AH, Burger D, Busatto S, Buschmann D, Bussolati B, Buzás EI, Byrd JB, Camussi G, Carter DRF, Caruso S, Chamley LW, Chang YT, Chaudhuri AD, Chen C, Chen S, Cheng L, Chin AR, Clayton A, Clerici SP, Cocks A, Cocucci E, Coffey RJ, Cordeiro-da-Silva A, Couch Y, Coumans FAW, Coyle B, Crescitelli R, Criado MF, D'Souza-Schorey C, Das S, de Candia P, De Santana EF, De Wever O, del Portillo HA, Demaret T, Deville S, Devitt A, Dhondt B, Di Vizio D, Dieterich LC, Dolo V, Dominguez Rubio AP, Dominici M, Dourado MR, Driedonks TAP, Duarte FV, Duncan HM, Eichenberger RM, Ekström K, Andaloussi ELS, Elie-Caille C, Erdbrügger U, Falcón-Pérez JM, Fatima F, Fish JE, Flores-Bellver M, Försönits A, Frelet-Barrand A, Fricke F, Fuhrmann G, Gabrielsson S, Gámez-Valero A, Gardiner C, Gärtner K, Gaudin R, Gho YS, Giebel B, Gilbert C, Gimona M, Giusti I, Goberdhan DCI, Görgens A, Gorski SM, Greening DW, Gross JC, Gualerzi A, Gupta GN, Gustafson D, Handberg A, Haraszti RA, Harrison P, Hegyesi H, Hendrix A, Hill AF, Hochberg FH, Hoffmann KF, Holder B, Holthofer H, Hosseinkhani B, Hu G, Huang Y, Huber V, Hunt S, AGE I, Ikezu T, Inal JM, Isin M, Ivanova A, Jackson HK, Jacobsen S, Jay SM, Jayachandran M, Jenster G, Jiang L, Johnson SM, Jones JC, Jong A, Jovanovic-Talisman T, Jung S, Kalluri R, Ichi KS, Kaur S, Kawamura Y, Keller ET, Khamari D, Khomyakova E, Khvorova A, Kierulf P, Kim KP, Kislinger T, Klingeborn M, Klinke DJ, Kornek M, Kosanović MM, Kovács ÁF, Krämer-Albers EM, Krasemann S, Krause M, Kurochkin IV, Kusuma GD, Kuypers S, Laitinen S, Langevin SM, Languino LR, Lannigan J, Lässer C, Laurent LC, Lavieu G, Lázaro-Ibáñez E, Le Lay S, Lee MS, Lee YXF, Lemos DS, Lenassi M, Leszczynska A, ITS L, Liao K, Libregts SF, Ligeti E, Lim R, Lim SK, Linē A, Linnemannstöns K, Llorente A, Lombard CA, Lorenowicz MJ, Lörincz ÁM, Lötvall J, Lovett J, Lowry MC, Loyer X, Lu Q, Lukomska B, Lunavat TR, SLN M, Malhi H, Marcilla A, Mariani J, Mariscal J, Martens-Uzunova ES, Martin-Jaular L, Martinez MC, Martins VR, Mathieu M, Mathivanan S, Maugeri M, LK MG, MJ MV, Meckes DG, Meehan KL, Mertens I, Minciacchi VR, Möller A, Møller Jørgensen M, Morales-Kastresana A, Morhayim J, Mullier F, Muraca M, Musante L, Mussack V, Muth DC, Myburgh KH, Najrana T, Nawaz M, Nazarenko I, Nejsum P, Neri C, Neri T,

Nieuwland R, Nimrichter L, Nolan JP, Nolte-'t Hoen ENM, Noren Hooten N, O'Driscoll L, O'Grady T, O'Loghlen A, Ochiya T, Olivier M, Ortiz A, Ortiz LA, Osteikoetxea X, Ostegaard O, Ostrowski M, Park J, Pegtel DM, Peinado H, Perut F, Pfaffl MW, Phinney DG, Pieters BCH, Pink RC, Pisetsky DS, Pogge von Strandmann E, Polakovicova I, Poon IKH, Powell BH, Prada I, Pulliam L, Quesenberry P, Radeghieri A, Raffai RL, Raimondo S, Rak J, Ramirez MI, Raposo G, Rayyan MS, Regev-Rudzki N, Ricklefs FL, Robbins PD, Roberts DD, Rodrigues SC, Rohde E, Rome S, Rouschop KMA, Rughetti A, Russell AE, Saá P, Sahoo S, Salas-Huenuleo E, Sánchez C, Saugstad JA, Saul MJ, Schiffelers RM, Schneider R, Schøyen TH, Scott A, Shahaj E, Sharma S, Shatnyeva O, Shekari F, Shelke GV, Shetty AK, Shiba K, PRM S, Silva AM, Skowronek A, Snyder OL, Soares RP, Sódar BW, Soekmadji C, Sotillo J, Stahl PD, Stoorvogel W, Stott SL, Strasser EF, Swift S, Tahara H, Tewari M, Timms K, Tiwari S, Tixeira R, Tkach M, Toh WS, Tomasini R, Torrecilhas AC, Tosar JP, Toxavidis V, Urbanelli L, Vader P, van Balkom BWM, Van der Grein SG, van Deun J, van Herwijnen MJC, Van Keuren-Jensen K, van Niel G, van Royen ME, van Wijnen AJ, Vasconcelos MH, Vechetti IJ, Veit TD, Vella LJ, Velot É, Verweij FJ, Vestad B, Viñas JL, Visnovitz T, Vukman KV, Wahlgren J, Watson DC, Wauben MHM, Weaver A, Webber JP, Weber V, Wehman AM, Weiss DJ, Welsh JA, Wendt S, Wheelock AM, Wiener Z, Witte L, Wolfram J, Xagorari A, Xander P, Xu J, Yan X, Yáñez-Mó M, Yin H, Yuana Y, Zappulli V, Zarubova J, Žėkas V, Ye ZJ, Zhao Z, Zheng L, Zheutlin AR, Zickler AM, Zimmermann P, Zivkovic AM, Zocco D, Zuba-Surma EK (2018) Minimal information for studies of extracellular vesicles 2018 (MISEV2018): a position statement of the International Society for Extracellular Vesicles and update of the MISEV2014 guidelines. J Extracell Vesicles 7. https://doi.org/10.1080/20013078.2018.1535750

80. Toupet K, Maumus M, Luz-Crawford P, Lombardo E, Lopez-Belmonte J, Van Lent P, Garin MI, Van Den Berg W, Dalemans W, Jorgensen C, Noël D (2015) Survival and biodistribution of xenogenic adipose mesenchymal stem cells is not affected by the degree of inflammation in arthritis. PLoS One 10:1–13. https://doi.org/10.1371/journal.pone.0114962
81. Toupet K, Maumus M, Peyrafitte JA, Bourin P, Van Lent PLEM, Ferreira R, Orsetti B, Pirot N, Casteilla L, Jorgensen C, Noël D (2013) Long-term detection of human adipose-derived mesenchymal stem cells after intraarticular injection in SCID mice. Arthritis Rheum 65:1786–1794. https://doi.org/10.1002/art.37960
82. Di Trapani M, Bassi G, Midolo M, Gatti A, Kamga PT, Cassaro A, Carusone R, Adamo A, Krampera M (2016) Differential and transferable modulatory effects of mesenchymal stromal cell-derived extracellular vesicles on T, B and NK cell functions. Sci Rep 6:1–13. https://doi.org/10.1038/srep24120
83. Vainieri ML, Lolli A, Kops N, D'Atri D, Eglin D, Yayon A, Alini M, Grad S, Sivasubramaniyan K, van Osch GJVM (2020) Evaluation of biomimetic hyaluronic-based hydrogels with enhanced endogenous cell recruitment and cartilage matrix formation. Acta Biomater 101:293–303. https://doi.org/10.1016/j.actbio.2019.11.015
84. Valencia J, Blanco B, Yáñez R, Vázquez M, Herrero Sánchez C, Fernández-García M, Rodríguez Serrano C, Pescador D, Blanco JF, Hernando-Rodríguez M, Sánchez-Guijo F, Lamana ML, Segovia JC, Vicente Á, Del Cañizo C, Zapata AG (2016) Comparative analysis of the immunomodulatory capacities of human bone marrow– and adipose tissue–derived mesenchymal stromal cells from the same donor. Cytotherapy 18:1297–1311. https://doi.org/10.1016/j.jcyt.2016.07.006
85. Vanlauwe J, Saris DBF, Victor J, Almqvist KF, Bellemans J (2011) Five-year outcome of characterized chondrocyte implantation versus microfracture for symptomatic cartilage defects of the knee early treatment matters. Am J Sports Med 39:2566–2574. https://doi.org/10.1177/0363546511422220
86. Varderidou-Minasian S, Lorenowicz MJ (2020) Mesenchymal stromal/stem cell-derived extracellular vesicles in tissue repair: challenges and opportunities. Theranostics 10:5979–5997. https://doi.org/10.7150/thno.40122
87. Vega A, Martín-Ferrero MA, Del CF, Alberca M, García V, Munar A, Orozco L, Soler R, Fuertes JJ, Huguet M, Sánchez A, García-Sancho J (2015) Treatment of knee osteoarthritis with allogeneic bone marrow mesenchymal stem cells: a randomized controlled trial. Transplantation 99:1681–1690. https://doi.org/10.1097/TP.0000000000000678
88. Vonk LA, van Dooremalen SFJ, Liv N, Klumperman J, Coffer PJ, Saris DBF, Lorenowicz MJ (2018) Mesenchymal stromal/stem cell-derived extracellular vesicles promote human cartilage regeneration in vitro. Theranostics 8:906–920. https://doi.org/10.7150/thno.20746
89. Vonk LA, De Windt TS, Slaper-Cortenbach ICM, Saris DBF (2015) Autologous, allogeneic, induced pluripotent stem cell or a combination stem cell therapy? Where are we headed in cartilage repair and why: a concise review. Stem Cell Res Ther 6:94
90. De Vries-Van Melle ML, Narcisi R, Kops N, Koevoet WJLM, Bos PK, Murphy JM, Verhaar JAN, Van Der Kraan PM, Van Osch GJVM (2014) Chondrogenesis of mesenchymal stem cells in an osteochondral environment is mediated by the subchondral bone. Tissue Eng Part A 20:23–33. https://doi.org/10.1089/ten.tea.2013.0080
91. Wakitani S, Mitsuoka T, Nakamura N, Toritsuka Y, Nakamura Y, Horibe S (2004) Autologous bone marrow stromal cell transplantation for repair of full-thickness articular cartilage defects in human patellae: two case reports. Cell Transplant 13:595–600. https://doi.org/10.3727/000000004783983747

92. Wang R, Maimaitijuma T, Ma Y, Jiao Y, Cao Y (2021) Mitochondrial transfer from bone-marrow-derived mesenchymal stromal cells to chondrocytes protects against cartilage degenerative mitochondrial dysfunction in rats chondrocytes. Chin Med J 134:212–218. https://doi.org/10.1097/CM9.0000000000001057
93. Weber AE, Bolia IK, Trasolini NA (2021) Biological strategies for osteoarthritis: from early diagnosis to treatment. Int Orthop 45:335–344. https://doi.org/10.1007/s00264-020-04838-w
94. Willis GR, Fernandez-Gonzalez A, Anastas J, Vitali SH, Liu X, Ericsson M, Kwong A, Mitsialis SA, Kourembanas S (2018) Mesenchymal stromal cell exosomes ameliorate experimental bronchopulmonary dysplasia and restore lung function through macrophage immunomodulation. Am J Respir Crit Care Med 197:104–116. https://doi.org/10.1164/rccm.201705-0925OC
95. de Windt TS, Hendriks JAA, Zhao X, Vonk LA, Creemers LB, Dhert WJA, Randolph MA, Saris DBF (2014) Concise review: unraveling stem cell cocultures in regenerative medicine: which cell interactions steer cartilage regeneration and how? Stem Cells Transl Med 3:723–733. https://doi.org/10.5966/sctm.2013-0207
96. De Windt TS, Saris DBF, Slaper-Cortenbach ICM, Van Rijen MHP, Gawlitta D, Creemers LB, De Weger RA, Dhert WJA, Vonk LA (2015) Direct cell–cell contact with chondrocytes is a key mechanism in multipotent mesenchymal stromal cell-mediated chondrogenesis. Tissue Eng Part A 21:2536–2547. https://doi.org/10.1089/ten.tea.2014.0673
97. de Windt TS, Vonk LA, Slaper-Cortenbach ICM, van den Broek MPH, Nizak R, van Rijen MHPP, De Weger RA, Dhert WJA, Saris DBF (2017) Allogeneic mesenchymal stem cells stimulate cartilage regeneration and are safe for single-stage cartilage repair in humans upon mixture with recycled autologous Chondrons. Stem Cells 35:256–264. https://doi.org/10.1002/stem.2475
98. de Windt TS, Vonk LA, Slaper-Cortenbach ICM, Nizak R, van Rijen MHP, Saris DBF (2017) Allogeneic MSCs and recycled autologous Chondrons mixed in a one-stage cartilage cell transplantion: a first-in-man trial in 35 patients. Stem Cells 35:1984–1993. https://doi.org/10.1002/stem.2657
99. Witwer KW, Van Balkom BWM, Bruno S, Choo A, Dominici M, Gimona M, Hill AF, De Kleijn D, Koh M, Lai RC, Mitsialis SA, Ortiz LA, Rohde E, Asada T, Toh WS, Weiss DJ, Zheng L, Giebel B, Lim SK (2019) Defining mesenchymal stromal cell (MSC)-derived small extracellular vesicles for therapeutic applications. J Extracell Vesicles 8. https://doi.org/10.1080/20013078.2019.1609206
100. Witwer KW, Buzás EI, Bemis LT, Bora A, Lässer C, Lötvall J, Nolte-'t Hoen EN, Piper MG, Sivaraman S, Skog J, Théry C, Wauben MH, Hochberg F (2013) Standardization of sample collection, isolation and analysis methods in extracellular vesicle research. J Extracell Vesicles 2. https://doi.org/10.3402/jev.v2i0.20360
101. Witwer KW, Théry C (2019) Extracellular vesicles or exosomes? On primacy, precision, and popularity influencing a choice of nomenclature. J Extracell Vesicles 8. https://doi.org/10.1080/20013078.2019.1648167
102. de Wolf C, van de Bovenkamp M, Hoefnagel M (2017) Regulatory perspective on in vitro potency assays for human mesenchymal stromal cells used in immunotherapy. Cytotherapy 19:784–797. https://doi.org/10.1016/j.jcyt.2017.03.076
103. Woodell-May JE, Sommerfeld SD (2020) Role of inflammation and the immune system in the progression of osteoarthritis. J Orthop Res 38:253–257. https://doi.org/10.1002/jor.24457
104. Wu L, Leijten J, Van Blitterswijk CA, Karperien M (2013) Fibroblast growth factor-1 is a mesenchymal stromal cell-secreted factor stimulating proliferation of osteoarthritic chondrocytes in co-culture. Stem Cells Dev 22:2356–2367. https://doi.org/10.1089/scd.2013.0118
105. Wu L, Leijten JCH, Georgi N, Post JN, van Blitterswijk CA, Karperien M (2011) Trophic effects of mesenchymal stem cells increase chondrocyte proliferation and matrix formation. Tissue Eng Part A 17:1425–1436. https://doi.org/10.1089/ten.tea.2010.0517
106. Wu L, Prins H-JJ, Helder MN, van Blitterswijk CA, Karperien M (2012) Trophic effects of mesenchymal stem cells in chondrocyte co-cultures are independent of culture conditions and cell sources. Tissue Eng Part A 18:1542–1551. https://doi.org/10.1089/ten.tea.2011.0715
107. Xia H, Liang C, Luo P, Huang J, He J, Wang Z, Cao X, Peng C, Wu S (2018) Pericellular collagen i coating for enhanced homing and chondrogenic differentiation of mesenchymal stem cells in direct intra-articular injection. Stem Cell Res Ther 9:1–12. https://doi.org/10.1186/s13287-018-0916-z
108. Yasui Y, Hart DA, Sugita N, Chijimatsu R, Koizumi K, Ando W, Moriguchi Y, Shimomura K, Myoui A, Yoshikawa H, Nakamura N (2018) Time-dependent recovery of human synovial membrane mesenchymal stem cell function after high-dose steroid therapy: case report and laboratory study. Am J Sports Med 46:695–701. https://doi.org/10.1177/0363546517741307
109. Zha K, Sun Z, Yang Y, Chen M, Gao C, Fu L, Li H, Sui X, Guo Q, Liu S (2021) Recent developed strategies for enhancing chondrogenic differentiation of MSC: impact on MSC-based therapy for cartilage regeneration. Stem Cells Int 2021:1–15. https://doi.org/10.1155/2021/8830834
110. Zhang H, Lin C, Zeng C, Wang Z, Wang H, Lu J, Liu X, Shao Y, Zhao C, Pan J, Xu S, Zhang Y, Xie D, Cai D, Bai X (2018) Synovial macrophage M1 polarisation exacerbates experimental osteoarthritis partially through R-spondin-2. Ann Rheum Dis 77:1524–1534. https://doi.org/10.1136/annrheumdis-2018-213450

Advanced Technologies for Potency Assay Measurement

6

Raghavan Chinnadurai

6.1 Introduction

Cell therapy is an emerging treatment platform of regenerative medicine that shows promising clinical efficacy. In the cell therapy approach, regenerative and immunomodulatory cells are isolated from human tissue/organs and expanded in a sterile manufacturing facility according to current good manufacturing practice (cGMP) regulations. Subsequently, these cells represent a medicinal product with the aim of mitigating inflammation, tissue injury and degeneration to improve healing. Crucial for application, cell products, like any other chemical/biological medication, need to be well-characterised in the cell manufacturing facilities and conform to regulatory approval criteria before infusion into the patients. Characterisation of advanced therapy medicinal products (ATMP) derived in cell manufacturing laboratories aims to ensure safety and promote efficacy/potency in patients upon infusion. Mesenchymal Stromal Cells (MSCs) are the leading cell therapy candidate in clinical trials worldwide [26]. Of importance, MSCs have been approved for the treatment of complex perianal fistulas in patients with Crohn's Disease, acute Graft versus Host Disease (GvHD) and critical limb ischemia associated with Buerger's disease [17, 68]. Early phase clinical trials have demonstrated that MSCs display an excellent safety profile and are well tolerated in the patients [44]. Despite initial enthusiasm and regulatory approval for the above-mentioned clinical conditions, MSCs have also exhibited contradictory efficacy in later-phase clinical trials [51]. There are multiple reasons for this discrepancy that include variability in patients, disease severity, involvement of other treatment regimens, poorly understood mechanism of action of MSCs and also variability of MSC therapeutics [51]. Despite these challenges, potency assays for infused MSCs need to be defined in order to obtain more consistent efficacy and clinical benefit [9]. Early phase clinical trials do not require potency assays that predict efficacy of MSCs since the primary endpoint of the early phase clinical trials is safety. In advanced-phase clinical trials and for marketing approval, regulatory authorities require the deployment of potency assays that quantitatively measure functional attributes that encompass mechanism of action and potentially predict efficacy as part of the release criteria of MSC.

6.2 Variability of MSC Therapeutics

Variability of MSC therapeutics is the biggest confounder for achieving sustainable clinical efficacy and potency measurements. These

R. Chinnadurai (✉)
Department of Biomedical Sciences, Mercer University School of Medicine, Savannah, GA, USA
e-mail: chinnadurai_r@mercer.edu

J. S. Burns (ed.), *Potency Assays for Advanced Stem Cell Therapy Medicinal Products*, Advances in Experimental Medicine and Biology 1420, https://doi.org/10.1007/978-3-031-30040-0_6

variabilities originate from the cell source to methodologies of cell manufacturing and practice [22]. MSCs were initially considered as an immune privileged cell type, but subsequent animal model studies have demonstrated that allogeneic MSCs can be immune-rejected [5]. Thus, autologous MSCs could be a best-choice epitome for personalised cell therapy for chronic disorders. However, the use of autologous MSCs in cell therapy is not always feasible. Although studies have shown that autologous whole-cell MSC are fit and can be used in cell therapy for certain chronic clinical conditions, they are not a pragmatic option for acute ailments. For acute clinical disorders, autologous cell therapy is not pragmatically feasible due to the typical need for a minimal timeline of two weeks for MSC isolation and expansion to a clinical dose. When very prompt timing of cell therapy is crucial for mitigating and reversing acute disorders, readily available allogeneic random-donor MSCs ('off-the-shelf') or MSC derivatives (e.g. matrix or extracellular vesicles) present the most feasible option. In allogeneic therapy, choice of the donor to isolate MSCs is random and expansion of MSCs from multiple donors may be required to make enough cell doses for multidose treatment strategies. Thus, it becomes very challenging to define potency assays that measure consistent and equal functionality of these cell therapeutics from independent donors. The family of cell types designated 'MSCs' can be isolated from various tissue sources including bone marrow, adipose tissue, umbilical cord and placenta [33]. Although these MSC populations from different tissue sources share mesenchymal phenotype similarity, they diverge in their more detailed, granular characteristics [62]. Thus, MSC populations isolated from differential tissue source need to be analysed rigorously to provide accurately informed potency assays that may predict their functionality in vivo. Differential methodologies of cell manufacturing and delivery also contribute to the variability of MSC therapeutics. Random donor MSCs that are readily available ('off-the-shelf') in the cryopreserved state would represent a feasible cell therapy option for clinical facilities that do not have cell manufacturing capacity. For such cell therapy practice, MSCs would be expanded and cryopreserved at the cell manufacturing facilities, and subsequently transported to the hospitals as a cryopreserved cellular product, where they would be promptly thawed and infused in to the patients within a few hours post-thaw. Although such a protocol would seem very feasible and viable, several studies have demonstrated that freshly thawed MSCs from cryopreservation are dysfunctional and not equivalent to the actively growing counterparts [55, 85]. Hence, potency assays may need to consider logistic factors such as transportation between manufacture and point of care [77] and the impact of freeze-thawing on MSC therapeutic quality and functionality.

6.3 General Considerations for MSC Release Criteria

United States Food and Drug Administration (FDA) and European Medicines Agency (EMA) classify MSC as a more than minimally manipulated Cellular and Gene Therapy (CGT) product for which an Investigational New Drug (IND) or Clinical Trial Application (CTA) is required for human clinical trials. Investigational cellular products are regulated with a well-defined cell manufacturing procedures and characterisation assays [53]. These assays ensure that the cellular product is safe and functional to infuse into humans. The three major determinants of the release criteria are identity, viability and sterility in early phase clinical trials [53]. The International Society for Cell and Gene Therapy (ISCT) has recommended minimal criteria to define bone marrowBone marrow derived MSC identity with (1) adherence to plastic culture plates, (2) trilineage (adipocyte, osteocyte and chondrocyte-Chondrocytes) differentiation potential and (3) combination of positive (CD105, CD73, CD90) and negative (CD45, CD34, CD14, CD11b, CD79a, CD19, HLA-DR) cell surface marker expression [19]. Widely accepted cell viability release criteria for fresh and frozen-thawed (cryopreserved) MSC products are above 90% and 70%, respectively. Trypan Blue exclusion assays are performed at the cell manufacturing facility to define the cell viability. Moreover, flow cytometry-based technologies that capture early apoptotic cells using Annexin V and Propidium Iodide staining allow the percentage quantification of live, apoptotic and necrotic cells. Sterility of the cell product is currently evaluated at the cell manufacturing stage using microbial culture analysis, automated microbial detection systems and Polymerase Chain Reaction (PCR) assays [65]. Endotoxin levels in MSC products using a limulus amoebocyte lysate (LAL) assay are also worthy of inclusion as part of the release criteria [27], since bacterial endotoxins can influence cell proliferation and differentiation [57, 58]. Malignant transformation or tumorigenesis of infused MSCs has never been reported in a patient. Analysis of autopsy tissue from patients who earlier received MSC therapy has demonstrated a lack of long term MSC engraftment, ectopic tissue formation or tumorigenic transformation [79]. Long-term MSC expansion may lead to cellular senescence and dysfunctionality, but not malignant transformation [14]. Some studies have demonstrated that cell culture expanded MSCs develop genomic mutations and aneuploidy but did not undergo malignant transformation [75, 81]. G-banding karyotype analysis, comparative genomic hybridisation (CGH) assay, fluorescence in situ hybridisation (FISH) are recommended assays to assess for chromosomal abnormalities [6] although a normal karyotype is not necessarily an incontrovertible indicator that the cells lack tumorigenic potential [11]. Another consideration in the product release criteria is the purity of the MSC product released as a cellular pharmaceutical. Purity measurements can be evaluated along with the identity characterisation as part of the release criteria. MSC products are evaluated for the percentage of total cells expressing positive and negative phenotypic markers which minimises the contamination from other cell populations such as hematopoietic cells and endothelial cells in the final preparations. All these considerations for basic release criteria of MSCs are for early phase clinical trials which assures safety while the advanced phase clinical trials/marketing approvals require both safety and potency analysis of the cellular products.

6.4 Key Aspects of Potency Assays

Potency assays should reflect the putative mechanism of action (MoA) of the MSC product that is being used in a given clinical condition. Hence these assays quantify the attributes of specific cell products and their functionality that are assumed to confer clinical benefit. The beneficial applicability of MSC is being explored for various clinical conditions although the precise mechanism of action that provides clinical benefit is not fully understood and will vary from one clinical condition to another. Hence, potency assays need to be developed that define the product characteristics suitable for the particular clinical condition being used. Considering a lack of understanding the MoA responsible for the MSCs' clinical benefit, moreover likelihood that a number of attributes are involved, an assay that measures a single property or characteristic of MSCs and its functionality may not adequately represent the potency of the product. Alternatively, a combination of bioassays and analytical assays collectively called 'assay matrix' can be used to measure more than one property of MSCs thereby defining the potency of MSCs more adequately (Fig. 6.1) [25]. For instance, MSC's ability to secrete CXCL5, IL-8, VEGF coupled with a functional angiogenic assay were considered as a surrogate assay matrix that defined the angiogenic potency of MSCs [47]. The ISCT has recommended that at least three analytic methods should be considered in matrix assay approaches: (i) quantitative RNA analysis of selected gene products, (ii) flow cytometry analysis of functionally relevant surface markers and (iii) Secretome analysis of bioactive molecules [25]. Another aspect of potency measurement is the inclusion of cellular reference standards in assays with direct comparison to the potency of test product. It has been proposed to utilise universal cellular reference standards in the potency assays [78]. However, an MSC-derived therapeutic

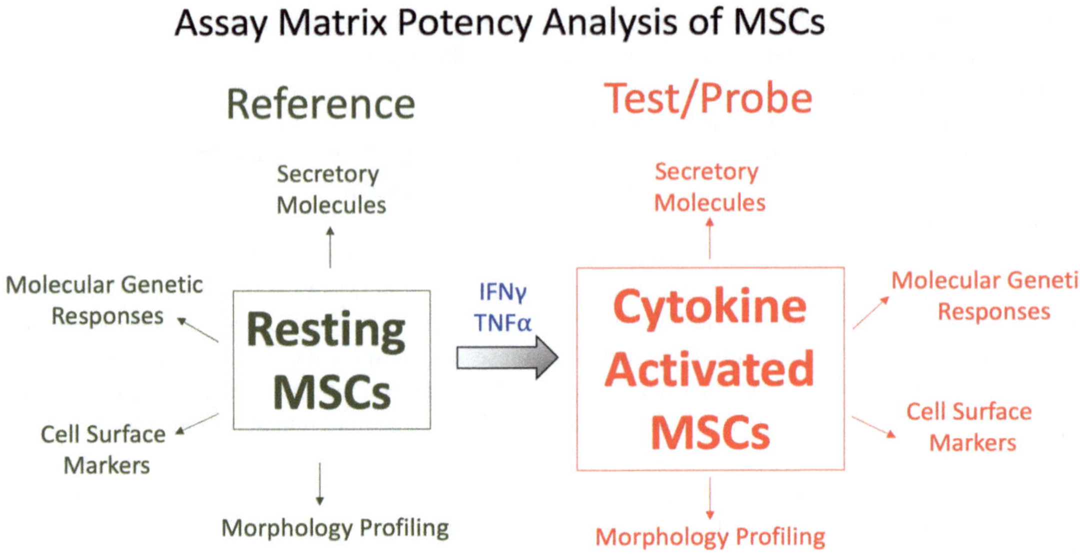

Fig. 6.1 Assay matrix strategy in defining the potency of human MSCs. In this strategy, resting and cytokine activated MSCs are compared for more than one effector pathways, collectively known as the assay matrix. Resting MSCs may serve as the reference standard, avoiding the need for a universal cellular reference/standard

effect may involve multiple effector molecules and pathways that may synergistically modulate inflammation and tissue injury via overlapping and non-overlapping mechanisms which are yet to be understood. Utilisation of cellular reference standards in defining the potency of autologous cell therapeutics further complicates the interpretation since the autologous MSC populations are distinct and patient-specific. Thus, identification of a universal cellular reference standards that meets this criterion is a challenge. Alternatively, another approach is the utilisation of internal reference controls to serve as the cellular reference standards. MSCs are responsive to environmental cues that activate the expression of several effector molecules important for tissue regeneration and immune modulation [80]. It is entirely possible to utilise this phenomenon of physiological shift from naïve to activated state in deploying reference standards. In this scenario, MSCs that are in naïve state are considered as a cellular reference standard and are compared with a cytokine activated counterpart for which effector molecules are modulated. Enumeration and quantification of the effector molecules between naïve and cytokine activated MSCs obviate the need of universal cellular reference standards [41]. Another strategy for defining the cellular reference standard is the inclusion of the inactive form of the cellular product that is being investigated. For instance, active MSC products can be compared with the heat inactivated counterparts. Thus, the quantitative difference in the expression of effector molecules between active and heat inactivated counterparts defines the potency.

6.5 Potency Assay Technologies

6.5.1 Immunological Assays

Assays to assess the immunosuppressive properties of MSCs have been widely considered as a surrogate measure of potency [41]. In these assays, MSCs ability to inhibit the proliferation of T cells is quantified predominantly by flow cytometry. Random donor derived Peripheral Blood Mononuclear Cells (PBMCs) serve as the source of T cells in these assays. T cells in the PBMCs may be activated by several methods such as classic Mixed Lymphocyte Reaction (MLR), engagement of T cell receptor (TCR) complex with anti-CD3 and anti-CD28 antibodies and TCR independent activation with PHA or

PMA and Ionomycin. In all these assays, T cell proliferation was measured by evaluating the dilution of proliferation dyes such as CFSE dye, expression of proliferation marker Ki67 and incorporation of nucleoside analogue bromodeoxyuridine (BRDU). The difference in the percentage of T cell proliferation between absence and presence of MSCs was calculated as the potency value of the MSCs. Unfractionated PBMCs contain lymphomyeloid populations and this complexity of heterogenous immune cell populations remains useful in potency assays, since infused MSCs encounter a similar multicellular environment in vivo. However, the lymphomyeloid populations can vary for each human subject and using the unfractionated PBMCs from a random donor in the potency assays would complicate the reproducibility of these assays from one PBMC donor to the next. One approach to minimise this complexity is to use a purified lymphoid population in the immunosuppressive potency assays. In these assays, purified T cells can be used in place of unfractionated PBMCs which minimise the issue related to the reproducibility in potency assays [24]. Other confounders of the potency assays are PBMC culture duration with MSCs and total reaction volume of the assay, which need to be considered and standardised in developing in vitro assays [7]. In a flow cytometry-based immunoassay, the expression of IFNγ-induced intracellular enzyme IDO1 and cell surface protein PD-L1 were evaluated within MSCs. These biomarker expressions were correlated with MSC-mediated suppression of T cell proliferation [28]. Thus, analysing IFNγ-induced IDO1 and PDL1 on MSCs could serve as a rapid potency assay for release criteria. MSC's interaction with macrophages are also being considered in potency analysis. One such example is the MSCs ability to inhibit LPS-induced TNFα expression on monocytes determined by intracellular flow cytometry and this assay system can be used as a surrogate measure of potency [67]. Similarly, MSCs can polarise macrophages from classic proinflammatory M1 into an immunosuppressive M2 subtype [23]. Although further studies are required, these metrics on macrophages are likely to become useful measures of potency.

6.5.2 Genomic Assays

MSCs possess a plurality of effector molecules that are important for immunomodulation and regeneration and thus quantifying their expression at RNA levels will likely provide a surrogate measure of potency. Next generation sequencing such as bulk *RNAseq* or single cell RNA sequencing provide information about the total transcriptome of individual MSC populations. These assays are expensive and their routine usage in small scale cell manufacturing laboratories are difficult. Nevertheless, these investigations identify target genes that can influence mechanisms of action and potentially serve as potency assay biomarkers. Transcriptome analysis of resting MSC also yielded candidate genes that may potentially predict their function and be applicable in potency assays. For example, it has been shown that gene expression levels of *TWIST1* predict intrinsic differences in the functionality of MSCs from independent donors. *TWIST1* expression also predicts MSC potency both in vitro and in vivo, and can be incorporated in potency testing [8]. Similarly, TNFα-stimulated gene 6 (*TSG-6/TNFAIP6*) expression predicted MSCs efficacy in sterile inflammation models for corneal injury, sterile peritonitis, and bleomycin-induced lung injury demonstrating a broad applicability for potency assays [46, 64]. Instead of quantifying a single effector gene expression, specific sets of genes that are significant for MSCs' function may also be used as a surrogate measure of potency [25]. For example, genomic cluster analysis of hMSC stimulated with osteogenic medium in vitro identified that a signature pattern of expression of 5 genes (*ALPL, COL1A2, DCN, ELN* and *RUNX2*), but not individual genes, correlated well with subsequent MSC bone forming osteogenic potential [56]. Reproduction and enhancement of the study revealed *TGFB2* expression was a highly indicative biomarker within an osteogenic potency assay gene cluster [59], reflecting that potency assays may be best regarded as continuously open to improvement and maturation. In another approach, MSC's fitness to respond to host inflammatory cues (cytokines and chemokines) that evoked effector molecules of significance to immunomodulation and regeneration were quantified at the RNA level to define potency. Activated PBMCs produce cytokines and chemokines influencing immunomodulatory genes on MSCs. Thus, MSC's fitness to upregulate immunomodulatory and regenerative genes upon coculture with the inflammatory cues produced by activated PBMCs could serve as the potency assay. However, there are challenges in using this assay system for reproducible analysis. Activated PBMCs could produce varying quantities of cytokines and chemokines that differ from donor to donor and hence this potency assay system may generate assay variabilities independent of MSC attributes. Alternatively, from a reductionist perspective recombinant cytokine or chemokine with a quantitative bioactivity could be used in the potency assay system. IFNγ is one such pro-inflammatory cytokine evoking immunosuppressive properties on MSCs. Thus, measuring MSC's responsiveness to IFNγ by quantitative PCR could be considered a surrogate measure of potency [41]. One notable such example is Indoleamine 2,3 Dioxygenase (IDO/IDO1) that is robustly induced in MSCs by IFNγ and plays a significant role on the immunosuppressive properties of MSCs. IDO1 catabolises the conversion of tryptophan into kynurenine which induces apoptosis of T cells. Blockade of IDO1 activity on MSCs completely abolished their in vitro suppressive properties on T cell proliferation [52]. The magnitude of *IDO1* gene induction by IFNγ also correlated with MSCs' immunosuppressive properties [21]. The MSC Committee of the ISCT has suggested that a standardized immune assay quantitative measurement of IFNγ-induced IDO and/or analysis of its transcriptional modulation in MSCs, could be deployed in predictive potency analysis [41]. As an alternative strategy, an ISCT guidance article has also recommended that the IFNγ-stimulated array of genes significant to MSC immunobiology and regenerative biology could be investigated in a small-scale quantitative RNA-based array as an 'assay matrix' to define their potency [25]. One such example is the Fluidigm™ nanoscale quantitative PCR array in which samples and targets can be probed in a 48X48 or 96X96 chip format. Utilisation of such platform has identified that IFNγ upregulates genes such as *IDO1, CXCL9, CXCL10, CXCL11, CIITA, HLADR, PD-L1* and *ICAM-1* [15]. In this matrix assay system, expression levels of these genes in the resting MSCs (prior to IFNγ stimulation) is compared with IFNγ stimulated counterparts. This approach obviates the need of universal reference standards/rulers in potency assays. The data from resting MSCs can serve as cellular reference standards/rulers and the magnitude of the differences versus stimulated MSC represent potency values. MicroRNA expression in MSCs are modulated by Toll-Like receptor molecules that regulate MSCs' immunomodulatory functions [2]. In addition, signatures of microRNA expression and their critical significance in regulation of differentiation, paracrine activity, survival and migration have been defined in MSCs [16]. Expression, regulation and functionality of microRNA in MSCs can be further correlated to identify their utility in predictive potency assays. Altogether, genomic assays focusing on selective sets of gene expression are not only sensitive, reliable and cost-effective but also quantitate the molecular fitness of the cells and thus can be incorporated in the potency assay matrix analysis.

6.5.3 Secretome Assays

MSCs secrete bioactive molecules such as cytokines and chemokines in the resting stage and upon interaction with host inflammatory cues. MSCs' capacity to secrete these bioactive molecules can be quantified and used as a surrogate measure of potency. Human bone marrow derived MSC secretion of soluble TNF receptor-1 has been used as a surrogate measure of potency for the product release criterion in a phase 3 trial of an MSC product efficacious for first-line therapy after initial steroid failure in acute graft versus host disease patients [42]. Similarly, Prostaglandin E2 secretion in human MSC cultures predicted their in vivo therapeutic potential, hence quantitation of its secretion could be used in prospective potency assays [39]. Seeking to understand how intravenously infused MSC might confer tissue repair benefits without significant engraftment, the observation that MSC secretion of TNF-α-induced protein 6 (TNAIP6 or TSG-6) could enhance regenerative efficacy in animal models presented a valuable surrogate potency assay model [45]. Although these strategies aimed to correlate functionality of MSCs with their ability to secrete a single bioactive molecule, the ISCT has recommended an 'assay matrix' approach that captures an array of cytokines and chemokines secreted by MSCs as part of a potency assay. This approach is possible with multiplexing technologies such as Luminex™ xMAP technology or BD™ Cytometric Bead Array (CBA). Using Luminex™ xMAP technology, a large panel of cytokines and chemokines (secretome) were analysed in independent cultures and cocultures of MSCs and activated PBMCs. In this assay system, the secretome of resting MSCs provided the reference values for their corresponding counterparts upon MSC interaction with activated PBMCs. This analysis has identified that MSC-mediated suppression of T cell proliferation was associated with unique secretome modulation. MSC mediated suppression of T cell proliferation was correlated with the downregulation of TNFα, IFNγ, IL-13, IL-5, IL-2R, CCL3 and CCL4, and upregulation of VEGF, IFNα, CXCL10, GCSF, CXCL9, IL-7 and CCL2 bioactive molecules. This analysis also identified that MSC and PBMC interactions were bidirectional, since bioactive molecules were modulated upon mutual interaction between both cell populations [15]. The assay matrix approach that captures the secretome of MSC's interaction with PBMCs is also useful in identifying the fitness of MSCs; an important consideration since prolonged cell expansion in culture expansion causes replicative exhaustion/senescence of MSCs. Secretome analysis has identified that senescent MSCs are significantly different to their early passage counterparts in modulating the PBMC secretome [14]. Similarly, in contrast to active cell culture conditions, MSCs immediately thawed from cryopreservation are relatively defective in modulating the PBMC secretome [15]. These validations suggest that the secretome assay matrix can predict the functionality of MSCs and can be utilised in potency testing assays.

6.5.4 Phosphorylation Assays

Cytokines, chemokines and growth factors communicate with their responding target cells by activating signal transduction pathways that are often initiated by the phosphorylation of signalling molecules. Evaluation of an array of the phosphorylated signalling molecules on MSCs that are induced with the combined secretome as a result of the interaction of activated PBMCs could serve as another matrix approach, termed a 'phosphomatrix approach', in potency testing (Fig. 6.2). In this approach, instant phosphorylation of Signal Transducer and Activator of Transcription (STAT) proteins on MSCs are captured upon stimulation with the secretome of activated PBMCs with and without MSC coculture. This approach may also be regarded as a 'loop analytical approach' since the secretome derived from MSC and PBMC coculture is tested on the same MSC populations [13]. Thus, the probe MSC populations are both generator and sensor of the secretome which obviates the need of additional primary or immortalised reporter cell lines to evaluate the effect of the secretome. STAT phosphorylation levels on MSCs induced by the secretome of heat-inactivated (HI) MSCs' cocultured with activated PBMCs served as the internal reference. Thus, the relative quantitation of phosphorylation induced by the secretome of live cultures can serve to inform upon MSC innate functional potency with reference to their unstimulated MSC counterparts. This strategy was implemented to derive potency assays for both autologous and allogeneic MSC products. BD™ Phosflow technology was used in this phosphomatrix loop analytical approach whereby phosphorylation levels of STAT molecules were measured as Mean Fluorescence Intensity (MFI) in flow cytometry. As a matrix assay, the phosphorylation status of an array of STAT molecules such as STAT1, STAT3, STAT4, STAT5, and STAT6 was investigated on MSCs variously

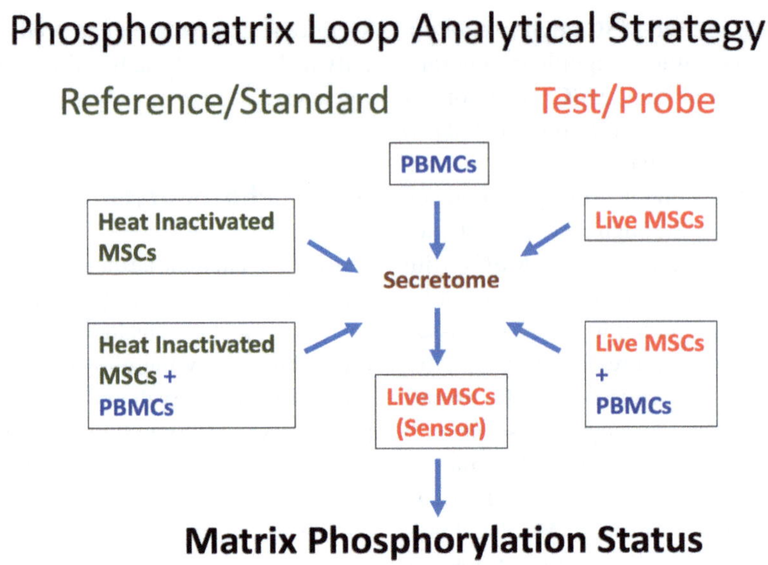

Fig. 6.2 Phosphomatrix Loop Analytical Strategy. The loop analytical potency testing approach investigates the fitness of the secretome of PBMCs cultured with and without live or heat-inactivated MSCs for inducing phosphorylation on MSCs. Secretome of heat inactivated MSCs serve as the reference standard. This strategy not only avoids the need for a universal standard, but also need for external reporter cells for testing the functionality of the secretome of MSC and PBMC interaction

derived from human bone marrow, adipose tissue, and umbilical cord. This study revealed that the secretome of activated PBMCs alone could induce STAT-1 and STAT-3 phosphorylation on MSCs. In contrast, the secretome of live MSC and PBMC coculture, but not heat inactivated MSC and PBMC coculture, failed to induce effective STAT-1 and STAT-3 phosphorylation. STAT1 and STAT3 phosphorylation levels on sensor MSCs correlated with and predicted allogeneic T-cell suppression mediated by the same MSC populations [13]. The phosphomatrix loop analytical approach was demonstrably a valuable strategy that could be incorporated into the potency testing of MSCs as a living immunoregulatory pharmaceutical.

6.5.5 Morphological Profiling Assays

Functionally-relevant morphological profiling (FRMP) denotes a strategy whereby the morphological attributes of cells that predict their functionality are quantified as cells respond to specific biological stimuli [49]. Methodologically, high content imaging is performed with automated microscopy that acquires cellular images in a rigorous high-throughput setting with the resulting high-dimensional morphological data then processed with computational approaches to obtain morphological signatures. These morphological signatures can be correlated with their functionality, thus FRMP can be deployed in potency assays [12, 40]. These two notable examples have highlighted the significance of FRMP for MSC potency analysis. Two studies, as mentioned below, used high-content imaging with automated high-dimensional morphological profiling software, cell profiler™, generating more than 90 morphological features of MSCs from multiple donors and passages that were then correlated with functions. In the first study, high-content imaging was performed on MSCs upon their induction for osteogenic differentiation with appropriate cues. Identified morphological signatures could correlate predictively with the MSCs' mineralisation ability [50]. In contrast, morphological signatures of control MSC cultures without osteogenic induction did not predict mineralisation, indicating that functional stimulation may be required for morphological profiling. In another study, unique morphological signatures of MSCs emerged upon stimulation with IFNγ and these phenotypic changes predicted their suppression of T cell activation. Morphological signatures of MSCs without IFNγ stimulation did not predict immunosuppression which again supported the significance of functional stimulation in these potency assays [38]. Both examples signified that specific morphological traits of MSC could serve as a predictor of their functionality, making FRMP an important new label-free tool with potential for predictive potency analysis.

6.5.6 Biomaterial-Based Assays

Biomaterials are natural or synthetic materials that possess and support biological functions with emerging applications in cell manufacturing and regenerative medicine [69]. Biomaterials can provide optimal culture conditions for in vitro cell growth that are more equivalent to the in vivo physiological conditions. For example, cells grown on matrices, scaffolds and hydrogels displayed closer resemblance to their native in vivo phenotype [1]. In addition, MSCs grown on planar surfaces and three-dimensional scaffold biomaterials display differential biological properties. For example, MSCs cultured on electro spun fibres produced significantly higher levels of secretory bioactive molecules than when cultured on microplates which suggested that the fibrous topography of the scaffolds influenced MSC functionality [72]. Similarly, MSCs grown in 3D spheroid scaffolds exhibited increased immunomodulatory potential and topological cues in three dimensional cultures played a significant role in promoting MSC differentiation [20, 34, 36]. These studies suggested that apical polarity and mechanical properties of conventional planar cell culture surfaces are different to the in vivo host microenvironment to which infused MSC home and perform regenerative functions. To better recapitulate physiological conditions three dimensional biomaterials that mimic the in vivo microenvironment can be incorporated in the potency assays to enhance predictive representation of MSC function in the host. Demonstrating the usefulness of biomaterials in potency testing, Williams et al. embedded MSCs in synthetic polyethylene glycol (PEG)-based hydrogels incorporated into microfluidic (tissue-on-a-chip) platforms [82]. Subsequently, stimuli were perfused within the chip and the resulting secretory molecules of MSCs were investigated. The results demonstrated that the IFNγ and TNFα-induced cytokine secretion profiles of MSCs in the hydrogels were different in comparison to profiles obtained from monolayer cells cultured on planar surfaces. In addition, this platform allowed investigation of the effect of gradient stimuli on MSC functionality and thus the bioengineered system provided a versatile tool for predicting MSC potency. Another important bioengineering intervention for improving potency assays involves the use of new biomaterials and/or engineered devices in the design of biosensors to measure analytes and molecules secreted by MSC. One of the more recent examples is the application of graphene as a biosensor nanomaterial in potency assays [4]. Functionalised graphene-based materials, graphene oxide (GO), reduced graphene oxide (RGO) and graphene quantum dot (GQD) can be utilised to sense biomolecules based on their charge interactions [73]. For example, VEGF-specific RNA aptamers in conjunction with GO in a Field-Effect Transistor (FET) electronic platform detected VEGF targets at very sensitive femtomolar concentrations [43, 48]. Anticipating future improvements in reliability and reproducibility, such engineered biosensors may introduce more cost-effective potency assays. To recapitulate the in vivo microenvironment that support MSC functionality and quantitate biomarkers using biosensors, predictive potency testing can be made more relevant by inclusion of biomaterials and engineered devices in the assay system.

6.5.7 Angiogenic Assays

MSCs' fitness to induce angiogenesis can be crucial for successful treatment and suitable potency assays can adopt two different approaches. In the first strategy, the MSC's ability to secrete proangiogenic factors is quantified to provide a surrogate measure of MSCs' potential to induce angiogenesis. VEGF levels of MSCs have been shown to be correlated with endothelial cell functions such as migration, proliferation and tube formation, justifying quantitation of VEGF secretion as a surrogate measure of angiogenic potency [76]. A similar approach has demonstrated that MSCs expressing high levels of aldehyde dehydrogenase (ALDH), an intracellular detoxification enzyme related to oxidative stress, display enhanced angiogenic properties compared to cells expressing ALDH at low levels [70] making evaluation of ALDH expression in MSCs a surrogate biomarker useful for angiogenic potency assays. In the second approach, proangiogenic properties of MSCs are tested using target-cell functional assays. For example, MSC condition media can be tested for the ability to induce endothelial cell tube formation, with quantities of MSC paracrine angiogenic factors, including C-X-C motif ligand 5 (CXCL5), interleukin 8 (IL-8) and vascular endothelial growth factor (VEGF) also evaluated in these assays. Depletion and supplementation of these cytokine levels in MSC-condition media established the indicative threshold values in the potency measurement [47]. Another functional approach termed 'in vitro aortic ring assay' has been developed, whereby Matrigel™-embedded thoracic segments of adult rat aortas were cultured with first trimester human umbilical cord-derived perivascular cells and bone marrow MSCs. Although bone marrow MSC did not show significant radial network growth and network loop formation, this assay system could be further investigated for its suitability in angiogenic potency assays [30]. Similarly, conditioned media from MSCs were tested in a three-dimensional fibrin matrix assay where human umbilical vein endothelial cells (HUVEC) were bound to gelatin-coated dextran beads and then embedded in a fibrin matrix that simulated the wound healing microenvironment. MSC conditioned media-induced sprouting and vessel formation was observed by microscopy [10]. MSC's angiogenic properties are highly regarded for their therapeutic and regenerative potential and hence appropriate angiogenic assays are a powerful tool in defining potency assays [66, 84].

6.5.8 Metabolic Assays

Human MSCs in their native quiescent state exhibit low proliferation but they undergo rapid proliferation upon cell culture adaptation. During this proliferation phase, energy metabolism is dependent on oxidative phosphorylation which leads to the accumulation of metabolic by-products such as reactive oxygen species that induce cellular senescence and reduce potency [60, 71]. Hence careful analysis of metabolic pathways and products of cell culture expanded MSCs can indicate their potency. In support of this, it has been shown that the immunosuppressive capacity of MSCs correlated with their glycolytic and respiratory activity [35]. A study has investigated MSC metabolism and measured immunomodulatory secreted factors from MSC seeded on the extraluminal side of hollow fibres in a longitudinally sampled bioreactor suitable for influencing human immune cells [3]. Such analysis of secreted metabolites and attributable functional factors can be used to define MSC potency. Mitochondrial metabolism is another target that can be deployed in potency assays. MSCs derived from obese individuals and atherosclerosis patients have been shown to display defective mitochondrial content and function that can lead to the loss of their function and stemness [37, 61]. Recently, it has been shown that mitochondrial transfer from MSCs to macrophages may play an important role in modulating macrophage function [31]. Further studies are necessary to define the applicability of mitochondrial function and metabolic fitness to MSC-derived mechanisms of action and potency assays [32, 54, 63, 74, 83].

6.6 Conclusion

MSCs present an apparently straightforward cell therapeutic product, amenable to cell manufacturing facilities since they can be expanded using standardised xenobiotic-free cell culture procedures complying with cGMP requirements. However, the mechanism of action of MSCs in mediating therapeutic benefit is complex and yet to be fully understood. In addition, heterogeneity and functional diversity can introduce variabilities that challenge the manufacture of MSC products and their application in diverse ailments [18, 29] making the development of potency assays more challenging. Nonetheless, sensitive, reliable, reproducible and economical assays coupled with advanced technologies are needed with developments underway to accurately characterise the potency of MSCs as part of their release criteria required for clinical translation.

References

1. Abdeen AA, Saha K (2017) Manufacturing cell therapies using engineered biomaterials. Trends Biotechnol 35(10):971–982. https://doi.org/10.1016/j.tibtech.2017.06.008
2. Abdi J, Rashedi I, Keating A (2018) Concise review: TLR pathway-miRNA interplay in mesenchymal stromal cells: regulatory roles and therapeutic directions. Stem Cells 36(11):1655–1662. https://doi.org/10.1002/stem.2902
3. Allen A, Vaninov N, Li M, Nguyen S, Singh M, Igo P, Tilles AW, O'Rourke B, Miller BLK, Parekkadan B, Barcia RN (2020) Mesenchymal stromal cell bioreactor for ex vivo reprogramming of human immune cells. Sci Rep 10(1):10142. https://doi.org/10.1038/s41598-020-67039-w
4. Amarandi RM, Becheru DF, Vlasceanu GM, Ionita M, Burns JS (2018) Advantages of graphene biosensors for human stem cell therapy potency assays. Biomed Res Int 2018:1676851. https://doi.org/10.1155/2018/1676851
5. Ankrum JA, Ong JF, Karp JM (2014) Mesenchymal stem cells: immune evasive, not immune privileged. Nat Biotechnol 32(3):252–260. https://doi.org/10.1038/nbt.2816
6. Barkholt L, Flory E, Jekerle V, Lucas-Samuel S, Ahnert P, Bisset L, Buscher D, Fibbe W, Foussat A, Kwa M, Lantz O, Maciulaitis R, Palomaki T, Schneider CK, Sensebe L, Tachdjian G, Tarte K, Tosca L, Salmikangas P (2013) Risk of tumorigenicity in mesenchymal stromal cell-based therapies--bridging scientific observations and regulatory viewpoints. Cytotherapy 15(7):753–759. https://doi.org/10.1016/j.jcyt.2013.03.005
7. Bloom DD, Centanni JM, Bhatia N, Emler CA, Drier D, Leverson GE, McKenna DH Jr, Gee AP, Lindblad R, Hei DJ, Hematti P (2015) A reproducible immunopotency assay to measure mesenchymal stromal cell-mediated T-cell suppression. Cytotherapy 17(2):140–151. https://doi.org/10.1016/j.jcyt.2014.10.002
8. Boregowda SV, Krishnappa V, Haga CL, Ortiz LA, Phinney DG (2016) A clinical indications prediction scale based on TWIST1 for human mesenchymal stem cells. EBioMedicine 4:62–73. https://doi.org/10.1016/j.ebiom.2015.12.020
9. Bravery CA, Carmen J, Fong T, Oprea W, Hoogendoorn KH, Woda J, Burger SR, Rowley JA, Bonyhadi ML, Van't Hof W (2013) Potency assay development for cellular therapy products: an ISCT review of the requirements and experiences in the industry. Cytotherapy 15(1):9–19. https://doi.org/10.1016/j.jcyt.2012.10.008
10. Brennen WN, Nguyen H, Dalrymple SL, Reppert-Gerber S, Kim J, Isaacs JT, Hammers H (2016) Assessing angiogenic responses induced by primary human prostate stromal cells in a three-dimensional fibrin matrix assay. Oncotarget 7(44):71298–71308. https://doi.org/10.18632/oncotarget.11347
11. Burns JS, Abdallah BM, Schrøder HD, Kassem M (2008) The histopathology of a human mesenchymal stem cell experimental tumor model: support for an hMSC origin for Ewing's sarcoma. Histol Histopathol 23:1229–1240.
12. Chen D, Dunkers JP, Losert W, Sarkar S (2021) Early time-point cell morphology classifiers successfully predict human bone marrow stromal cell differentiation modulated by fiber density in nanofiber scaffolds. Biomaterials 274:120812.
13. Chinnadurai R, Rajakumar A, Schneider AJ, Bushman WA, Hematti P, Galipeau J (2019) Potency analysis of mesenchymal stromal cells using a Phospho-STAT matrix loop analytical approach. Stem Cells 37(8):1119–1125. https://doi.org/10.1002/stem.3035
14. Chinnadurai R, Rajan D, Ng S, McCullough K, Arafat D, Waller EK, Anderson LJ, Gibson G, Galipeau J (2017) Immune dysfunctionality of replicative senescent mesenchymal stromal cells is corrected by IFNgamma priming. Blood Adv 1(11):628–643. https://doi.org/10.1182/bloodadvances.2017006205

15. Chinnadurai R, Rajan D, Qayed M, Arafat D, Garcia M, Liu Y, Kugathasan S, Anderson LJ, Gibson G, Galipeau J (2018) Potency analysis of mesenchymal stromal cells using a combinatorial assay matrix approach. Cell Rep 22(9):2504–2517. https://doi.org/10.1016/j.celrep.2018.02.013
16. Clark EA, Kalomoiris S, Nolta JA, Fierro FA (2014) Concise review: MicroRNA function in multipotent mesenchymal stromal cells. Stem Cells 32(5):1074–1082. https://doi.org/10.1002/stem.1623
17. Cuende N, Rasko JEJ, Koh MBC, Dominici M, Ikonomou L (2018) Cell, tissue and gene products with marketing authorization in 2018 worldwide. Cytotherapy 20(11):1401–1413. https://doi.org/10.1016/j.jcyt.2018.09.010
18. Dunn CM, Kameishi S, Grainger DW, Okano T (2021) Strategies to address mesenchymal stem/stromal cell heterogeneity in immunomodulatory profiles to improve cell-based therapies. Acta Biomater S1742–7061(21)00227
19. Dominici M, Le Blanc K, Mueller I, Slaper-Cortenbach I, Marini F, Krause D, Deans R, Keating A, Prockop D, Horwitz E (2006) Minimal criteria for defining multipotent mesenchymal stromal cells. The International Society for Cellular Therapy position statement. Cytotherapy 8(4):315–7. https://doi.org/10.1080/14653240600855905. PubMed PMID: 16923606
20. Follin B, Juhl M, Cohen S, Pedersen AE, Kastrup J, Ekblond A (2016) Increased paracrine immunomodulatory potential of mesenchymal stromal cells in three-dimensional culture. Tissue Eng Part B Rev 22(4):322–329. https://doi.org/10.1089/ten.TEB.2015.0532
21. Francois M, Romieu-Mourez R, Li M, Galipeau J (2012) Human MSC suppression correlates with cytokine induction of indoleamine 2,3-dioxygenase and bystander M2 macrophage differentiation. Mol Ther 20(1):187–195. https://doi.org/10.1038/mt.2011.189
22. Galipeau J (2013) The mesenchymal stromal cells dilemma--does a negative phase III trial of random donor mesenchymal stromal cells in steroid-resistant graft-versus-host disease represent a death knell or a bump in the road? Cytotherapy 15(1):2–8. https://doi.org/10.1016/j.jcyt.2012.10.002
23. Galipeau J (2021) Macrophages at the nexus of mesenchymal stromal cell potency: the emerging role of chemokine cooperativity. Stem Cells. https://doi.org/10.1002/stem.3380
24. Galipeau J, Krampera M (2015) The challenge of defining mesenchymal stromal cell potency assays and their potential use as release criteria. Cytotherapy 17(2):125–127. https://doi.org/10.1016/j.jcyt.2014.12.008
25. Galipeau J, Krampera M, Barrett J, Dazzi F, Deans RJ, DeBruijn J, Dominici M, Fibbe WE, Gee AP, Gimble JM, Hematti P, Koh MB, LeBlanc K, Martin I, McNiece IK, Mendicino M, Oh S, Ortiz L, Phinney DG, Planat V, Shi Y, Stroncek DF, Viswanathan S, Weiss DJ, Sensebe L (2016) International Society for Cellular Therapy perspective on immune functional assays for mesenchymal stromal cells as potency release criterion for advanced phase clinical trials. Cytotherapy 18(2):151–159. https://doi.org/10.1016/j.jcyt.2015.11.008
26. Galipeau J, Sensebe L (2018) Mesenchymal stromal cells: clinical challenges and therapeutic opportunities. Cell Stem Cell 22(6):824–833. https://doi.org/10.1016/j.stem.2018.05.004
27. Gee AP, Sumstad D, Stanson J, Watson P, Proctor J, Kadidlo D, Koch E, Sprague J, Wood D, Styers D, McKenna D, Gallelli J, Griffin D, Read EJ, Parish B, Lindblad R (2008) A multicenter comparison study between the Endosafe PTS rapid-release testing system and traditional methods for detecting endotoxin in cell-therapy products. Cytotherapy 10(4):427–435. https://doi.org/10.1080/14653240802075476
28. Guan Q, Li Y, Shpiruk T, Bhagwat S, Wall DA (2018) Inducible indoleamine 2,3-dioxygenase 1 and programmed death ligand 1 expression as the potency marker for mesenchymal stromal cells. Cytotherapy 20(5):639–649. https://doi.org/10.1016/j.jcyt.2018.02.003
29. Han ZC, Du WJ, Han ZB, Liang L (2017) New insights into the heterogeneity and functional diversity of human mesenchymal stem cells. Biomed Mater Eng 28:S29–S45.
30. Iqbal F, Szaraz P, Librach M, Gauthier-Fisher A, Librach CL (2017) Angiogenic potency evaluation of cell therapy candidates by a novel application of the in vitro aortic ring assay. Stem Cell Res Ther 8(1):184. https://doi.org/10.1186/s13287-017-0631-1
31. Jackson MV, Morrison TJ, Doherty DF, McAuley DF, Matthay MA, Kissenpfennig A, O'Kane CM, Krasnodembskaya AD (2016) Mitochondrial transfer via tunneling nanotubes is an important mechanism by which mesenchymal stem cells enhance macrophage phagocytosis in the in vitro and in vivo models of ARDS. Stem Cells 34(8):2210–2223. https://doi.org/10.1002/stem.2372
32. Jorgensen C, Khoury M (2021) Musculoskeletal Progenitor/Stromal Cell-Derived Mitochondria Modulate Cell Differentiation and Therapeutical Function. Front Immunol 12:606781.
33. Kabat M, Bobkov I, Kumar S, Grumet M (2020) Trends in mesenchymal stem cell clinical trials 2004-2018: is efficacy optimal in a narrow dose range? Stem Cells Transl Med 9(1):17–27. https://doi.org/10.1002/sctm.19-0202
34. Kilian KA, Bugarija B, Lahn BT, Mrksich M (2010) Geometric cues for directing the differentiation of mesenchymal stem cells. Proc Natl Acad Sci U S A 107(11):4872–4877. https://doi.org/10.1073/pnas.0903269107
35. Killer MC, Nold P, Henkenius K, Fritz L, Riedlinger T, Barckhausen C, Frech M, Hackstein H, Neubauer A, Brendel C (2017) Immunosuppressive capacity of mesenchymal stem cells correlates with metabolic activity and can be enhanced by valproic acid.

Stem Cell Res Ther 8(1):100. https://doi.org/10.1186/s13287-017-0553-y
36. Kim IL, Khetan S, Baker BM, Chen CS, Burdick JA (2013) Fibrous hyaluronic acid hydrogels that direct MSC chondrogenesis through mechanical and adhesive cues. Biomaterials 34(22):5571–5580. https://doi.org/10.1016/j.biomaterials.2013.04.004
37. Kizilay Mancini O, Lora M, Cuillerier A, Shum-Tim D, Hamdy R, Burelle Y, Servant MJ, Stochaj U, Colmegna I (2018) Mitochondrial oxidative stress reduces the immunopotency of mesenchymal stromal cells in adults with coronary artery disease. Circ Res 122(2):255–266. https://doi.org/10.1161/CIRCRESAHA.117.311400
38. Klinker MW, Marklein RA, Lo Surdo JL, Wei CH, Bauer SR (2017) Morphological features of IFN-gamma-stimulated mesenchymal stromal cells predict overall immunosuppressive capacity. Proc Natl Acad Sci U S A 114(13):E2598–E2607. https://doi.org/10.1073/pnas.1617933114
39. Kota DJ, Prabhakara KS, Toledano-Furman N, Bhattarai D, Chen Q, DiCarlo B, Smith P, Triolo F, Wenzel PL, Cox CS Jr, Olson SD (2017) Prostaglandin E2 indicates therapeutic efficacy of mesenchymal stem cells in experimental traumatic brain injury. Stem Cells 35(5):1416–1430. https://doi.org/10.1002/stem.2603
40. Kowal JM, Schmal H, Halekoh U, Hjelmborg JB, Kassem M (2020) Single-cell high-content imaging parameters predict functional phenotype of cultured human bone marrow stromal stem cells. Stem Cells Transl Med 9:189–202.
41. Krampera M, Galipeau J, Shi Y, Tarte K, Sensebe L, Therapy MSCCotISfC (2013) Immunological characterization of multipotent mesenchymal stromal cells--the International Society for Cellular Therapy (ISCT) working proposal. Cytotherapy 15(9):1054–1061. https://doi.org/10.1016/j.jcyt.2013.02.010
42. Kurtzberg J, Abdel-Azim H, Carpenter P, Chaudhury S, Horn B, Mahadeo K, Nemecek E, Neudorf S, Prasad V, Prockop S, Quigg T, Satwani P, Cheng A, Burke E, Hayes J, Skerrett D, Group M-GS (2020) A phase 3, single-arm, prospective study of Remestemcel-L, ex vivo culture-expanded adult human mesenchymal stromal cells for the treatment of pediatric patients who failed to respond to steroid treatment for acute graft-versus-host disease. Biol Blood Marrow Transplant 26(5):845–854. https://doi.org/10.1016/j.bbmt.2020.01.018
43. Kwon OS, Park SJ, Hong JY, Han AR, Lee JS, Lee JS, Oh JH, Jang J (2012) Flexible FET-type VEGF aptasensor based on nitrogen-doped graphene converted from conducting polymer. ACS Nano 6(2):1486–1493. https://doi.org/10.1021/nn204395n
44. Lalu MM, McIntyre L, Pugliese C, Fergusson D, Winston BW, Marshall JC, Granton J, Stewart DJ, Canadian Critical Care Trials G (2012) Safety of cell therapy with mesenchymal stromal cells (SafeCell): a systematic review and meta-analysis of clinical trials. PLoS One 7(10):e47559. https://doi.org/10.1371/journal.pone.0047559
45. Lee RH, Pulin AA, Seo MJ, Kota DJ, Ylostalo J, Larson BL, Semprun-Prieto L, Delafontaine P, Prockop DJ (2009) Intravenous hMSCs improve myocardial infarction in mice because cells embolized in lung are activated to secrete the anti-inflammatory protein TSG-6. Cell Stem Cell 5(1):54–63. https://doi.org/10.1016/j.stem.2009.05.003
46. Lee RH, Yu JM, Foskett AM, Peltier G, Reneau JC, Bazhanov N, Oh JY, Prockop DJ (2014) TSG-6 as a biomarker to predict efficacy of human mesenchymal stem/progenitor cells (hMSCs) in modulating sterile inflammation in vivo. Proc Natl Acad Sci U S A 111(47):16766–16771. https://doi.org/10.1073/pnas.1416121111
47. Lehman N, Cutrone R, Raber A, Perry R, Van't Hof W, Deans R, Ting AE, Woda J (2012) Development of a surrogate angiogenic potency assay for clinical-grade stem cell production. Cytotherapy 14(8):994–1004. https://doi.org/10.3109/14653249.2012.688945
48. Lin CW, Wei KC, Liao SS, Huang CY, Sun CL, Wu PJ, Lu YJ, Yang HW, Ma CC (2015) A reusable magnetic graphene oxide-modified biosensor for vascular endothelial growth factor detection in cancer diagnosis. Biosens Bioelectron 67:431–437. https://doi.org/10.1016/j.bios.2014.08.080
49. Marklein RA, Lam J, Guvendiren M, Sung KE, Bauer SR (2018) Functionally-relevant morphological profiling: a tool to assess cellular heterogeneity. Trends Biotechnol 36(1):105–118. https://doi.org/10.1016/j.tibtech.2017.10.007
50. Marklein RA, Lo Surdo JL, Bellayr IH, Godil SA, Puri RK, Bauer SR (2016) High content imaging of early morphological signatures predicts long term mineralization capacity of human mesenchymal stem cells upon osteogenic induction. Stem Cells 34(4):935–947. https://doi.org/10.1002/stem.2322
51. Martin I, Galipeau J, Kessler C, Le Blanc K, Dazzi F (2019) Challenges for mesenchymal stromal cell therapies. Sci Transl Med 11(480). https://doi.org/10.1126/scitranslmed.aat2189
52. Meisel R, Zibert A, Laryea M, Gobel U, Daubener W, Dilloo D (2004) Human bone marrow stromal cells inhibit allogeneic T-cell responses by indoleamine 2,3-dioxygenase-mediated tryptophan degradation. Blood 103(12):4619–4621. https://doi.org/10.1182/blood-2003-11-3909
53. Mendicino M, Bailey AM, Wonnacott K, Puri RK, Bauer SR (2014) MSC-based product characterization for clinical trials: an FDA perspective. Cell Stem Cell 14(2):141–145. https://doi.org/10.1016/j.stem.2014.01.013
54. Michaeloudes C, Li X, Mak JCW, Bhavsar PK (2021) Study of Mesenchymal Stem Cell-Mediated Mitochondrial Transfer in In Vitro Models of Oxidant-Mediated Airway Epithelial and Smooth Muscle Cell Injury. Methods Mol Biol 2269:93–105.
55. Moll G, Geissler S, Catar R, Ignatowicz L, Hoogduijn MJ, Strunk D, Bieback K, Ringden O (2016)

Cryopreserved or fresh mesenchymal stromal cells: only a matter of taste or key to unleash the full clinical potential of MSC therapy? Adv Exp Med Biol 951:77–98. https://doi.org/10.1007/978-3-319-45457-3_7
56. Murgia A, Veronesi E, Candini O, Caselli A, D'Souza N, Rasini V, Giorgini A, Catani F, Iughetti L, Dominici M, Burns JS (2016) Potency biomarker signature genes from multiparametric osteogenesis assays: will cGMP human bone marrow mesenchymal stromal cells make bone? PLoS One 11(10):e0163629. https://doi.org/10.1371/journal.pone.0163629
57. Nomura Y, Fukui C, Morishita Y, Haishima Y (2017) A biological study establishing the endotoxin limit for in vitro proliferation of human mesenchymal stem cells. Regenerative therapy 7:45–51.
58. Nomura Y, Fukui C, Morishita Y, Haishima Y (2018) A biological study establishing the endotoxin limit for osteoblast and adipocyte differentiation of human mesenchymal stem cells. Regen Ther 8:46–57.
59. Ofiteru AM, Becheru DF, Gharbia S, Balta C, Herman H, Mladin B, Ionita M, Hermenean A, Burns JS (2020) Qualifying osteogenic potency assay metrics for human multipotent stromal cells: TGF-beta2 a telling eligible biomarker. Cell 9(12). https://doi.org/10.3390/cells9122559
60. Pattappa G, Heywood HK, de Bruijn JD, Lee DA (2011) The metabolism of human mesenchymal stem cells during proliferation and differentiation. J Cell Physiol 226(10):2562–2570. https://doi.org/10.1002/jcp.22605
61. Perez LM, Bernal A, de Lucas B, San Martin N, Mastrangelo A, Garcia A, Barbas C, Galvez BG (2015) Altered metabolic and stemness capacity of adipose tissue-derived stem cells from obese mouse and human. PLoS One 10(4):e0123397. https://doi.org/10.1371/journal.pone.0123397
62. Phinney DG, Sensebe L (2013) Mesenchymal stromal cells: misconceptions and evolving concepts. Cytotherapy 15(2):140–145. https://doi.org/10.1016/j.jcyt.2012.11.005
63. Piekarska K, Urban-Wójciuk Z, Kurkowiak M, Pelikant-Małecka I, Schumacher A, Sakowska J, Spodnik JH, Arcimowicz Ł, Zielińska H, Tymoniuk B, Renkielska A, Siebert J, Słomińska E, Trzonkowski P, Hupp T, Marek-Trzonkowska NM (2022) Mesenchymal stem cells transfer mitochondria to allogeneic Tregs in an HLA-dependent manner improving their immunosuppressive activity. Nat Commun 13:856.
64. Prockop DJ (2016) Inflammation, fibrosis, and modulation of the process by mesenchymal stem/stromal cells. Matrix Biol 51:7–13. https://doi.org/10.1016/j.matbio.2016.01.010
65. Radrizzani M, Soncin S, Lo Cicero V, Andriolo G, Bolis S, Turchetto L (2016) Quality control assays for clinical-grade human mesenchymal stromal cells: methods for ATMP release. Methods Mol Biol 1416:313–337. https://doi.org/10.1007/978-1-4939-3584-0_19
66. Rezaie J, Heidarzadeh M, Hassanpour M, Amini H, Shokrollahi E, Ahmadi M, Rahbarghazi R (2019) The Angiogenic Paracrine Potential of Mesenchymal Stem Cells. Update on Mesenchymal and Induced Pluripotent Stem Cells.
67. Ribeiro A, Ritter T, Griffin M, Ceredig R (2016) Development of a flow cytometry-based potency assay for measuring the in vitro immunomodulatory properties of mesenchymal stromal cells. Immunol Lett 177:38–46. https://doi.org/10.1016/j.imlet.2016.07.010
68. Robb KP, Fitzgerald JC, Barry F, Viswanathan S (2019) Mesenchymal stromal cell therapy: progress in manufacturing and assessments of potency. Cytotherapy 21(3):289–306. https://doi.org/10.1016/j.jcyt.2018.10.014
69. Roh KH, Nerem RM, Roy K (2016) Biomanufacturing of therapeutic cells: state of the art, current challenges, and future perspectives. Annu Rev Chem Biomol Eng 7:455–478. https://doi.org/10.1146/annurev-chembioeng-080615-033559
70. Sherman SE, Kuljanin M, Cooper TT, Putman DM, Lajoie GA, Hess DA (2017) High aldehyde dehydrogenase activity identifies a subset of human mesenchymal stromal cells with vascular regenerative potential. Stem Cells 35(6):1542–1553. https://doi.org/10.1002/stem.2612
71. Sherr CJ, DePinho RA (2000) Cellular senescence: mitotic clock or culture shock? Cell 102(4):407–410. https://doi.org/10.1016/s0092-8674(00)00046-5
72. Su N, Gao PL, Wang K, Wang JY, Zhong Y, Luo Y (2017) Fibrous scaffolds potentiate the paracrine function of mesenchymal stem cells: a new dimension in cell-material interaction. Biomaterials 141:74–85. https://doi.org/10.1016/j.biomaterials.2017.06.028
73. Suvarnaphaet P, Pechprasarn S (2017) Graphene-based materials for biosensors: a review. Sensors (Basel) 17(10). https://doi.org/10.3390/s17102161
74. Tan YL, Eng SP, Hafez P, Abdul Karim N, Law JX, Ng MH (2022) Mesenchymal Stromal Cell Mitochondrial Transfer as a Cell Rescue Strategy in Regenerative Medicine: A Review of Evidence in Preclinical Models. Stem Cells Transl Med 11:814–827.
75. Tarte K, Gaillard J, Lataillade JJ, Fouillard L, Becker M, Mossafa H, Tchirkov A, Rouard H, Henry C, Splingard M, Dulong J, Monnier D, Gourmelon P, Gorin NC, Sensebe L, de Greffe SF, de Moelleet Therapie C (2010) Clinical-grade production of human mesenchymal stromal cells: occurrence of aneuploidy without transformation. Blood 115(8):1549–1553. https://doi.org/10.1182/blood-2009-05-219907
76. Thej C, Ramadasse B, Walvekar A, Majumdar AS, Balasubramanian S (2017) Development of a surrogate potency assay to determine the angiogenic activity of Stempeucel(R), a pooled, ex-vivo expanded, allogeneic human bone marrow mesenchymal stromal cell product. Stem Cell Res Ther 8(1):47. https://doi.org/10.1186/s13287-017-0488-3

77. Veronesi E, Murgia A, Caselli A, Grisendi G, Piccinno MS, Rasini V, Giordano R, Montemurro T, Bourin P, Sensebé L, Rojewski MT, Schrezenmeier H, Layrolle P, Ginebra MP, Panaitescu CB, Gómez-Barrena E, Catani F, Paolucci P, Burns JS, Dominici M (2014) Transportation conditions for prompt use of ex vivo expanded and freshly harvested clinical-grade bone marrow mesenchymal stromal/stem cells for bone regeneration. Tissue Eng Part C Methods 20:239–251.
78. Viswanathan S, Keating A, Deans R, Hematti P, Prockop D, Stroncek DF, Stacey G, Weiss DJ, Mason C, Rao MS (2014) Soliciting strategies for developing cell-based reference materials to advance mesenchymal stromal cell research and clinical translation. Stem Cells Dev 23(11):1157–1167. https://doi.org/10.1089/scd.2013.0591
79. von Bahr L, Batsis I, Moll G, Hagg M, Szakos A, Sundberg B, Uzunel M, Ringden O, Le Blanc K (2012) Analysis of tissues following mesenchymal stromal cell therapy in humans indicates limited long-term engraftment and no ectopic tissue formation. Stem Cells 30(7):1575–1578. https://doi.org/10.1002/stem.1118
80. Wang Y, Chen X, Cao W, Shi Y (2014) Plasticity of mesenchymal stem cells in immunomodulation: pathological and therapeutic implications. Nat Immunol 15(11):1009–1016. https://doi.org/10.1038/ni.3002
81. Wang Y, Zhang Z, Chi Y, Zhang Q, Xu F, Yang Z, Meng L, Yang S, Yan S, Mao A, Zhang J, Yang Y, Wang S, Cui J, Liang L, Ji Y, Han ZB, Fang X, Han ZC (2013) Long-term cultured mesenchymal stem cells frequently develop genomic mutations but do not undergo malignant transformation. Cell Death Dis 4:e950. https://doi.org/10.1038/cddis.2013.480
82. Williams EK, Garcia JR, Mannino RG, Schneider RS, Lam WA, Garcia AJ (2019) Enabling mesenchymal stromal cell immunomodulatory analysis using scalable platforms. Integr Biol (Camb) 11(4):154–162. https://doi.org/10.1093/intbio/zyz014
83. Yan W, Diao S, Fan Z (2021) The role and mechanism of mitochondrial functions and energy metabolism in the function regulation of the mesenchymal stem cells. Stem Cell Res Ther 12:140.
84. Yang G, Mahadik B, Choi JY, Fisher JP (2020) Vascularization in tissue engineering: fundamentals and state-of-art. Progress in Biomedical Engineering 2:012002.
85. Yin JQ, Zhu J, Ankrum JA (2019) Manufacturing of primed mesenchymal stromal cells for therapy. Nat Biomed Eng 3(2):90–104. https://doi.org/10.1038/s41551-018-0325-8

Innovative Quantification of Critical Quality Attributes

7

Sotirios Papamatthaiou and Despina Moschou

7.1 Introduction

Potency testing involves the quantitative measure of the biological activity of a product, with the main in vitro assays involving quantification of protein biomarkers via ELISA or ELISPOT, genetic sequences via RT-PCR or cell-based analysis via flow cytometry. Nonetheless, at the moment the respective assays usually take considerable time to develop. The main reasons can be traced to the cellular therapy's patient-specific nature, the limited amount of time available to perform quality testing and the limited stability of the products over time [1]. An ideal technological enabler in overcoming these existing on critical attribute accurate quantification can be found in Lab-on-Chip microsystems.

Lab-on-Chip (LoC) is a novel technology that promises democratisation of access to diagnosis through the miniaturisation of biochemical analysis. LoC technology involves devices that incorporate several laboratory processes on a single substrate that has a dimension of a few square millimeters or centimeters. In this perspective, the term 'chip' does not strictly refer to the widespread silicon chip but to any material that can serve as the substrate for the process integration (i.e. glass, polydimethylsiloxane (PDMS) and paper). It must be noted that the terms Lab-on-Chip and micro-Total Analysis System (μTAS) are used indiscriminately today. However, this was not always the case as μTAS was terminology first introduced by Manz et al. [2] in 1990 to describe the integration of the total sequence of lab processes performed on micro-devices and later the term LoC has been used to denote a more general purpose device that still uses the μTAS technology [3]. Apart from the promise for low-cost and upscalable manufacturing, some other major advantages include lower application cost due to the relative ease of operation that does not require specialised staff, reduced chance of human error and faster diagnosis and response times, since everything is done automatically in handheld microchips. Hence, LoC is the technology that has at its disposal strong characteristics which can advance Point of Need Testing (PONT) [4]: Improvement of PONT is an eagerly anticipated feature which will offer added value to the healthcare providers with evident gains as the transfer of specimens to laboratories ceases to be a requirement.

Although LoC is a well-proven laboratory technology, few products have managed to achieve a widespread commercial use, with the most well-known being the test strips known as lateral-flow tests introduced in the late 1980s (tests for cardiac markers, pregnancy and drug abuse) [5, 6]. Apart from the strong advantages

S. Papamatthaiou · D. Moschou (✉)
Centre for Biosensors, Bioelectronics and Biodevices (ToC3Bio) and Department of Electronic & Electrical Engineering, University of Bath, Bath, UK
e-mail: sp2216@bath.ac.uk; dm855@bath.ac.uk

J. S. Burns (ed.), *Potency Assays for Advanced Stem Cell Therapy Medicinal Products*, Advances in Experimental Medicine and Biology 1420, https://doi.org/10.1007/978-3-031-30040-0_7

that LoC exhibits, there are also challenges to be addressed by engineers. The most significant drawback that LoC currently faces, is the lack of a mass manufacturing standard, preventing it from tackling the ASSURED criteria bottleneck (Affordable, Sensitive, Specific, User-friendly, Rapid and robust, Equipment free and Deliverable to end-users) [7]. Currently, there is not a manufacturing method that is perceived to be clearly advantageous over others. Thus, most LoC devices are not presently ready for mass production at competitive costs and are still mainly relevant to research laboratory settings.

Lab-on-PCB is currently considered as a promising strategy addressing this issue, due to its intuitive compatibility with electronics and the extensive experience in industrial manufacturing processes. In this chapter, we will extensively present Lab-on-PCB components and systems quantifying a variety of different analytes relevant to potency testing and highlight the upscaling potential this technology can introduce to the LoC field.

7.2 Lab-on-PCB

7.2.1 PCB Technology Overview

LoC can handle fluids of very small volume (picolitres) transferring them across the chip's surface for analytical processes to perform functions such as pumping, mixing, filtering and sorting at the micro-scale. The delivery of the fluids to the specific sites on the chip is handled by microfluidics: a set of micro-channels etched or molded into a compatible material (glass, silicon or polymers) [8]. The microfluidic channels are connected to the outside macro-world via inlets and outlets pierced through the chip and connected to suitable tubing. A considerable handicap for the widespread use of LoC is the use of materials not ideally fitted for the mass production of high-performance devices. Silicon, glass, polydimethylsiloxane (PDMS) and paper are some of the most used materials in laboratories and have exhibited satisfactory results so far. However, silicon is too expensive for mass production when cm-scale LoC devices are needed, despite the advantage of a well-established manufacturing infrastructure. Glass is transparent (convenient for optical microfluidic testing) and biocompatible, but at the same time a relatively expensive material lagging in electronics integration. PDMS is cheap, transparent, biocompatible, flexible and versatile but similarly to glass, it lacks in electronics integration thus the cost becomes unviable for advanced quantification applications, such as potency assays. Paper is a fairly novel material for LoC [9] having exhibited moderate quantification sensitivity with more research required to unlock its full potential, especially in terms of microfluidic integration. In the near future, it is widely expected that LoCs will be routinely used in clinical practice. It is not surprising that governments and funding bodies are keen to support the rapid commercialisation of LoC: the global LoC market was valued at $4.23 billion in 2016 and is expected to reach $7.95 billion by 2022 [10]. That being said, cost-effective, scalable techniques have to be further explored to overcome the described LoC bottleneck.

PCB manufacturing is a mature industry, well-established for over 70 years. It has massively contributed to the evolution of consumer electronics by reducing the size and the cost of the circuitry. Contemporary PCB infrastructure is capable of a manufacturing precision and quality comparable to that of the micrometer-scale semiconductor industry. PCBs have evolved in complexity, capable of multilayering with up to 50 layers and a capacity to go beyond 100b [11, 12]. This technology can potentially extend the benefits it introduced to the electronics industry to the LoC field, thus promising a similar impact on the broadening of consumer access to bioelectronics. More specifically, Lab-on-PCB offers straightforward electronics integration, eliminating the need for deposition methods that require expensive clean-room facilities. Indeed, this applies not only to the electrical tracks and sensing electrodes but also to the uncomplicated customisation of the device with electronic components often required for improved sensitivity and reliability [13]. Commonly, this may include micro-heaters, amplifiers, filters, optoelectronics and control circuitry. As it will be extensively presented here, microfluidic integration is achievable with standard PCB industry equipment and practices (or newly developed PCB compatible approaches), facilitating production of devices ready to be used directly out of the factory. Interestingly, the usual dimensions of the microfluidic features incorporated in the bioelectronic devices are in the range of 50 μm–100 mm [14]. This characteristic perfectly matches the standard PCB machinery capabilities making redundant the highly precise and complex Si technology offering nm-scale features [15]. Another convenient asset of the PCB industry is the fabrication of flexible printed boards, equally useful in biosensing applications. Environmental concerns about the disposability of Lab-on-PCBs are alleviated by the already established recycling facilities and standardised processes of the PCB industry.

Hence, it is suggested that the sought-after radical change in the clinical analysis and diagnostic testing fields towards a non-laboratory scheme can be realised by the industrial-scale compatible technology of PCBs.

7.2.2 Early Prototypes

In 1996, Lammerink et al. demonstrated for the first time the Mixed Circuit Board (MCB), expanding the conventional PCB to a microfluidic platform [16]. Just 1 year later, Jobst et al. [17] promptly reported the implementation of the PCB in biosensing applications. The PCB was used as a platform to accommodate the counter electrode for the glucose-lactate three electrodes sensor and the electronic interface with the potentiostat. Photo-patterned spacers were necessary to create the flow through cell. The sensor array was made on a glass carrier by means of thermal evaporation and the latter was used to seal the channel. Expanding on this work, Petrou et al. fabricated a micro-device for continuous sampling and monitoring of glucose [18]. They stressed that the sensor was produced by thin film technology, whereas the fluidic paths were photo-patterned onto a thin film photoresist using conventional printed circuit equipment rather than relying on any silicon processing facility. In the same way, a glass chip was bonded with adhesive to seal the fluidic device. Similarly, Nguyen and Huang [14, 19] demonstrably introduced micro-machined peristaltic and diffuser/nozzle pumps on a PCB substrate by integrating piezoelectric discs on the PCB pump chamber, focusing on lowering the packaging cost but this included processing steps not related to the standard PCB manufacturing technology.

Pagel's group pioneered the proposal of microfluidic integration in the standard PCB manufacturing technology. In their 1999 work [20], the fluidic layers were inserted in the board with the same processes used to form the electronic layers. The basic principles were the creation of cavities between the copper tracks to form the channels and the attachment of a second board to cover the PCB using epoxies (Fig. 7.1). The introduction of micro-channels, valves, heaters and fluid reservoirs on one common PCB was proposed to minimise the analytical micro-system manufacturing costs, compared to the considerably more expensive option of silicon and LIGA technology. In one of their next articles, Wego and Pagel [21] demonstrated a more sophisticated PCB-based device that incorporated flexible parts by introducing a thin polymeric Kapton membrane layer between the PCB layers. This enabled the formation of pumps and actuators. In this direction, they further revealed a capacitance device capable of detecting gas bubbles in the microfluidic channel, a pH-regulation system keeping the physiological environment of cell cultures stable and a sensor for pressure differences [22]. It is important to note that all of the above was made feasible by following multi-layer PCB technology procedures, utilising conventional double-sided copper-plated rigid base material (FR4). The channels and the fluidic structures were developed in the 'sandwich' formation between the two individual PCBs, as Fig. 7.1d depicts.

Since then, the appealing relative low cost and upscaling prospects of the PCB industry have attracted more interest in developing microfluidic

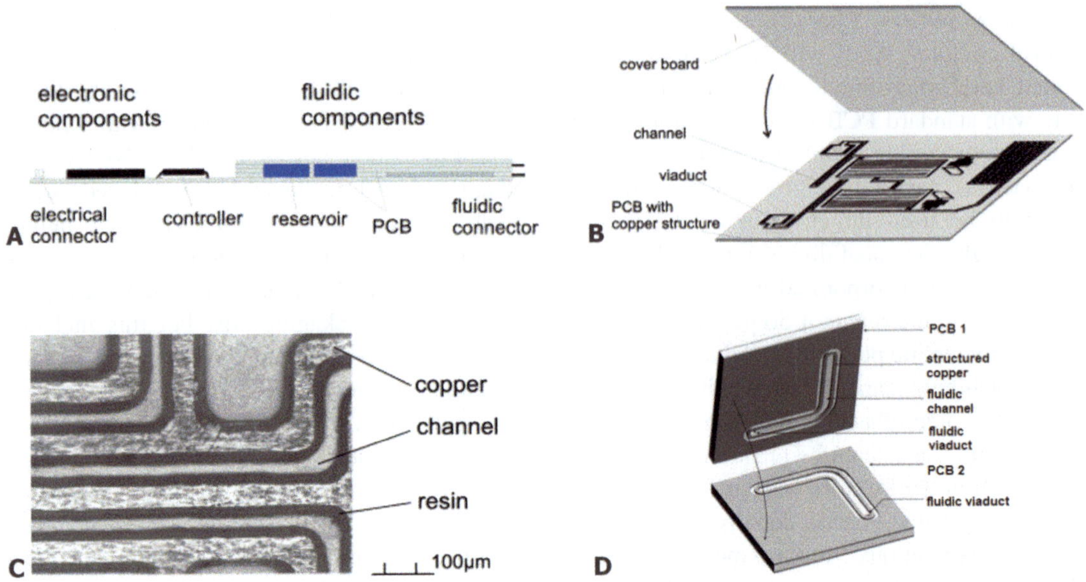

Fig. 7.1 Principles integrating PCB technology with fluidic microsystems. (**a**) Integration of electronic and fluidic components on one common PCB cross-section, (**b**) Fluidic system of different elements in PCB technology, cover board is shown lifted up, (**c**) Fluidic channels in PCB-technology. The cover board is in glass top view (Reprinted from Merkel et al., [20], copyright 1999, Elsevier publisher), (**d**) Principle of fluidic microsystems based on PCB technology. (Reprinted from Wego et al., [22], copyright 2001, IOP Publishing, Ltd)

sensing devices. For example, Gong and Kim [23] in 2008 reported the building of digital (i.e. drop-on-demand) microfluidic plates based on a PCB, dispensing picoliter to nanoliter drops on demand directly in the liquid-filled channels of the polymer chip. Their electrowetting-on-dielectric (EWOD) chip required a 2-D electrode pads pattern in which multilayer electrical access lines were created inexpensively using the mature PCB technology. They managed to create a chip with comparable performance to those on polished glass or Si substrates and due to its low production-cost it was suitable even for disposable applications. Most notably, the authors claimed the PCB manufacturing technology incurred fabrication costs that were 1000 times lower than the typical fabrication cost/cm^2 for the IC industry. Introducing further development, Pittet et al. [24] fabricated an electrochemiluminescence (ECL) microfluidic device with integrated PCB electrodes to sense H_2O_2, obtaining a 100 nM limit of detection. They used a pair of gold/Ag/AgCl electrodes as cathode and anode for the electrochemical reaction. The Ag/AgCl electrode was fabricated on top of the standard gold PCB electrode by depositing Ag and then oxidising in AgCl. Apart from the cost-effective advantage that the PCB technology offered, the authors highlighted the low resistance of these electrodes on insulator substrates, making them suitable for withstanding large currents compared to the thin-film electrodes. They also acknowledged the benefit of etching the solder mask to create the fluidic channels, concurrently within the same phase of the PCB manufacturing process as the electrodes.

7.2.3 Materials and Processes for Microfluidic Integration

7.2.3.1 Hybrid Polymer/Si – PCB Integration Approaches

More recently, an alternative technique for fluidic channel formation was proposed by Gassmann et al. [25] by making the channel from a thick (2 mm) polycarbonate (PC) layer adhered to the PCB by an acrylic glue transfer tape. This design was specifically selected to satisfy the requirement of the thermal treatment of seawater to totally isolate the sample from the copper layer (with the acrylic glue). There are a plethora of approaches that combine the PCB substrates and processes, mainly as the host for the electronic connections, with materials that require non-compatible processes to the PCB industry for the fluidic channels/components construction. For example, Ortiz et al. [26] provided a proof of concept assay utilising one of the first hybrid systems to combine a PCB packaged silicon microelectromechanical system (MEMS) with polymer microfluidics for cancer diagnosis. Particularly, the core sensing element of the device was a silicon MEMS mass sensor employing a circular diaphragm resonator (CDR), with suitable surface functionalisation converting it into a label-free BioMEMS analyte sensor. The MEMS devices were mounted onto a rigid-flex PCB to establish electrical connections and a biocompatible epoxy layer encapsulated the CDR loaded PCB, leaving uncovered only the functionalised sensing diaphragm area. The packaging process was finalised when the chip was inserted in a disposable microfluidic cartridge. The final device is shown in Fig. 7.2. It is noteworthy that this work was funded by the European Commission as part of the SmartHEALTH Integrated Project consortium to address the high-cost issues of healthcare.

Some research groups have reported that the adoption of non-standard PCB industry materials such as SU-8, PDMS and PMMA enabled them to develop the required characteristics for their devices. For instance, Kontakis et al. [27] formed fluidic micro-channels with polymer walls on top of a PCB-based chip to develop a thermal flow sensor. For this application, the thermal isolation of the sensing electrodes was crucial for high sensitivity and extending the sensor working range. Apart from the superior thermal resistivity of the PCB compared to the typical Si-based MEMS thermal flow sensors, the usage of polymer materials further enhanced the thermal isolation. The process steps are summarised in Fig. 7.3a and the micro-channel with the sensing electrodes are shown in Fig. 7.3b. An SU-8 layer was lithographically created on top of the PCB substrate, alleviating the height inconsistencies of the PCB surface. The Pt resistors were then sputtered and vias were made to connect them with the copper tracks under the previously formed SU-8 film. At this point, a thicker SU-8 film (100 μm) was spin-coated and the fluidic channel was defined by performing lithography. Finally, insulation of the channel was performed by spin coating a thin PMMA layer (1 μm) and a thick PMMA plate (3 mm) was thermally bonded on both sides of the chip to seal it. Similarly, Wu et al., from the University of California, Berkeley [28], devised a modular chip for the integration of Hall effect sensors into a programmable microfluidic format for the automated detection of magnetically labeled serum protein–PAH adducts. A thin, flexible PCB was used for the electronics part of the device and multiple layers of PDMS and glass comprised the fluidic channels, the valve membrane and the pneumatic valve actuation. Several non-PCB compatible

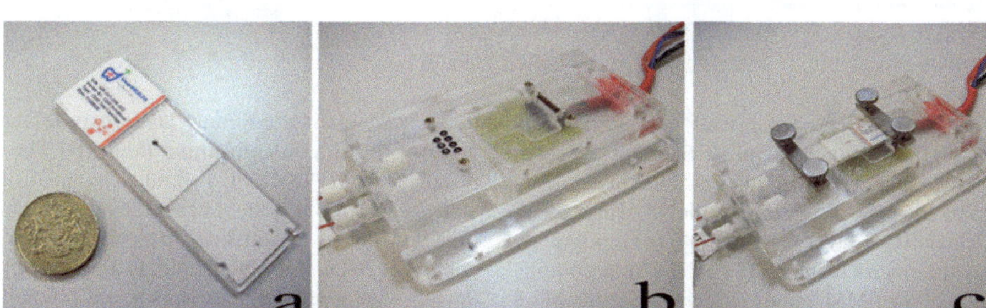

Fig. 7.2 A disposable microfluidic cartridge device. (**a**) Microfluidic cartridge containing the CDR loaded PCB. (**b**) The instrument manifold: electrical and fluidic interfaces can be observed. (**c**) Microfluidic cartridge clamped on manifold prototype. (Reprinted from Ref. [26], copyright 2008, SPIE)

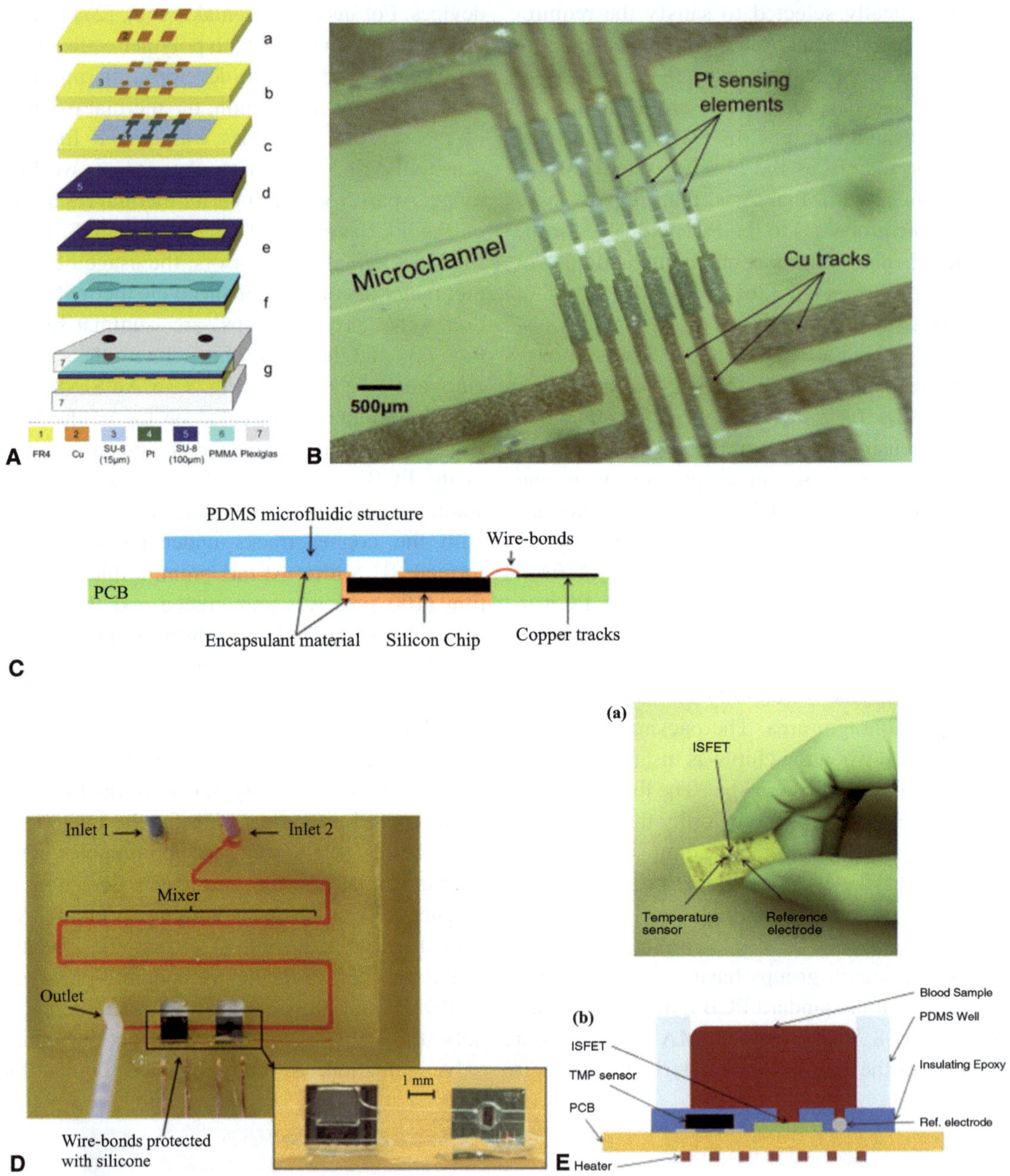

Fig. 7.3 Flow microsensor fabrication incorporating PCB. (**A**) Main fabrication steps of PCB-based flow sensor: (**a**) patterned PCB, (**b**) SU-8 planarisation layer, (**c**) Pt sputtering and lift-off. Formation of the Pt sensing elements, (**d**) Thick SU-8-layer spin coating, (**e**) lithography definition of the microchannel, (**f**) PMMA spin coating, (**g**) lamination of the PMMA plate (Plexiglas) on the SU-8 surface. Sealing of the microchannel's upper surface. (**B**) A photograph of microfluidic integration of the microsensor. (Reprinted from Ref. [27], copyright 2009, Elsevier publisher). (**C**) Cross-sectional diagram (not to scale) of the integration of microfluidic structures with small silicon chips. Photolithographically patterned openings in the thin encapsulating polymer layer expose the sensitive areas of the silicon chip to the fluid. (**D**) The assembled prototype with PDMS microfluidic channel full of red ink solution. Detail of the channel over the chips before injecting ink in the channel. (Reprinted from Ref. [31], copyright 2012, IOP Publishing, Ltd). (**E**) Top: Integrated device with a temperature sensor, ISFET and reference electrode on a printed circuit board with a heater on the back. Bottom: Cross-sectional diagram of the overall design of the device from the front view (the figure is not to scale). (Reprinted from Tseng et al. [32], copyright 2015, Elsevier publisher)

processes were employed here again, such as spin-coating and chemical vapor deposition, rendering this a hybrid approach too.

Similar achievements on microfluidic fabrication from SU-8 or analogous materials on the PCB platform include the development of an electroosmotic micropump [29], a flow sensor based on a paddle wheel [30] and an array of electrochemical sensors along with provision for sample preparation [30]. Particularly, Gassmann et al. [30] highlighted the added benefit of higher resolution (down to 1 μm) for fluidic structures when using SU-8 in comparison to the plain copper etching method (around 50 μm).

Burdallo et al. [31] adopted a more radical approach when hybridising the PCB field with the established silicon microtechnology. They envisaged the integration of solid-state sensors and actuators fabricated on silicon with molded microfluidic structures on the PCB. The PCB acted as a hosting plate for two silicon detection chips, including one Inter-Digitated Electrode (IDE) chip for conductivity and an Ion Sensitive Field-Effect Transistor (ISFET) for pH measurement. Figure 7.3c, d describes the encapsulation of the silicon in PCB chip. Diacrylate bisphenol A (DABA) photocurable polymer was used for the encapsulation. Consequently, a perfectly flat surface was generated onto which the microfluidic network was tightly sealed. They reported that this setup facilitated laminar flow on top of the sensors due to the minimised volume of the measurement chambers that in turn was made possible because of the shallow openings (~44 μm deep) over the chips. In addition, they reported IDE and ISFET response results that were satisfactorily on par with the contemporary literature. Another successful integration of PCB and standard IC technology was presented by Tseng et al. [32]. The integrated device (Fig. 7.3e) included an ISFET for rapid (2 min) Glucose 6 Phosphate Dehydrogenase (G6PD) deficiency screening, employing pH-based detection that incorporated a heater and a temperature control unit to ensure stable working temperatures. A PDMS well was attached on top of the ISFET to hold the sample.

7.2.3.2 Dry Film Photoresist Seamless Integration

The dry film photoresist is a PCB adopted technology as it is characterised by high yield and superior quality (uniform thickness) to the liquid photoresists [33]. The efficiency of this technology is also better suited to the mass production philosophy of the PCB industry than the material wasting deposition processes (i.e. spin-coating) of the liquid state SU-8 and PDMS. A group from the University of California, Irvine [15] explored two alternative polymers, polyurethane and 1002F, to construct the microfluidic channels. Their main argument was that the material within which the microfluidic channel was patterned, i.e. in the metal or solder mask layer, might raise an issue on biocompatibility. Avoiding any surface treatment that risked contamination and to promote biocompatibility, they supported the planarisation with a polymer layer that provided a flat, biocompatible surface for microfluidics. Both 1002F and polyurethane were applied with standard PCB techniques and notably, polyurethane did not require lithography. The authors performed electroosmotic flow measurement in micro-channels, demonstrating the suitability of these materials for biochemical and electrochemical applications. During the same period, they also reported [34] development of a microfluidic PCB device employing the 1002F dry photoresist. This same approach allowed implementation of sample lysis and a capillary-based separation termed isotachophoresis (ITP) of the target nucleic acid for detection of malaria. Interestingly, Guijt et al. [35] laminated Ordyl dry film photoresist with an office laminator to obtain a smooth surface on top of the detection electrodes and then produced two versions of LoC devices capable of detecting capacitively coupled contactless conductivity. The first option provided reusable detectors to be bonded to a separate microfluidic network formed in a PDMS chip, whilst the second option provided integrated detectors whereby the microfluidic network, created by subsequent lamination and lithographic steps of the dry film photoresist, was irreversibly sealed to the detector. Both designs demonstrated comparable performance to previously reported detectors fabricated by more costly advanced and sophisticated fabrication processes.

From the above-described studies, it is evident that Lab-on-PCB may be considered a promising platform for biosensing applications. It is thus logical that more effort has been focused recently on further optimisation of a PCB industry-compatible integration of the different device components. Franco et al. [36] developed a PCB compatible technique for bonding the PCB substrate to a polymeric solid material for microfluidic integration. Particularly, they used PMMA but this can be easily extended to PC, polyethylene terephthalate (PET) and cyclic olefin copolymer (COC). Instead of using a glue or/and adhesive tape for the bonding, they developed a thermal method as they argued that this technique was more pragmatically oriented towards mass production. A copper microheater provided the energy to temporarily melt the plastic and the bonding was completed after it solidified again. They further applied this technique on fabricating the first reported normally open PCB-based microvalve [37]. This was done by incorporating an additional copper microheater, placed under the channel. The generated heat melted the PMMA which in return blocked the channel.

7.2.4 Advanced Quantification Diagnostic Device Examples

7.2.4.1 PCR Modules

Adding to the proven benefits of PCB implementation and the experience/knowledge regarding successful manufacturing methods, there has progress in the development of self-sufficient PCB-based diagnostic devices incorporating the previously described individual modules (e.g. mixers, pumps, etc.) with increasingly reported benchmarks matching standard, non-PCB devices or traditional benchtop methods. At Stanford University, Marshall et al. [38] integrated mixing, thermal lysis of whole blood and nucleic acid isotachophoresis extraction-purification on a single PCB chip with microfluidic structure made from polyurethane. Their results were comparable to those obtained using standard off-chip lysis and a glass capillary for ITP [39]. The validation of their on-chip lysis and extraction was performed with off-chip quantitative polymerase chain reaction (qPCR). Given the powerful relevance of amplifying specific regions of DNA by PCR for biosensor applications, miniaturisation of the underlying technology has been explored. The integration of micro PCR (μPCR) in a PCB-based device that is also capable of sample preparation and subsequent DNA detection was first reported in 2004 by Liu et al. [40]. The device consisted of three modules (Fig. 7.4b): (i) the plastic chip which included a mixing unit for cell capture using immunomagnetic separation, (ii) a cell pre-concentration/purification/lysis/PCR unit and (iii) a DNA microarray chamber; comprising a PCB with Peltier heaters and control circuitry and a second PCB chip with 4 × 4 gold electrodes, where the target DNA hybridisation took place (Motorola eSensor). The plastic chip was micromachined in PC and was sealed by another (500 μm thick) PC cover via solvent assistant thermal bonding. The valves were made by melting and re-solidifying paraffin while the three boards were attached together by means of double-sided adhesive tape.

On a similar quest, Moschou et al. [41, 42] presented a more PCB industry compatible μPCR device, whereby the fluidic compartments were made by laminating polyimide films (Dupont PC1015) on the PCB board and the micro-heaters were made on the copper layer (Fig. 7.4b). The μPCR module fed a label-free, silicon-based, capacitive DNA-sensor for mutations of the KRAS gene, of diagnostic significance for colon cancer. Compared to earlier studies this device was simpler, requiring an external instrument for fluid pumping and temperature control of the integrated micro-heaters for the μPCR steps. Subsequent development focused on deriving an efficient array of micro-heaters with a combined temperature sensing/heating feature, employing also simulations to ensure uniform temperature across each PCR zone (denaturation, extension, annealing) with no thermal cross-talk between the zones while achieving comparable DNA amplification results to commercial bench-top thermocycler in a shorter time [43]. This continu-

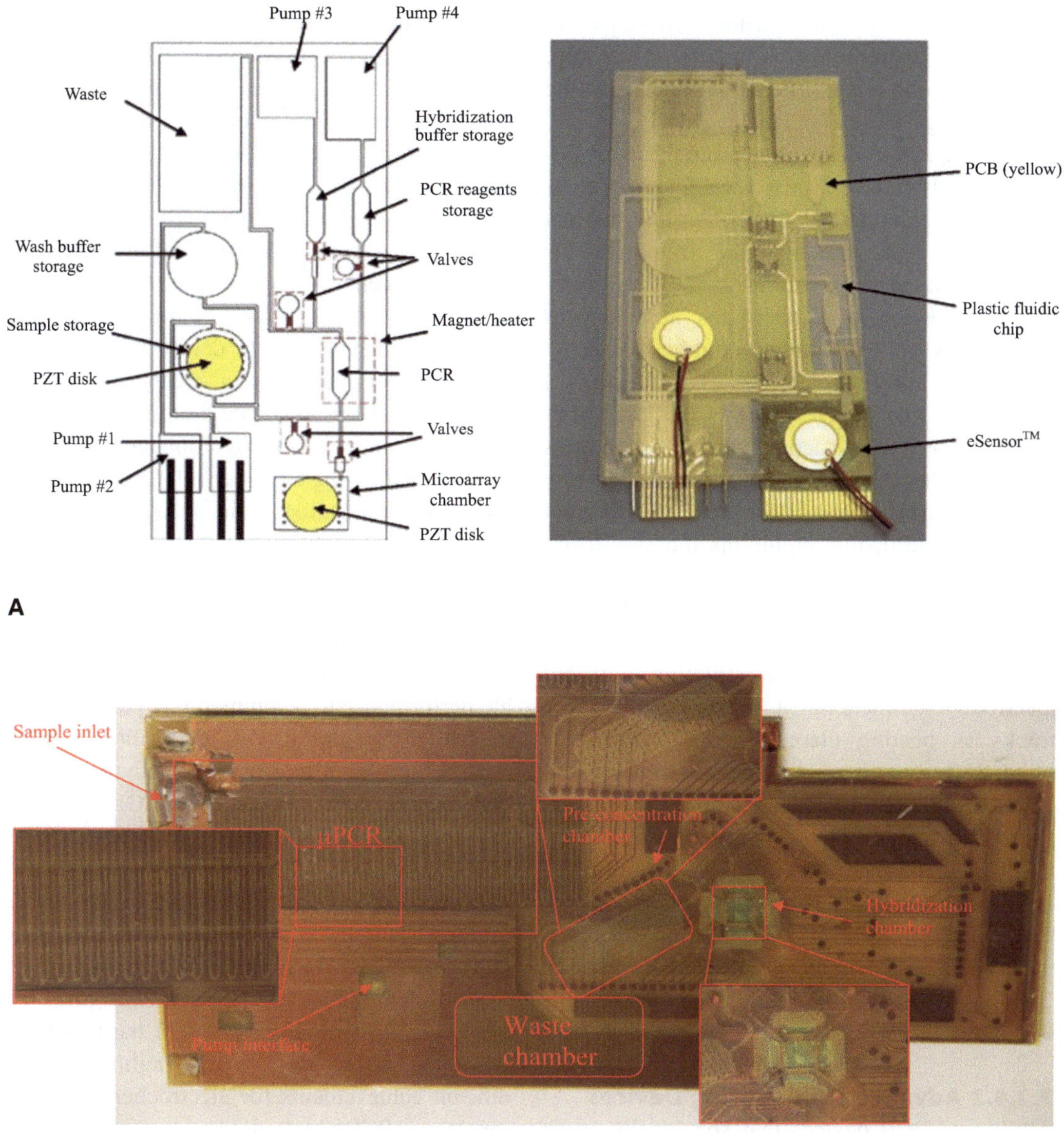

Fig. 7.4 Integrated lab-on-PCB device. (**a**) Left: Schematic of the plastic fluidic chip. Pumps 1–3 are electrochemical pumps, and pump 4 is a thermopneumatic pump, Right: Photograph of the integrated device that consists of a plastic fluidic chip, a printed circuit board (PCB), and a Motorola eSensor microarray chip. (Reprinted from Ref. [40], copyright 2004, American Chemical Society). (**b**) Lab-on-PCB chip featuring μPCR and DNA-silicon sensor with laminated polyimide films. (Reprinted from Moschou et al. [42], copyright 2013, SPIE)

ous flow μPCR device was further improved by the same group in terms of amplification speed and power consumption, providing a means for developing portable, battery-operated μPCR [44, 45]. The achievement of robust sealing (withstanding 12 bars), retaining complete PCB manufacturing processes compatibility, enabled the group to increase the channel length and the flow velocity (15 mL/min) further, consequently decreasing the amplification time to only 2 min, rendering it one of the fastest PCR devices in the literature regardless of the material [45].

Although the concept of creating the microfluidic channels in the PCB was ideal for continuous flow μPCR applications, due to the low thermal conductivity of the material (required for consistent fixed temperatures at the three individual μPCR areas), an alternative approach for static μPCR has involved microfluidics in a separate PMMA formation on the PCB chip [46, 47] introducing interesting solutions (ranging from added copper layer to active fan cooling) to mitigate the higher thermal mass and the need for low thermal inertia (static PCR requires thermal cycling). Indeed, this highlights the significant growing interest in the Lab-on-PCB approach.

Tseng et al. [48] followed a different approach for microfluidic integration of a qPCR device on PCB. After constructing a three-electrode electrochemical sensor with copper tracks as heating elements, they adhesively bonded a commercially supplied sterilised chamber on top of the board to hold the fluid for the qPCR. In addition, the droplet-based microfluidics on PCB could be highly suitable for DNA amplification as the copper electrodes used for electrowetting could be simultaneously used as heaters for the amplification process. This idea was recently applied to human papillomavirus (HPV) diagnosis [49].

7.2.4.2 Advanced Bio-Sensing Devices

Advancing PCB devices that specifically focus on biological sensing (as shown in Table 7.1), a carbon nanotube-based PCB electrode array demonstrated a state-of-the-art sensing performance, achieving simultaneous amperometric detection of lactate and glucose [50], just one of several increasingly sophisticated carbon-based sensors made possible by adopting the latest manufacturing technologies (Fig. 7.5a). Inkjet-printing was employed to deposit graphene on the working electrode of a flexible PCB electrochemical sensor for wearable bio-electronics designed for continuous glucose sensing [51]. Gold nanoparticles have been electro-deposited on graphene to enhance the sensitivity, reaching a 0.3 mg/dL limit of detection (LOD).

Ultrasensitive protein detection by PCB devices has also been accomplished. Jacobs et al. [52] sputtered ZnO on PCB chips to detect a protein biomarker for cardiovascular diseases. Their aim was to exploit the inherently nano-textured ZnO surfaces for electrochemical biosensing on the cost-effective PCB platform. Screen-printed silver electrodes were made by means of conventional PCB technology to detect carcinoembryonic antigen (CEA) protein, a cancer biomarker [53]. A novel antibody-like biomimetic material has been used as a biorecognition element resulting in sensitive (pg CEA per mL), rapid (15-min maximum incubation period) and precise (5% signal change) performance at a tenth of the manufacturing cost of traditional commercial devices.

Adopting alternative approaches, the coulter principle has been applied to enumerate tumor cells on a PCB chip [54, 55], showing comparable performance to a commercial cytometer with the added benefit of PoC capability. Sanchez et al. achieved an impressive selectivity of seven breast cancer gene markers with LOD of 25 pM by multiplex amplification and detection of mRNA on gold PCB electrode-arrays [56]. Furthermore, Jolly et al. developed a DNA microfluidic sensor by immobilising PNA probes on PCB gold micro-electrodes [57]. They researched two different industrially-applied PCB gold electroplating technologies (soft and hard plating), reporting LOD as low as 57 fM, highlighting significant achievements for electrochemical DNA sensing on PCB electrodes.

Paving the way for high-quality commercial products, Moschou et al. [58] incorporated a commercially available assay for IFN-gamma immunosensing into a double-layer PCB chip, that consisted of a reference electrode layer (silver plated) and a sensing electrode layer (gold plated). The first layer also included cylindrical, gold-plated micro-chambers for solution handling. The described chip is shown in Fig. 7.5b. This platform was later upgraded to a microfluidic one, optimised for microfluidic diffusion kinetics [59]. These studies were two of several assisted by partnerships between academia and

Table 7.1 PCB-based devices for µTAS applications

Application/feature	Analyte (where applicable)	Sensor technology (where applicable)	Microfluidic/fluid handling technology	Reference
Micropump			Copper etching	Wego and Pagel [21]
Peristaltic/diffuser, nozzle pumps			Copper etching	Nguyen and Huang [14]/[19]
Micropump			Spin-coated SU-8	Luque et al. [29]
Automated control of micropumps			SU-8	Flores et al. [69]
Electrolytic pump			PMMA	Kim et al. [70]
pH-regulation system, pressure, bubble sensor		Optical, capacitive	Copper etching	Wego et al. [22]
Volume actuator, bubble detector		Capacitive	Copper etching	Merkel et al. [20]
Bubble detector		Capacitive	Plastic pipes	Quoc et al. [71]
Single-use, unidirectional microvalve			Copper etching, spin-coated SU-8	Flores et al. [72]
Normally open microvalve			Copper etching, PMMA	Perdigones and Quero [37]
Thermopneumatically actuated microvalve			SU-8	Aracil et al. [66]
Temperature regulation				Haci et al., [73]
Cell culture (microheater and micromixer)			PDMS	Cabello et al. [74]
Emulsion generation, flow monitor		Capacitive	Copper etching, glass sealed	Dong et al. [68]
Electroosmotic flow measurement			Dry film photoresist	Wu et al. [15]
Flow sensor		Thermal dispersion	Spin-coated SU-8, PMMA	Kontakis et al. [27]
Passive microfluidics			Dry film photoresist	Vasilakis et al. [61]
Thermal method for PMMA-PCB bonding for microfluidics			Copper etching, PMMA	Franco et al. [36]
Paddle wheel flow sensor, sensor cell array with heaters for future sensing applications		Electrochemical (Chronopotentiometry)	SU-8 spin coating	Gassmann et al. [30]
SAW-based acoustofluidics			PDMS	Mikhaylov et al. [67]
Droplet-based microfluidics (EWOD)				Gong and Kim [23]
Microelectrode arrays (MEA)				Cabello and Aracil [64]

(continued)

Table 7.1 (continued)

Application/feature	Analyte (where applicable)	Sensor technology (where applicable)	Microfluidic/fluid handling technology	Reference
Continuous sampling	Glucose	Enzyme	Dry film photoresist, glass cover	Petrou et al. [18]
	Glucose-lactate	Electrochemical (Amperometric)	Dry film photoresist, glass wafer	Jobst et al. [17]
	Glucose	Electrochemical (Amperometric)	Dry film photoresist/PDMS	Morgan et al. [60]; Pu et al. [51]
	Glucose	Electrochemical (Amperometric, CV and EIS)		Alhans et al. [75]
	Glucose-lactate	Electrochemical (Amperometric)		Li et al. [50]; Kassanos et al. 76]
	H_2O_2	Electrochemiluminescence	Dry film photoresist and glass cover	Pittet et al. [24]
Seawater thermal treatment			Machined polycarbonate (PC)	Gassmann et al. [25]
Conductivity and pH		Electrochemical (IDE & ISFET)	PDMS	Burdallo and Fern [31]
pH sensing		Electrochemical (OCP, extended gate ISFET)	PMMA	Moschou et al. [4]
pH sensing for G6PD deficiency detection		Electrochemical (ISFET)	PDMS	Tseng et al. [32]
Capacitively coupled contactless conductivity detection	Inorganic cations	Electrochemical	PDMS and dry film photoresist	Guijt et al. [35]
Lysis and isotachophoresis	Nucleic acids (Malaria detection)	Optical	Dry film photoresist	Wu et al. [34]
On-chip mixing, thermal lysis, PCR-compatible nucleic acid extraction			Polyurethane pouring with PDMS mold and PMMA supporting frame	Marshall et al. [38]
Continuous flow μPCR module	KRAS gene (colon cancer)	Capacitive	Laminated polyimide films	Moschou et al. [41]; Mavraki et al. 44]
Sample preparation, μPCR	Esherichia coli K12 detection	Electrochemical (eSensor™ microarray)	Machined polycarbonate sealed by a thin PC layer	Liu et al. [40]
Three electrode sensors integrated with qPCR	Quantification of amplification product of qPCR	Electrochemical (CV)	SU-8, commercial sterilised chambers adhesively bonded	Tseng et al. [48]
Ultrafast continuous flow μPCR			Laminated polyimide films	Kaprou et al. [45]

Static μPCR			PMMA/Glass	Kaprou et al. [47]; Shen et al. [77]
Droplet microfluidics for isothermal DNA amplification			Plastic	Diaz-Diaz and Campos-Canton [49]
	Metal compounds/circulating tumor cell	Electrochemical (EIS)	PDMS	Narakathu et al. [78]; Ren et al. [65]
	Tumor cell	Electrochemical (impedance)	PDMS	Guo et al. [13]
	Enumeration of tumor cells	Electrochemical (impedance)	PDMS	Shi et al. [54; Fu et al. 55]
Dielectrophoresis-based cell sorting			Copper etching, olefin-sheet and pressure sensitive adhesive sealing/PDMS	Leiterer et al. [63]; Luo et al. [79]
Microfluidic active control diluter	Tetramethybenzadine (TMB)	Electrochemical (Amperometric)	Dry film photoresist	Vasilakis et al. [80]
	CEA antigen/HPV DNA target recognition	Circular diaphragm resonator	Machined PMMA	Ortiz et al. [26]
	DNA separation & detection	Electrochemical (Amperometric)	PDMS	Ghanim et al. [81]
	IFN-γ sensor for tuberculosis	Electrochemical (Amperometric)	PMMA	Moschou et al. [58]
	DNA	Electrochemical (EIS)	PMMA sealed by thin FR-4	Jolly et al. [57]
	Serum protein–PAH adducts	Hall effect	Glass	Wu et al. [28]
	Protein biomarker for cardiovascular diseases (Troponin-T)	Electrochemical (EIS)	PDMS	Jacobs et al. [52]
	CEA protein (cancer biomarker)	Electrochemical (DPV)		Moreira et al. [53]
	mRNA	Electrochemical (Amperometric)	PMMA	Acero Sánchez et al. [56]

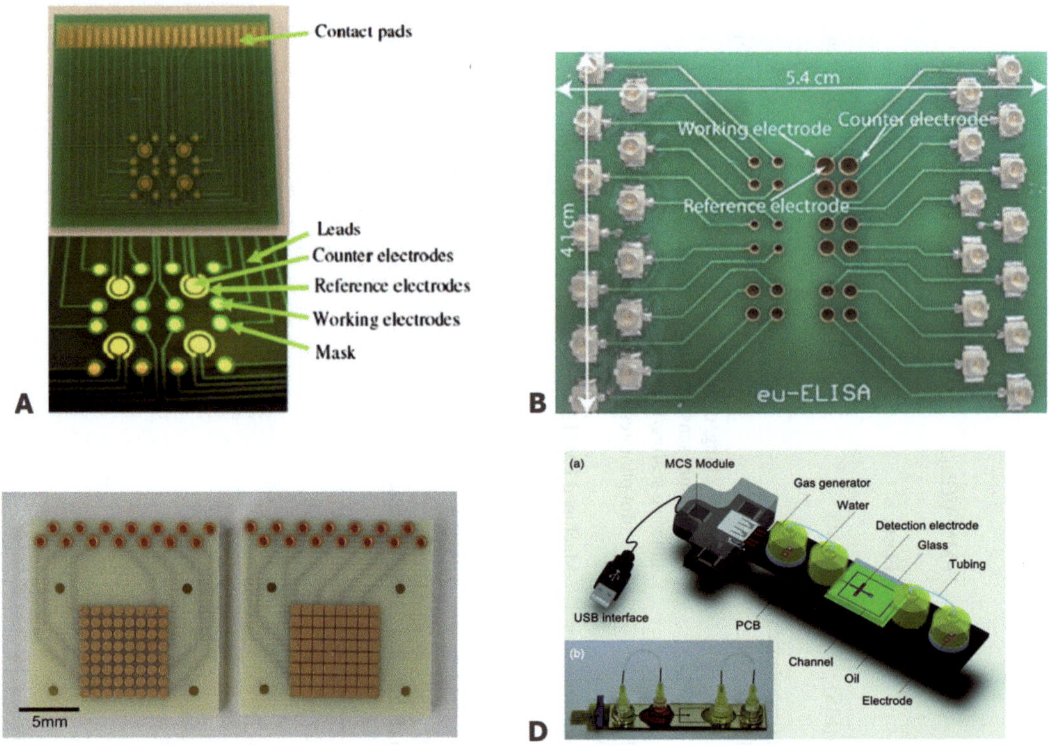

Fig. 7.5 Examples of integrated microfluidic channel applications on PCB. (**a**) Photograph of the array chip used for multi-biosensors base. The four gold rings are to be modified to work as Ag/AgCl reference electrodes, the four larger disk electrodes within the ring electrodes work as counter electrodes, the 16 smaller disk electrodes are to be fabricated as lactate, glucose sensors and sensor layers without enzyme as interference detection sensors. (Reprinted from Ref. [50], copyright 2013, Elsevier publisher). (**b**) Commercially fabricated micro-chambers used for IFN-γ detection. (Reprinted from Ref. [58], copyright 2016, Elsevier publisher). (**c**) PCB-chips for dielectrophoresis. Circular and square-shaped copper electrodes plated with nickel and gold to minimise electrochemical oxidation on the electrodes. (Reprinted from Ref. [63], copyright 2015, SPIE). (**d**) 3D schematics of the USB-driven microfluidic device on a PCB (U-Chip). The inset shows the fabricated U-chip on a PCB with a standard USB interface. (Reprinted from Dong et al., [68], copyright 2013, Royal Society of Chemistry publisher)

the PCB industry [4, 33, 47, 60–62] showing the up-scaling potential of the PCB platform for LoC applications.

PCB devices in the LoC field have also been purposed for dielectrophoresis using nickel and gold plated electrodes [63] (Fig. 7.5c), 3D micro-electrode arrays (MEA) for the detection of electrical signals from cells or tissues [64], electrical impedance spectroscopy (EIS) for tumor cells detection [65], thermopneumatical actuation of single use microvalve [66] and a surface acoustic wave (SAW)-based acoustofluidic PCB device [67]. A particularly interesting application of integrated microfluidic channels on the PCB (although sealed with a glass chip) was developed by Li et al. [68] whereby the copper electrodes were used to generate oil-water emulsions by electrolysis (Fig. 7.5d).

7.2.5 Recent Developments in Lab-on-PCB Commercially Relevant Issues

The engagement of the PCB platform for use as an integral part of biosensing applications, advancing the Lab-on-PCB concept, was evidenced by detailed studies that focused on PCB material characterisation and quality control of properties, aiming to render more reliable and

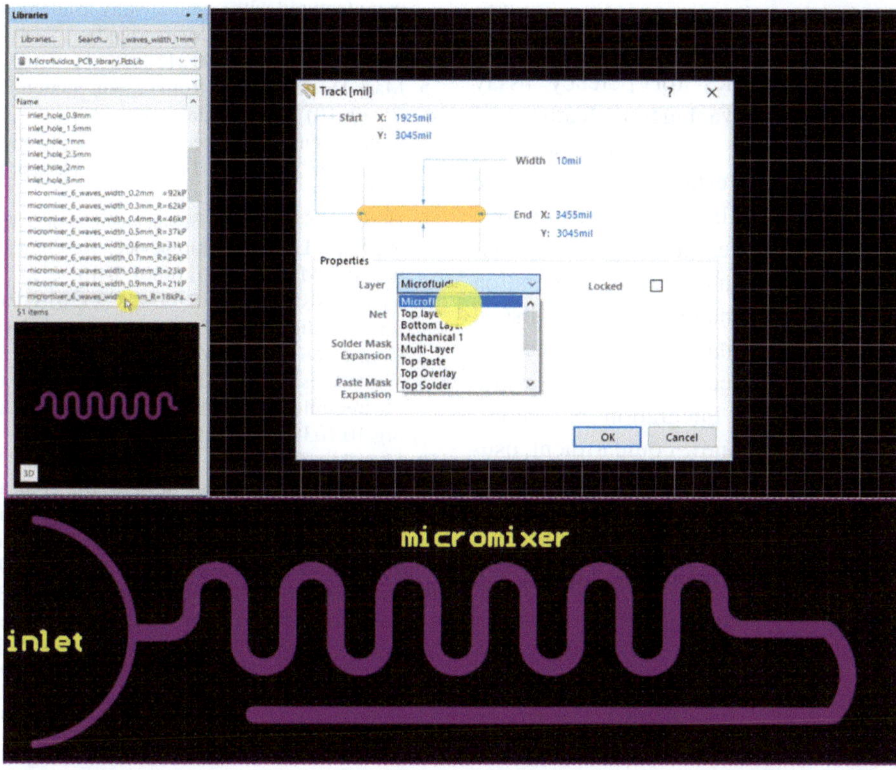

Fig. 7.6 Computer assisted design software facilitates microfluidic design. A passive micro-mixer component and its inlet designed in Altium® and a microfluidic footprint library that includes various components

robust sensors, suited for commercial use. Stable PCB Ag/AgCl electrodes could serve as biosensor integral reference electrodes [4, 82] and long-term hydrophilisation stability of FR-4 would be advantageous for passive microfluidics [61, 83]. More specifically, the PCB reference electrodes demonstrated stable open circuit potential behavior under continuous buffer flow of various pH values and the suitably treated FR-4 surfaces retained their hydrophilic properties for at least 26 days.

Furthermore, unification of electronic and microfluidic manufacturing processes in the PCB industry mandates the same unifying practice in the design phase. Hence, adoption of the PCB industry standard CAD software to design the microfluidic structures of the Lab-on-PCB platform represents a very welcome recent ambition [84, 85]. Essentially, this achieves merger of electronic and microfluidic design within a single computer assisted design (CAD) platform resulting in improved communication with the factory and unhindered implementation of the design during the manufacturing phase. Key steps in the realisation of this idea involve the layer stack manager configuration and the design rule check (DRC) set of the CAD software to facilitate the microfluidics design along with the creation of libraries dedicated solely to the microfluidic and sensing components (Fig. 7.6).

7.3 Conclusion

An extensive overview of PCB-based LoC prototype development by the research community, reveals an evolution from mostly individual components present in pioneering devices, to two decades of enhanced integration, establishing more complex and self-sufficient platforms serving the μTAS approach. The Lab-on-PCB platform can successfully accommodate most diagnostic related application, enabling highly accurate analyte quantification at the point of need, ideal characteristics for potency assay applications. Highly advantageous features of Lab-on-PCB technology are the long-standing industrial infrastructure, established appropriate micro-fabrication capabilities and the intuitive electronics integration. Research focus on manufacturing techniques and materials for integration of microfluidics with mature and established PCB industry practices, has achieved successful proofs of principle in recent years. In several cases, the microfluidic integration is accomplished by bonding the fluidic compartment, usually made from glass/PMMA/PDMS, onto the PCB chip housing the electronics. Alternatively, channel formation directly on the PCB, i.e. by metal etching, essentially using the metallic layer's thickness as the channel's walls, introduces a holistic integration simplicity that distinguishes of the Lab-on-PCB LoC field. These hallmarks lay foundation for the full exploitation of the upscaling advantages that the PCB platform offers with promise for cost-effective potency assays. Use of graphene ink drop-casted to form a transistor channel helped establish the first example of an electrolyte gated field-effect transistor (FET)-based PCB biosensor [86] introduces versatility for sensitive measurement of a broad range of potential biomarker types [87].

References

1. Stroncek DF, Jin P, Wang E, Jett B (2007) Potency analysis of cellular therapies: the emerging role of molecular assays. J Transl Med 10:1–10. https://doi.org/10.1186/1479-5876-5-24
2. Manz A, Graber N, Widmer HM (1990) Miniaturized total chemical analysis systems: a novel concept for chemical sensing. Sens Actuators B Chem 1:244–248
3. Nikolelis DP, Varzakas T, Erdem A, Nikoleli G-P (2013) Portable biosensing of food toxicants and environmental pollutants, 1st edn. CRC Press. https://www.routledge.com/Portable-Biosensing-of-Food-Toxicants-and-Environmental-Pollutants/Nikolelis-Varzakas-Erdem-Nikoleli/p/book/9781466576322
4. Moschou D, Trantidou T, Regoutz A, Carta D, Morgan H, Prodromakis T (2015) Surface and electrical characterization of Ag/AgCL pseudo-reference electrodes manufactured with commercially available PCB technologies. Sensors (Switzerland) 15:18102–18113. https://doi.org/10.3390/s150818102
5. Mark D, Haeberle S, Roth G, Von Stetten F, Zengerle R (2010) Microfluidic lab-on-a-chip platforms: requirements, characteristics and applications. Chem Soc Rev 39:1153–1182. https://doi.org/10.1039/b820557b
6. Hermsen SA, Roszek B, van Drongelen AW, Geertsma RE (n.d.) Lab-on-a-chip devices for clinical diagnostics. RIVM report 080116001/2013
7. Yetisen AK, Akram MS, Lowe CR (2013) Paper-based microfluidic point-of-care diagnostic devices. Lab Chip 13:8–15. https://doi.org/10.1039/c3lc50169h
8. Whitesides GM (2006) The origins and the future of microfluidics. Nature 442:368–373. https://doi.org/10.1038/nature05058
9. Sher M, Zhuang R, Demirci U, Asghar W (2017) Paper-based analytical devices for clinical diagnosis: recent advances in the fabrication techniques and sensing mechanisms. Expert Rev Mol Diagn 17:351–366. https://doi.org/10.1080/14737159.2017.1285228
10. Pal K, Kraatz H-B, Khasnobish A, Bag S, Banerjee I, Kuruganti U (eds) (2019) Bioelectronics and medical devices, from materials to devices – fabrication, applications and reliability, 1st edn. Woodhead Publishing
11. Asia Circuits (n.d.). https://www.asiacircuits.com/. Accessed 10 May 2020
12. PCBOnestop (n.d.). https://www.pcbonestop.com/custom-pcb/100-layers-dobolue-side-crimping-backplane.html. Accessed 10 May 2020
13. Guo J, Li CM, Kang Y (2014) PDMS-film coated on PCB for AC impedance sensing of biological cells. Biomed Microdevices 16:681–686. https://doi.org/10.1007/s10544-014-9872-2
14. Nguyen X, Huang N-T (2000) High-performance micropumps based on printed circuit board technology. Proc SPIE 4177:249–256. https://doi.org/10.1117/12.395669
15. Wu LL, Babikian S, Li G, Bachman M (2011) Microfluidic printed circuit boards. In: Proceedings – electronic components and technology conference. IEEE, pp 1576–1581. https://doi.org/10.1109/ECTC.2011.5898721
16. Lammerink TS, Spiering V, Elwenspoek M, Fluitman JH, van den Berg A (1996) Modular concept for fluid handling systems. In: Proceedings of Ninth International Workshop on Micro Electromechanical Systems. IEEE, pp 389–394. https://doi.org/10.1109/MEMSYS.1996.494013
17. Jobst G, Moser I, Svasek P, Varahram M, Trajanoski Z, Wach P (1997) Mass producible miniaturized flow through a device with a biosensor array. Sens Actuators B Chem 43:121–125
18. Petrou PS, Moser I, Jobst G (2002) BioMEMS device with integrated microdialysis probe and biosensor array. Biosens Bioelectron 17:859–865
19. Nguyen N, Huang X (2001) Miniature valveless pumps based on printed circuit board technique. Sens Actuators A Phys 88:104–111

20. Merkel T, Graeber M, Pagel L (1999) A new technology for fluidic microsystems based on PCB technology. Sens Actuators A Phys 77:98–105
21. Wego A, Pagel L (2001) A self-flling micropump based on PCB technology. Sens Actuators A Phys 88:220–226
22. Wego A, Richter S, Pagel L (2001) Fluidic microsystems based on printed circuit board technology. J Micromech Microeng 11:528
23. Gong J, Kim C (2008) Direct-referencing two-dimensional-array digital microfluidics using multilayer printed circuit board. J Microelectromech Syst 17:257–264
24. Pittet P, Lu G, Galvan J, Ferrigno R, Blum LJ, Leca-bouvier BD (2008) PCB technology-based electrochemiluminescence analytical systems. IEEE Sens J 8:565–571
25. Gassmann S, Trozjuk A, Singhal J, Miranda ML, Zielinski O (2015) PCB based micro fluidic system for thermal cycling of seawater samples. In: 2015 IEEE International Conference on Industrial Technology (ICIT). IEEE, pp 3365–3369
26. Ortiz P, Keegan N, Spoors J, Hedley J, Harris A, Burdess J, Velten T, Biehl M, Knoll T, Haberer W, Solomon M (2008) A hybrid MEMS-based microfluidic system for cancer diagnosis. IEEE Biomed Circuits Syst Conf 7270:1–8. https://doi.org/10.1117/12.810010
27. Kontakis K, Petropoulos A, Kaltsas G (2009) A novel microfluidic integration technology for PCB-based devices: application to microflow sensing. Microelectron Eng 86:1382–1384. https://doi.org/10.1016/j.mee.2009.01.088
28. Wu A, Wang L, Jensen E, Boser B (2010) Modular integration of electronics and microfluidic systems using flexible printed circuit boards. Lab Chip 10:519–521. https://doi.org/10.1039/b922830f
29. Luque A, Perdigones F, Aracil C (2012) Fabrication of electroosmotic micropump using PCB and SU-8. In: IEEE Industrial Electronics Society. IEEE, pp 3958–3961
30. Gassmann S, Luque A, Perdigones F, Quero JM (2013) Sensor structures generated with combination of SU8 and PCBMEMS. In: Proceedings of the 39th annual conference of the IEEE Industrial Electronics Society, IECON 2013. IEEE, pp 108–112
31. Burdallo I, Fern C (2012) Integration of microelectronic chips in microfluidic systems on printed circuit board. J Micromech Microeng 22:105022. https://doi.org/10.1088/0960-1317/22/10/105022
32. Tseng H, Lum J, Malfesi S, Gray BL (2015) Development of rapid screening for glucose-6-phosphate dehydrogenase deficiency prior to malaria treatment utilizing on-board pH-based electrochemical assay. Measurement 73:158–161. https://doi.org/10.1016/j.measurement.2015.05.012
33. Moschou D, Tserepi A (2017) The lab-on-PCB approach: tackling the μTAS commercial upscaling bottleneck. Lab Chip 17:1388–1405. https://doi.org/10.1039/c7lc00121e
34. Wu LL, Marshall LA, Babikian S, Han CM, Santiago JG, Bachman M (2011) A printed circuit board based micrfluidic system for pint-of-care diagnostics applications. In: Proceedings of the 15th international conference on miniaturized systems for chemistry and life sciences. Chemical and Biological Microsystems Society, pp 1819–1821
35. Guijt RM, Armstrong JP, Candish E, Lefleur V, Percey WJ, Shabala S, Hauser PC, Breadmore MC (2011) Microfluidic chips for capillary electrophoresis with integrated electrodes for capacitively coupled conductivity detection based on printed circuit board technology. Sensors Actuators B Chem 159:307–313. https://doi.org/10.1016/j.snb.2011.06.023
36. Franco E, Salvador B, Perdigones F, Cabello M, Quero JM (2018) Fabrication method of lab-on-PCB devices using a microheater with a thermo-mechanical barrier. Microelectron Eng 194:31–39. https://doi.org/10.1016/j.mee.2018.02.019
37. Perdigones F, Quero JM (2019) Physical highly integrable and normally open microvalve for industrial thermoplastic-based lab on PCB. Sensors Actuators A Phys 300:111639. https://doi.org/10.1016/j.sna.2019.111639
38. Marshall LA, Wu LL, Babikian S, Bachman M, Santiago JG (2012) Integrated printed circuit board device for cell lysis and nucleic acid extraction. Anal Chem 84:9640–9645
39. Marshall LA, Han CM, Santiago JG (2011) Extraction of DNA from malaria-infected erythrocytes using isotachophoresis. Anal Chem 83:9715–9718
40. Liu RH, Yang J, Lenigk R, Bonanno J, Grodzinski P (2004) Self-contained, fully integrated biochip for sample preparation, polymerase chain reaction amplification, and DNA microarray detection. Anal Chem 76:1824–1831
41. Moschou D, Vourdas N, Kokkoris G, Tsekenis G, Tsouti V, Zergioti I, Tserepi A (2012) Fabrication of a label-free micromechanical capacitive biosensor and integration with μPCR towards a LoC for disease diagnosis. In: Proceedings of the 16th international conference on miniaturized systems for chemistry and life sciences (MicroTAS), 2012. Chemical and Biological Microsystems Society, pp 1804–1806
42. Moschou D, Vourdas N, Filippidou MK, Tsouti V, Kokkoris G, Tsekenis G, Zergioti I (2013) Integrated biochip for PCR-based DNA amplification and detection on capacitive biosensors. Proc SPIE 8765:1–9. https://doi.org/10.1117/12.2017690
43. Mavraki E, Moschou D, Kokkoris G, Vourdas N, Chatzandroulis S (2011) A continuous flow μPCR device with integrated microheaters on a flexible polyimide substrate. Procedia Eng 25:1245–1248. https://doi.org/10.1016/j.proeng.2011.12.307
44. Moschou D, Vourdas N, Kokkoris G, Papadakis G, Parthenios J, Chatzandroulis S, Tserepi A (2014) All-plastic, low-power, disposable, continuous-flow PCR chip with integrated microheaters for rapid DNA amplification. Sensors Actuators B Chem 199:470–478. https://doi.org/10.1016/j.snb.2014.04.007

45. Kaprou GD, Papadopoulos V, Papageorgiou DP, Kefala I, Papadakis G, Gizeli E, Chatzandroulis S, Kokkoris G, Tserepi A (2019) Ultrafast, low-power, PCB manufacturable, continuous-flow microdevice for DNA amplification. Anal Bioanal Chem 411:5297–5307
46. Kaprou G, Papadakis G, Kokkoris G, Papadopoulos V, Kefala I, Papageorgiou D, Gizeli E, Tserepi A (2015) Miniaturized devices towards an integrated lab-on-a-chip platform for DNA diagnostics. Proc SPIE 9518:1–8. https://doi.org/10.1117/12.2181953
47. Kaprou GD, Papadopoulos V, Loukas CM, Kokkoris G (2020) Towards PCB-based miniaturized thermocyclers for DNA amplification. Micromachines (Basel) 11:258. https://doi.org/10.3390/mi11030258
48. Tseng AH, Adamik V, Parsons J, Scott SL, Jenny M, Lesley L, Gray B (2014) Development of an electrochemical biosensor array for quantitative polymerase chain reaction utilizing three-metal printed circuit board technology. Sensors Actuators B Chem 204:459–466. https://doi.org/10.1016/j.snb.2014.07.123
49. Diaz-Diaz IA, Campos-Canton E (2018) Design of an electrowetting biosensor prototype controlling microfluidic droplet movement for isothermal nucleic acid amplification assays. In: 2018 IEEE international autumn meeting on power, electronics and computing. IEEE, pp 1–4
50. Li X, Zang J, Liu Y, Lu Z, Li Q, Ming C (2013) Simultaneous detection of lactate and glucose by integrated printed circuit board based array sensing chip. Anal Chim Acta 771:102–107. https://doi.org/10.1016/j.aca.2013.02.011
51. Pu Z, Wang R, Wu J, Yu H, Xu K, Li D (2016) A flexible electrochemical glucose sensor with composite nanostructured surface of the working electrode. Elsevier B.V. https://doi.org/10.1016/j.snb.2016.02.115
52. Jacobs M, Muthukumar S, Panneer A, Engel J, Prasad S (2014) Ultra-sensitive electrical immunoassay biosensors using nanotextured zinc oxide thin fi lms on printed circuit board platforms. Biosens Bioelectron 55:7–13. https://doi.org/10.1016/j.bios.2013.11.022
53. Moreira FTC, Ferreira MJMS, Puga JRT, Sales MGF (2016) Screen-printed electrode produced by printed-circuit board technology. Application to cancer biomarker detection by means of plastic antibody as sensing material. Sens Actuators B Chem 223:927–935. https://doi.org/10.1016/j.snb.2015.09.157
54. Shi D, Guo J, Chen L, Xia C, Yu Z, Ai Y (2015) Differential microfluidic sensor on PCB for biological cells analysis. Electrophoresis 36:1854–1858. https://doi.org/10.1002/elps.201400524
55. Fu Y, Yuan Q, Guo J (2017) Lab-on-PCB-based micro-cytometer for circulating tumor cells detection and enumeration. In: Microfluidics and nanofluidics. Springer-Verlag, Berlin, Heidelberg, pp 1–4. https://doi.org/10.1007/s10404-017-1854-2
56. Acero Sánchez JL, Henry OYF, Joda H, Werne Solnestam B, Kvastad L, Johansson E, Akan P, Lundeberg J, Lladach N, Ramakrishnan D, Riley I (2016) Multiplex PCB-based electrochemical detection of cancer biomarkers using MLPA-barcode approach. Biosens Bioelectron 82:224–232. https://doi.org/10.1016/j.bios.2016.04.018
57. Jolly P, Rainbow J, Regoutz A, Estrela P, Moschou D (2018) A PNA-based Lab-on-PCB diagnostic platform for rapid and high sensitivity DNA quantification. Biosens Bioelectron 123:244–250. https://doi.org/10.1016/j.bios.2018.09.006
58. Moschou D, Greathead L, Pantelidis P, Kelleher P, Morgan H, Prodromakis T (2016) Amperometric IFN-γ immunosensors with commercially fabricated PCB sensing electrodes. Biosens Bioelectron 86:805–810. https://doi.org/10.1016/j.bios.2016.07.075
59. Evans D, Papadimitriou KI, Vasilakis N, Pantelidis P, Kelleher P, Morgan H, Prodromakis T (2018) A novel microfluidic point-of-care biosensor system on printed circuit board for cytokine detection. Sensors (Basel) 18:1–14. https://doi.org/10.3390/s18114011
60. Morgan H, Prodromakis T, Moschou D (2016) A PCB-based electrochemical glucose biosensing platform. In: MTAS 2016, pp 3–4. https://doi.org/10.1089/jpm.2007.9828
61. Vasilakis N, Moschou D, Carta D, Morgan H (2016) Long-lasting FR-4 surface hydrophilisation towards commercial PCB passive microfluidics. Appl Surf Sci 368:69–75. https://doi.org/10.1016/j.apsusc.2015.12.123
62. Evans D, Papadimitriou KI, Greathead L, Vasilakis N, Pantelidis P, Kelleher P, Morgan H, Prodromakis T (2017) An assay system for point-of-care diagnosis of tuberculosis using commercially manufactured PCB technology. Sci Rep 7:1–10. https://doi.org/10.1038/s41598-017-00783-8
63. Leiterer C, Urban M, Fritzsche W (2015) Printed circuit boards as platform for disposable lab-on-a-chip applications. Proc SPIE 9668:8–13. https://doi.org/10.1117/12.2202413
64. Cabello M, Aracil C (2018) Lab-on-PCB: low cost 3D microelectrode array device for extracellular recordings. In: 2018 Spanish conference on electron devices (CDE). IEEE, pp 8–11
65. Ren C, Zhang S, Song D, Guo J (2016) Lab on dielectric film deposited PCB device for characterization of electrical property of biological cells. IEEE Trans Dielectr Electr Insul 23:1895–1897. https://doi.org/10.1109/TDEI.2016.005284
66. Aracil C, Perdigones F, Moreno JM, Luque A, Quero JM (2015) Portable Lab-on-PCB platform for autonomous micromixing. Microelectron Eng 131:13–18. https://doi.org/10.1016/j.mee.2014.10.018
67. Mikhaylov R, Wu F, Wang H, Clayton A, Sun C (2020) Development and characterisation of acoustofluidic devices using detachable electrodes made from PCB†. Lab Chip 20:1807–1814. https://doi.org/10.1039/c9lc01192g
68. Dong E, Chen H, Li J, Wang Y (2014) USB-driven microfluidic chips on printed circuit boards. Lab Chip 14:860–864. https://doi.org/10.1039/c3lc51155c

69. Flores G, Aracil C, Perdigones F, Quero JM (2018) Lab-protocol-on-PCB: prototype of a laboratory protocol on printed circuit board using MEMS technologies. Microelectron Eng 200:26–31. https://doi.org/10.1016/j.mee.2018.08.003
70. Kim AH, Hwang H, Baek S, Kim D (2018) Design, fabrication, and performance evaluation of a printed-circuit-board microfluidic electrolytic pump for lab-on-a-chip devices. Sensors Actuators A Phys 277:73–84. https://doi.org/10.1016/j.sna.2018.04.042
71. Quoc TV, Dac HN, Quoc TP (2014) A printed circuit board capacitive sensor for air bubble inside fluidic flow detection. Microsyst Technol 21:1–8. https://doi.org/10.1007/s00542-014-2141-8
72. Flores G, Aracil C, Perdigones F, Quero JM (2014) Low consumption single-use microvalve for microfluidic PCB-based platforms. J Micromech Microeng 24:065013. https://doi.org/10.1088/0960-1317/24/6/065013
73. Haci D, Liu Y, Nikolic K, Demarchi D, Constandinou TG, Georgiou P (2018) Thermally controlled Lab-on-PCB for biomedical applications. In: 2018 IEEE Biomedical Circuits and Systems Conference. IEEE, pp 1–4. https://doi.org/10.1109/BIOCAS.2018.8584664
74. Cabello M, Aracil C, Perdigones F, Quero JM, Member S (2017) Conditioning lab on PCB to control temperature and mix fluids at the microscale for biomedical applications. In: 2017 Spanish conference on electron devices (CDE). IEEE, pp 1–4
75. Alhans R, Singh A, Singhal C, Narang J, Wadhwa S (2018) Comparative analysis of single-walled and multi-walled carbon nanotubes for electrochemical sensing of glucose on gold printed circuit boards. Mater Sci Eng C 90:273–279. https://doi.org/10.1016/j.msec.2018.04.072
76. Kassanos P, Anastasova S, Yang G (2018) A low-cost amperometric glucose sensor based on PCB technology. In: 2018 IEEE SENSORS. IEEE, pp 1–4
77. Shen K, Chen X, Guo M, Cheng J (2005) A microchip-based PCR device using flexible printed circuit technology. Sensors Actuators B Chem 105:251–258. https://doi.org/10.1016/j.snb.2004.05.069
78. Narakathu BB, Member S, Guruva S, Avuthu R, Member S (2015) Development of a microfluidic sensing platform by integrating PCB technology and inkjet printing process. IEEE Sens J 15:6374–6380
79. Luo J, Simon MG, Jiang AYL, Nelson EL, Lee AP, Li G-P, Bachman M (2016) 3-D In-Bi-Sn electrodes for Lab-on-PCB cell sorting. IEEE Trans Compon Packag Manuf Technol 6:1295–1300
80. Vasilakis KIP, Evans ND, Morgan H, Prodromakis T (2016) The Lab-on-PCB framework for affordable, electronic-based point-of-care diagnostics: from design to manufacturing. In: 2016 IEEE healthcare innovation point-of-care technologies conference (HI-POCT). IEEE, pp 126–129
81. Ghanim MH, Abdullah MZ (2013) Design of disposable DNA biosensor microchip with amperometric detection featuring PCB substrate. BioChip J 7:51–56. https://doi.org/10.1007/s13206-013-7108-9
82. Papamatthaiou S, Estrela P, Moschou D (2019) PCB-implemented graphene electrolyte-gated field-effect transistors for biosensing applications. In: Proceedings of the 23rd international conference on miniaturized systems for chemistry and life sciences (µTAS 2019). Chemical and Biological Microsystems Society, pp 1172–1173
83. Vasilakis N, Papadimitriou KI, Morgan H, Prodromakis T (2017) High-performance PCB-based capillary pumps for affordable point-of-care diagnostics. Microfluid Nanofluidics 21:1–11. https://doi.org/10.1007/s10404-017-1935-2
84. C3Bio, C3Bio YouTube Channel (2019). https://www.youtube.com/channel/UCB91_SEjJv7swM0TmlcgmgA. Accessed 6 May 2020
85. Moschou D (2019) Challenges in the design of Lab-on-PCB platforms. https://www.altium.com/live-conference/munich
86. Papamatthaiou S, Estrela P, Moschou D (2021) Printable graphene BioFETs for DNA quantification in Lab-on-PCB microsystems. Sci Rep 11:1–9. https://doi.org/10.1038/s41598-021-89367-1
87. Amărandi RM, Becheru DF, Vlăsceanu GM, Ioniă M, Burns JS (2018) Advantages of graphene biosensors for human stem cell therapy potency assays. Biomed Res Int 2018:1676851. https://doi.org/10.1155/2018/1676851

Release Assays and Potency Assays for CAR T-Cell Interventions

8

Juliana Dias, Amaia Cadiñanos-Garai, and Claire Roddie

8.1 Introduction

Chimeric antigen receptor (CAR) T-cells are considered "living drugs" and offer a compelling alternative to conventional anticancer therapies. Briefly, T-cells are redirected, using gene engineering technology, toward a specific cancer cell surface target antigen via a synthetic CAR protein. CARs have a modular design comprising four main structures: an antigen-binding domain, a hinge region, a transmembrane domain, and one or more intracellular signaling domains for T-cell activation (Fig. 8.1) [71, 81]. The antigen-binding domain is typically composed of a single-chain variable fragment, derived from a monoclonal antibody, providing specificity against the desired antigen.

CD19-targeting is at the forefront of CAR T-cell technology development. This antigen is highly expressed across different types of B cell malignancies, but virtually absent outside the B-cell compartment and its expression is confined to the B cell development stages but lost upon terminal differentiation into plasma cells. These characteristics confer a high specificity and high coverage that is ideal for CAR T-cell therapy targets [93]. The unprecedented responses observed in clinical trials using CD19-targeting CAR T-cells have led to U.S. Food and Drug Administration (FDA) approvals for four different CAR T-cell products for relapsed/refractory (r/r) B-cell malignancies: YESCARTA™ (*axicabtagene ciloleucel*), KYMRIAH™ (*tisagenlecleucel*), TECARTUS™ (*brexucabtagene autoleucel*), and most recently, BREYANZI® (*lisocabtagene maraleucel*) [62].

The CAR T-cell field is rapidly evolving: a growing number of new targets and indications are under development, such as B-cell maturation antigen (BCMA) for multiple myeloma, CD30 for Hodgkin's lymphoma, and CD20/CD22 for B-cell malignancies [94], with the first BCMA-targeting therapy, ABECMA (*idecabtagene vicleucel*), recently approved by the FDA [69]. CAR T-cells for application in solid tumor oncology are also the subject of intense investigation, posing additional challenges in overcoming the immunosuppressive tumor microenvironment, and low-expression/promiscuous target antigens. Despite this, encouraging results have been

J. Dias (✉)
UCL Cancer Institute, University College London, London, UK

Royal Free Hospital London, NHS Foundation Trust, London, UK
e-mail: j.pinto@ucl.ac.uk

A. Cadiñanos-Garai
USC/CHLA Cell Therapy Program, Keck School of Medicine of USC, University of Southern California (USC), Los Angeles, CA, USA

C. Roddie
UCL Cancer Institute, University College London, London, UK

Department of Haematology, UCL Hospital, London, UK

J. S. Burns (ed.), *Potency Assays for Advanced Stem Cell Therapy Medicinal Products*, Advances in Experimental Medicine and Biology 1420, https://doi.org/10.1007/978-3-031-30040-0_8

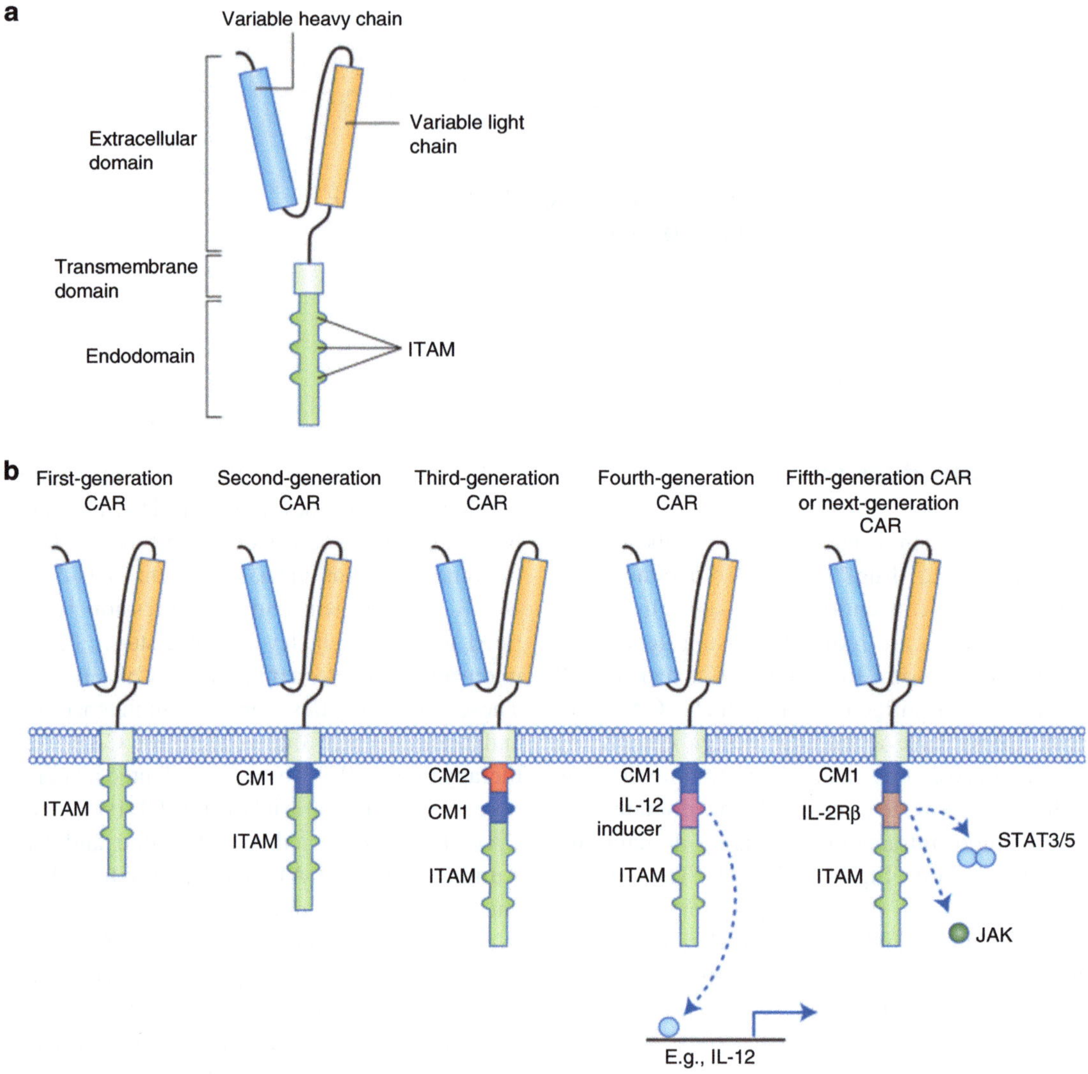

Fig. 8.1 Structure of chimeric antigen receptors (CAR). (**a**) The core structure of a CAR, highlighting its main components: the extracellular domain (responsible for antigen recognition), the transmembrane domain, and the intracellular domain (endodomain). The antigen-recognition domain is a single-chain fragment variant (scFV) generally composed of the variable light and heavy chain regions of an antigen-specific immunoglobulin separated by a flexible linker. This is linked to the transmembrane domain through the hinge. This spacers region generally supplies stability and flexibility for efficient CAR expression and activity, and it is often derived from the structure of immunoglobulins. The endodomain contains the intracellular motifs that enable downstream signaling proteins to be recruited and phosphorylated upon antigen binding for T-cell activation. Most CARs contain the intracellular domain of CD3ζ, which contains three immunoreceptor tyrosine-based activation motifs (ITAMs), as well as different co-stimulatory domains (e.g., CD28 and 41BB). (**b**) Evolution of the development of CARs from the first generation, which contained only ITAM motifs in the intracellular domain. Introduction of one (second generation) or more (third generation) co-stimulatory domains were crucial for the success of CAR T-cell therapies. New CARs are now under development to further improve efficacy by introduction of constitutive or inducible chemokines (e.g., IL-12) (fourth generation) or intracellular domains of cytokine receptors (fifth or next generation). (Image reproduced from Tokarew et al. [82] under the terms of the Creative Commons CC BY license (http://creativecommons.org/licenses/by/4.0/))

observed with EGFR, HER2, mesothelin, MUC1, and EpCAM CAR targeting for a broad range of indications [54].

Manufacturing protocols for CAR T-cell products vary between products and institutions but are always governed by the principles of good manufacturing practice (GMP). Briefly, patient T-cells are harvested (using apheresis) followed by enrichment, activation, and transduction steps ex vivo, typically using a viral vector as a transgene delivery system. Transduced T-cells are expanded for 6–22 days ex vivo to obtain the target therapeutic dose and subsequently cryopreserved while awaiting completion of quality control testing, batch certification, and release to the patient [71].

A major challenge in the CAR T-cell manufacturing field is balancing product quality with scalability and cost-effectiveness, especially when transitioning from an academic clinical trial into a marketed product, to be implemented across many collection, manufacturing, and treatment sites. Achieving product consistency while circumnavigating the intrinsic variability associated with autologous products is an additional barrier. To overcome these limitations, a robust understanding of the product and its biological actions is crucial to establish a target product profile with a defined list of critical quality attributes to be assessed for each batch prior to product certification. Additional challenges arise as the field progresses, such as new safety considerations associated with the use of allogenic T-cells and genome-editing tools.

In this chapter, we will discuss the release and potency assays required for CAR T-cell manufacturing, covering their relevance, current challenges, and future perspectives.

8.2 Regulations and Requirements for Quality Control Testing and Batch Release

CAR T-cell therapies are considered advanced therapy medicinal products (ATMPs) in Europe, under the scope of the European Commission (EC) regulation 1394/2007 (as amended) and must be manufactured following the Guidelines on Good Manufacturing Practice specific to Advanced Therapy Medicinal Products (EudraLex, Volume 4, Part IV, 2017) [24]. In the United States, these therapies are regulated by the FDA Office of Tissues and Advanced Therapies of the Center for Biologics Evaluation and Research [53].

Both the FDA [53] and the European Medicines Agency (EMA) [16] have historically published guidelines for cell and gene therapy that are applicable, although not specific, to CAR T-cell products. The FDA's "Considerations for the Development of Chimeric Antigen Receptor (CAR) T-Cell Therapies" was released on March 2022 and a revised version of the EMA Guideline on quality, nonclinical, and clinical aspects of medicinal products containing genetically modified cells came into effect in June 2021 and covers more details pertinent to CAR T-cell therapies [26].

Due to the rapidly evolving nature of the field, diversity of manufacturing practice, product complexity, and critical knowledge gaps concerning the biological action and the determinant features associated with clinical efficacy and safety of CAR T-cell therapies, it is challenging to establish harmonized and consolidated guidelines that apply to the entire industry. Furthermore, compendial testing methods are often not suitable for analysis of this type of product, so alternative assays should be validated.

In general, rigorous characterization studies throughout the earlier stages of development for each individual CAR T-cell product are essential to identify critical quality attributes, i.e., molecular and biological characteristics found to be necessary to ensure product safety and efficacy. These should cover the determinants of product safety, identity, purity, and potency that will form the requirements for final batch release (Table 8.1). A certificate that summarizes the test methods used, the corresponding test results, and the acceptable range must be provided for release of each batch. Specifications should be appropriate to the stage of product development and should be refined and tightened as product development progresses toward

Table 8.1 Typical release assays for CAR T-cell products

Category	Quality attribute	Usual assays
Safety	Sterility	Compendial culture methods (USP <71> and Ph. Eur. 2.6.1) BACTEC and BacT/ALERT (Ph. Eur. 2.6.27) qPCR methods under validation Combination of rapid and in-process testing in case of conditional product release (short shelf-life)
	Mycoplasma	Compendial culture method (USP <63> and Ph. Eur. 2.6.7) qPCR assays
	Replication competent virus (RCR/RCL) (when using viral vectors)	Amplification using permissive cell line and cell-based indicator assay Alternative PCR assays may be considered appropriate for release testing
	Vector copy number	qPCR Droplet digital PCR for single-cell resolution
Identity	CAR expression	Flow cytometry (anti-idiotype antibodies, anti-Fab antibodies, Protein L, antigen-Fc detection reagents, marker gene expression) qPCR
	CD3 expression (or relevant target cell population)	Flow cytometry
Purity	Endotoxin	Limulus Amebocyte Lysate (LAL) assay (Ph. Eur. 2.6.14 and USP <85>)
	Transduction efficiency	Determination of percentual CAR$^+$ cells as described above
	Viability	Flow cytometry Vital dyes
	Residual ancillary materials	Residual beads counted by validated morphology assay (microscopy) Quantitative assays when required Risk assessment to determine clearance of each substance during manufacturing
	Residual infectious particles (when using viral vectors)	Transduction of permissive cell lines
	Residual nontarget cells/ tumor burden	Flow cytometry panels, to be defined and validated according to manufacturing method and patient's characteristics (e.g., residual CD14$^+$ monocytes, CD19$^+$ blasts, etc.)
Potency/ quantity	Dose level	Flow cytometry determination of CAR$^+$ cells detection and absolute counting methods (e.g., BD Trucount™)
Potency	Cytotoxic potential	Cytotoxicity against target expressing cell lines (usually assessed by ^{51}Cr release) INFγ secretion upon target exposure (flow cytometry, ELISpot)
	Other determinants of product efficacy	Dependent on product characterization for identification of critical parameters that determine long-term response May include analysis of surrogate markers, proliferative and migratory capacities, polyfunctionality, etc.

licensing. For Phase 1 trials, it is generally understood by the regulatory agencies that few specifications are finalized and that assays may still be under development. However, as a minimum, specifications and acceptance criteria for product safety and quantity (cell doses) should be defined and an appropriate testing plan for characterization defined. It is also generally accepted that validation of analytical procedures will not be complete at this stage. Nevertheless, methods should be appropriately controlled, specific, sensitive, and reproducible and, whenever possible, compendial methods should be used. Furthermore, safety-related assays should be qualified prior to initiation of clinical trials [26, 86]. The quality attributes, most commonly assessed for CAR T-cell product batch release, are discussed below.

8.3 Safety

Safety testing must be in place for all Phase 1 trials and usually includes assays to ensure products are free from microbial contamination, adventitious agents, and replication competent virus. These are outlined in detail below. Specifications with established acceptance criteria must be defined based on the quality attributes of each specific CAR T-cell product and details of manufacturing methods and transgene delivery strategy should be considered.

Due to the rapidly evolving nature of CAR technology and clinical trials, the risks associated with novel products must be accordingly mapped to define critical quality attributes and appropriate testing/assays to ensure product safety. Examples include fourth-generation CAR designs that combine direct tumor targeting with programmed cytokine secretion [13]; genome-editing tools that have the potential to induce (off target) double-stranded DNA breaks; and the immunological implications of off-the-shelf, allogeneic CAR T-cell products [41].

8.3.1 Sterility Assessment

In Europe and the United States, sterility testing of biopharmaceutical products is historically performed as defined by USP <71> [88] and Ph. Eur. 2.6.1 [15], with detection of microbial or fungal growth in test samples by turbidity assessment after 14 days of incubation. These time-consuming assays greatly increase the turnaround time of CAR T-cell products. Over recent years, automation and introduction of colorimetric and fluorescence-based CO_2 measurements of metabolic activity (e.g., BacT/Alert 3D® and BD BACTEC™ systems) or adenosine triphosphate (ATP) detection by bioluminescence (Rapid Milliflex® Detection System) have increased the sensitivity of culture-based methods and permit faster detection of contamination when compared with standard methods. Further, the use of enriched aerobic or anaerobic media and incubation at 35–37 °C for a minimum of 7 days is an approach formally recognized by Ph. Eur. 2.6.27 [34].

The use of rapid and nonculture methods is also covered by the American legislation, under the Code of Federal Regulation (CFR) on "General Biological Product Standards" (21 CFR 610) [85]. Any alternative methods must be validated (as covered by Ph. Eur. 5.1.6) [15] and results must be demonstrated to be equal or superior to the compendial references. PCR-based approaches for bacteria and fungi detection through amplification of highly conserved sequences, such as the bacterial 16S rRNA, are currently under development [83]. Although optimization and comparability in sensitivity to the gold standard compendial tests are still to be determined, these methods are promising and have the potential to generate results within a few hours.

For products with a short shelf-life, product release prior to final sterility results can be accepted as part of a risk-based approach. The FDA mandate a combination of in-process sterility controls, a rapid detection test (such as Gram staining) and final sterility assessment based on 21 CFR 610 compliant methods, with a clear management strategy for positive results detected following product administration. This is not currently common practice for CAR T-cell therapies where products are cryopreserved prior to infusion. However, disease progression (and in some cases patient death) during the manufacturing period represents a significant challenge, affecting up to 13% of the patients in pivotal trials [36]. Shorter vein-to-vein times and the use of closed, automated manufacturing platforms (where the risks of in-process product contamination are considerably reduced) are of critical importance to the field.

8.3.2 Mycoplasma Detection

Mycoplasma contamination can arise from the use of cell culture reagents of animal origin, from the starting donor material, or the environment and personnel. Contrary to most bacterial contaminations, the presence of Mycoplasma does not always result in noticeable changes to cell culture turbidity or cell morphology and may go undetected for several cell passages. To address this risk, a compendial culture-based assay using indicator cell lines and multiple cell passages is described by USP <63> [88] and Ph. Eur. 2.6.7 [15], but this laborious and time-consuming testing method is not well suited for release testing of single-batch cell products. Indeed, the use of alternative detection methods such as PCR-based assays is supported by the FDA and EMA for use in the CAR T-cell therapy space [84].

8.3.3 Replication Competent Lentivirus (RCL) or Retrovirus (RCR)

Retroviral and lentiviral vectors are commonly used in CAR T-cell manufacture as efficient tools for delivery of transgene to target cells. Viral vector design has improved significantly over the last decade, with safety features designed to reduce the likelihood of generating replication competent viral vector during the manufacture process. However, exposure to replication-competent lentivirus (RCL) or retrovirus (RCR) remains a theoretical safety concern for patients treated with CAR T-cell therapies. Recombination events could lead to the generation of novel, replicating viruses during CAR T-cell manufacture or post-infusion, posing a risk of genotoxicity and malignant cell transformation. To mitigate for this, recommendations for RCL/RCR testing include assessment of all viral vector lots, manufactured cell products, and monitoring patients post-infusion. Assays for RCL/RCR detection in the viral vector batch rely on the use of permissive cell lines such as the PG4 cell line, which assumes a transformed phenotype in the presence of RCR, cultivated during multiple passages with the test material to support virus entry, amplification, and particle production [14]. The amplified material is then detected with a bespoke indicator assay, developed specifically for the vector under investigation.

For analysis of ex vivo transduced cells for batch release, PCR-based assays may be considered appropriate, particularly when time constraints are present. The use of alternative assays should be defined based on a risk assessment and should be validated, with an appropriate limit of detection. Recent guidelines from both European and American regulatory bodies have introduced flexibility to the requirement for RCL/RCR testing as part of final batch release. Indeed, RCR/RCL testing can be omitted once sufficient manufacturing and clinical experience is obtained to demonstrate that transduced cell products are consistently RCL/RCR-negative [87], or if the absence of RCL/RCR is demonstrated for each viral vector batch and generation of replicating virus during manufacture is ruled out by appropriate risk assessments [26]. Reassuringly, long-term safety data from multiple clinical trials using genetically modified cell products continues to accumulate, without evidence of RCR/RCL, indicating that any associated risks are low [36, 49, 51].

8.3.4 Vector Copy Number (VCN) per Transduced Cell

When cells are transduced with integrating vectors, the risk of insertional mutagenesis needs to be carefully considered. The risks are determined by several factors, including the insertion profile of the vector used, the vector design including the choice of enhancer and promoter sequences, the transgene product, and the vector copy number (VCN) per transduced cell.

Gammaretroviral vectors confer a risk of leukemogenesis due to their pattern of integration near transcription start sites and proto-oncogenes. This is also a potential (lesser) risk for lentiviral vectors [26]. Available clinical data suggest that newer generation vectors strongly reduce the risks of insertional mutagenesis [57], nevertheless as the total number of transduced cell infusions increases, the likelihood of infusing cells bearing at-risk insertions also increases.

Regulatory agencies require characterization of integration profile and integration sites to support marketing authorization applications. Analysis of VCN per transduced cell is a critical quality attribute determining product safety. Since VCN also has a direct impact on transgene expression, products must be carefully designed to achieve a balance between safety and efficacy. Less than five copies per transduced cell is usually considered a safe limit [99].

VCN assessment of CAR T-cell products and patient peripheral blood during follow-up is usually performed by quantitative PCR (qPCR), although a recommended, standardized assay is yet to be defined [40]. Methods employing single-cell level analysis such as droplet digital PCR (ddPCR) have the advantage of allowing detection of cell-to-cell variability in the distribution of vector copies rather than an average of the whole cell population, thus allowing identification of clones with a high number of integrations that could pose a higher risk [73].

8.3.5 Identity

Identity testing is required to identify a product and distinguish it from other products in the same facility. Most CAR T-cell therapies are patient-specific; autologous products and efficient traceability systems must be in place from apheresis to the final cell product, such that the correct product is infused to the correct patient.

For CAR T-cell products, identity assays include an assessment of specific cell populations such as $CD3^+/CD4^+/CD8^+$ T-cells in addition to the intended genetic modification(s) by qPCR or flow cytometry. Transduction efficiency can be defined based on CAR expression or on the expression of marker genes, using polyclonal anti-mouse Fab reagents for CARs derived from murine scFv, anti-idiotype antibodies generated against specific binders or antigen-Fc detection reagents [40].

Immunophenotyping by flow cytometry is widely used in the clinical setting, but a lack of assay standardization remains. Promoting standardization is a priority for the field and efforts to address this include the definition of standard panels for evaluation of major immune cell subsets, the availability of internal controls, the development of automated analysis strategies, the definition of proficiency assessment programs, and the requirement for accreditation of flow facilities by external agencies such as the UK National External Quality Assessment Scheme (UK NEQAS) system. The EuroFlow consortium [63] and the Human ImmunoPhenotyping Consortium [27] are examples of initiatives to streamline and standardize immunophenotyping assays, so that data can be compared across different sites and studies. However, each CAR T-cell product has unique characteristics such that there may be a requirement to develop and validate new transduction efficiency assays for each new construct.

8.3.6 Purity

Purity is defined as the relative freedom from extraneous materials in the final product, excepting the drug substance and excipients. Purity criteria should be defined according to the nature and intended use of the cell product, the manufacturing method used, and the consistency of the production process. Assays to demonstrate product purity should be adequate to the phase of development and adjusted as data accumulates or whenever the manufacture process changes.

Process-related impurities may include media and supplements, growth factors and cytokines, antibiotics, activation and enrichment reagents, and vectors. These should be kept to a minimum in the final formulation. Risk assessment should consider the clearance of each substance throughout the manufacturing process and the risk to the patient upon infusion, setting quantitative limits for the final product as appropriate. An example of the CAR T-cell space is residual activation beads, generally quantified by microscopy [84, 90].

Viral vectors require particular consideration. Calculations based on initial vector volumes added alongside the reduction ratio achieved (defined by the vector half-life, inactivation steps, and final dilution) can help to ensure that free infectious vector particles in the final product are reduced to negligible concentrations.

There are significant technical challenges in demonstrating absence of infectious viral particles in the final product and this is acknowledged by EMA [26]. Residual infectious particle concentration can be roughly detected using permissible cell lines (HEK 293T cells), but for the purposes of environmental risk determination, theoretical calculations are generally accepted by the regulatory authorities. For lentiviral vectors, the Dutch Commission on Genetic Modification (COGEM) proposed a formula based on available experimental data that can estimate residual free infectious viral particles in the cell product. However, variability between vectors, products, and processes means that this should be used with caution [18].

Product-related impurities can include nontarget cells, unmodified target cells, and nonviable cells, which may be present after selection or enrichment. For CAR T-cell products, a minimum of 70% viability is recommended by the FDA [83]. Release criteria for CAR T-cell products often include % $CD3^+$ T-cells, but a full characterization of final cell composition (including residual tumor burden) is desirable, especially when the manufacturing method does not include an enrichment step.

With regard to safety, evaluation of bacterial endotoxin level is mandatory. The FDA may require in vivo rabbit pyrogen tests for some licensed products. More often, the Limulus Amoebocyte Lysate (LAL) method is used, as defined by Ph. Eur. 2.6.14 [15] and USP <85> [88]. This test uses hemolymph extracted from the Limulus Polyphemus crab, which clots in the presence of bacterial endotoxins. The FDA recommends that the upper limit acceptance criterion for endotoxin should be set at 5 Endotoxin Units (EU)/kg body weight/hour for intravenous infusion. For intrathecal and/or intraocular administration, the recommendation is 0.2 EU/kg body weight/hour and 2.0 EU/dose/eye, respectively [86]. Although most CAR T-cell therapies are intravenously infused, local administration in the tumor or at the resection site are being evaluated for solid tumors [74].

Dimethyl sulfoxide (DMSO) is considered an excipient for cryopreserved cells rather than an impurity, but a safety limit for infusion is defined as 1 mL/kg/day, and this should be taken into account for high volume CAR T-cell products.

8.3.7 Quantity

To ensure consistent dosing throughout clinical investigation, specification of methods to measure dosing should also be defined at Phase 1. For CAR T-cell products, methods to determine absolute cell counts and flow cytometry assays for detection of CAR expression are usually used.

Image-based automated counting methods, such as the NucleoCounter® and Vi-CELL™ are useful for determination of total cell numbers and viability. Alongside automated hematological analyzers, these can be useful as quick tools for in-process controls. Precise assessment of final product dose is usually performed by flow cytometry, as this permits determination of CAR expression in viable $CD3^+$ T-cells/other target cells. It also gives additional information on the expression of other proteins such as memory and exhaustion markers, which may be relevant features for potency assessment and allows enumeration of cells using counting beads.

The FDA recommends that assays to determine dose should be qualified prior to initiation of clinical studies and a detailed description of the qualification protocol submitted in the original Investigational New Drug (IND) application, along with data supporting the accuracy, reproducibility, sensitivity, and specificity of the method [86].

8.4 Potency

Potency assessment is an essential aspect of the quality control system to evaluate biological function of cellular products and to ensure batch-to-batch consistency. These assays should be defined according to the products' mechanisms of action and critical attributes assessed by well-controlled investigations throughout the development stages and conducted with consistently manufactured products (Table 8.2). In the CAR T-cell arena, development of potency assays is challenging, due to the intrinsic batch-to-batch variability associated with the use of autologous cells.

Table 8.2 U.S. Food and Drug Administration (FDA) requirements for potency testing and relevance of these assays during the complete cycle of product development[a]

FDA guideline requirements for potency assays	Relevant for potency testing
• Indicate product-specific biologic activity • Measure specific activity of active component • Provide test results for product release	*Development and characterization* • Demonstrate key biological activities • Correlate product attributes and activity
• Provide data to establish stability specifications • Meets the mechanism of action for intended product use • Comply with biologics regulations and good manufacturing practice	*Manufacture and batch release* • Assess product batches against set criteria • Identification of sub-competent batches • Inter-batch comparability • Stability assessment
• Have predefined acceptance or rejection criteria • Include appropriate reference materials, standards, and controls • Be amenable to validation • Have established and documented accuracy, sensitivity, specificity, and reproducibility • Provide quantitative data	*Clinic* • Measure of product efficacy • Definition of required/adjusted product doses

[a]FDA Guidance for Industry Potency Tests for Cellular and Gene Therapy Products

Potency assessment for CAR T-cell batch release generally involves analysis of target-specific cytotoxicity, but this does not give insights into many aspects of their biological potential such as the ability to deliver long-term responses and persistence. It is unlikely that the immense complexity of these products can be captured by a single assay. This is discussed in more detail in Sect. 8.3.

8.4.1 Potency Assessment for CAR T-Cell Therapies

Potency assessment of CAR T-cells is a quantitative measurement of their biological activity and should ensure the quality and consistency of released batches [29]. These assays should be capable of identifying subpotent batches and used as a measure of drug product quality and consistency. Traditional approaches to potency testing are based on the development of in vivo and in vitro assays that measure the product's mechanism of action (MoA). Assays should be developed, optimized, and validated to characterize product attributes/biological activity that reflects or predicts clinical outcome and that could be defined as a critical quality feature [7].

Potency assay development for cellular therapy products poses several challenges [67]. First, each drug product is manufactured using patient-specific starting material, such that there is limited QC material available for potency assessment(s). Second, autologous products can be highly variable, making it challenging to define and validate a consistent assay. Furthermore, CAR T-cells exert their action through multiple, complementary mechanisms and it is difficult to capture this complexity in a single, accurate assay. Potency testing can be time-consuming, and development of rapid assays should be the priority so as not to delay final batch certification, particularly for patients with rapidly progressive disease. For the reasons outlined, standardized potency assays for CAR T-cell products are not yet defined for widespread use.

Although potency testing is not a prerequisite for early-stage clinical studies and is only essential for product release from Phase III onward, implementation of potency assays in earlier phase clinical studies may facilitate the development of more sophisticated and well-defined assays for use in Phase III via continuous optimization.

8.5 Regulatory Aspects

According to American and EU pharmaceutical legislation, cellular therapeutic products (and therefore CAR T-cell products) require potency evaluation prior to market entry [25, 28]. Both regulators stress the complexity of potency assay development and adopt a flexible regulatory approach, albeit the FDA specifies certain requirements for potency testing, outlined in Table 8.2 [25].

The analytical method used for potency assessment should ideally be quantitative, with appropriate controls and standards. However, regulatory agencies acknowledge that quantitative methods are not always feasible and will accept semiquantitative assays in this setting.

Although not required at early stages, a progressive developmental approach to potency assays is suggested and acceptance criteria specifications set for Phases 1 and 2 should be adjusted throughout product development stages to reflect manufacturing and clinical experience. The presentation of early-stage results as "information only" is also valid.

Both the FDA and EMA accept the use of in vivo and in vitro functional biological assays for product characterization. Due to the time constraints for product release, both agencies recognize the usage of nonfunctional, surrogate assays if correlation has been previously established. Most importantly, accuracy, sensitivity, specific, and reproducibility of all types of assays should be established [29].

In some cases, assay development may require the generation of novel standards or procedures that are not yet covered in the guidelines. In which case, manufacturers are responsible for the assessment and development of suitable assays for their specific product.

8.6 Methods for Potency Assessment of CAR T-Cell Products

Potency assessment should be defined based on the proposed mode of action of the product and on how and why it is expected to give therapeutic benefit. This can be based on scientific literature around nonclinical studies (animal or in vitro), or preliminary clinical data from early-stage trials. If the MoA is not fully defined, an assay matrix approach can permit assessment of biological activity, but when the MoA is known, the assay should focus on that attribute. For simplicity, most CAR T-cell potency assays are designed to measure cytotoxic activity in vitro against target-expressing cell lines. However, this simplified assessment does not account for the complex factors that determine product efficacy and influence response in vivo, such as the interaction with other immune cells, the role of the tumor microenvironment, the effects of chronic activation, and other factors not yet defined (Fig. 8.2). Extended potency assays can be helpful in this regard and are essential during the development of novel therapeutics, to permit full comprehension of the product and to inform potential product improvements.

In vivo assays are often central to product development, as in process controls or to evaluate the effect of manufacture process change [79], but even then, the lack of appropriate animal models, difficult standardization, technical complexity, and experimental duration limit their use [25]. Efforts have been made into the development, characterization, and standardization of xenograft mice models for anti-CD19 CAR T-cell therapies [1] and such comprehensive studies of tumor behavior and kinetics could be critical in widening the application of these models for the characterization of product potency, prediction of clinical outcomes, and particularly applicable in the field of solid tumors.

Simplified in vitro assays have the advantage of allowing a higher degree of standardization and are likely to remain the preferred choice for potency evaluation for batch release. For instance, the use of target-expressing cell line banks, although not fully representative of patient's target tumor cells, provides a standardized model that allows a batch-to-batch comparison of product activity. As long as supported by efficiency data, simple in vitro assays are easier to qualify, allowing definition of a numeric acceptance criteria and providing invaluable comparative information on product quality. The most common in vitro methods for potency assessment are described and summarized in Fig. 8.3.

In the future, several assays may be required for full product characterization, but we acknowledge that more complex/advanced assays for product release may have an adverse impact on time to release and overall costs. As an example, CRISPR screening has recently emerged as a

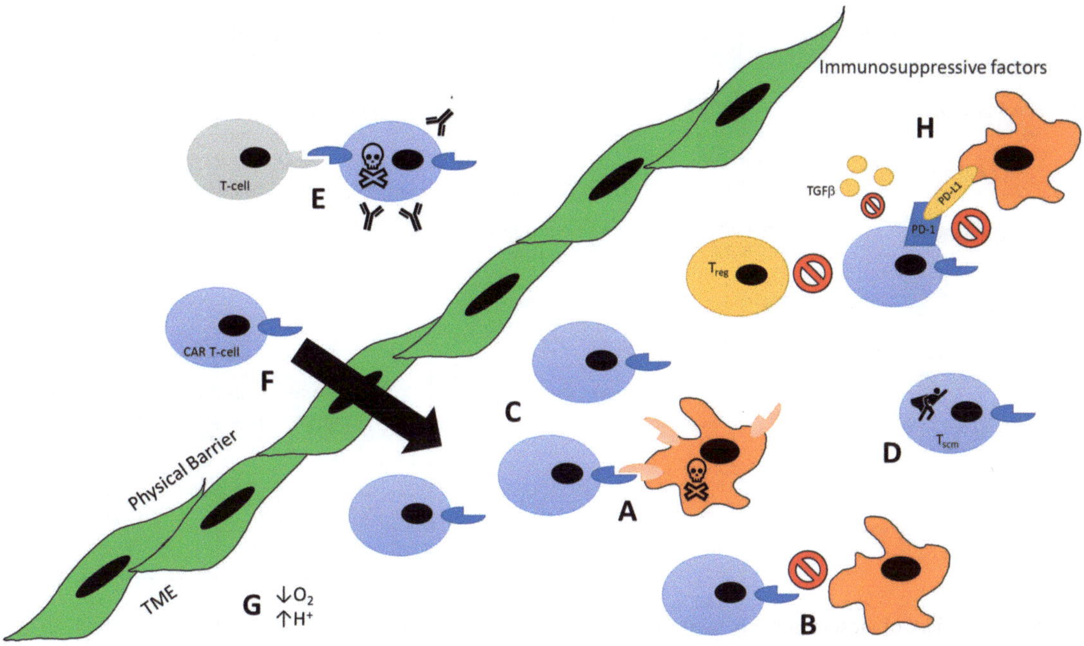

Fig. 8.2 In vivo determinants of CAR T-cell therapy effectiveness to be considered for the design of potency assays. (A) The main feature associated with CAR T-cell function is target-directed cytotoxicity. (B) Target-directed cytotoxic function is directly dependent on antigen expression. Therefore, antigen escape, low antigen density, and heterogenous antigen expression are common concerns. (C) Robust in vivo expansion and (D) sustained persistence are key features related with long-lasting clinical responses and are often associated with the T stem cell-memory compartment (Tscm), defined as CD45RA−/CCR7+/CD62L+/CD95+ CAR T-cells, and the resistance to exhaustion and senescence upon prolonged antigen exposure. (E) Lack of immunogenicity is also determinant in therapy success. (F) Particularly in the case of solid tumors, the capacity to reach tumor site and bypass the physical barriers of the tumor stroma (migration/invasion) are critical. (G) CAR T-cells designed for such application must also be resistant to the hostile tumor microenvironment (TME), often hypoxic and acidic and (H) to the many immunosuppressive factors expressed by the tumor cells (such as PD-L1), secreted into the TME (such as TGFb), and the presence of suppressive immune cells, such as regulatory T (Treg)

valuable tool for the identification of genes that are determinants for CAR T-cell function and clinical efficacy [91]. Although impractical for product release, such screening approaches, if implemented in the course of product development and characterization, have the potential to reveal critical quality attributes that can be used as biomarkers for an efficient product release assay.

8.6.1 Target-Directed Cytotoxic Activity

Cytotoxicity assays measure CAR T-cell tumor targeting with the use of methods such as flow cytometry, radioactive labeling, and impedance analysis. Direct assays aim to quantify effector activity and target cell lysis, whereas indirect assays measure a by-product of the effector–target interaction (e.g., measurement of cytokines).

8.6.2 Direct Assays

During cytotoxicity assay development, it is essential to optimize read-outs, incubation times, and effector to target cell ratios. Controls should be included to demonstrate antigen-specific cytotoxicity (e.g., antigen-negative targets) and to ensure that measured cytotoxicity is effector-specific (e.g., by culturing targets without effectors and with non-CAR T-cells to account for background signal).

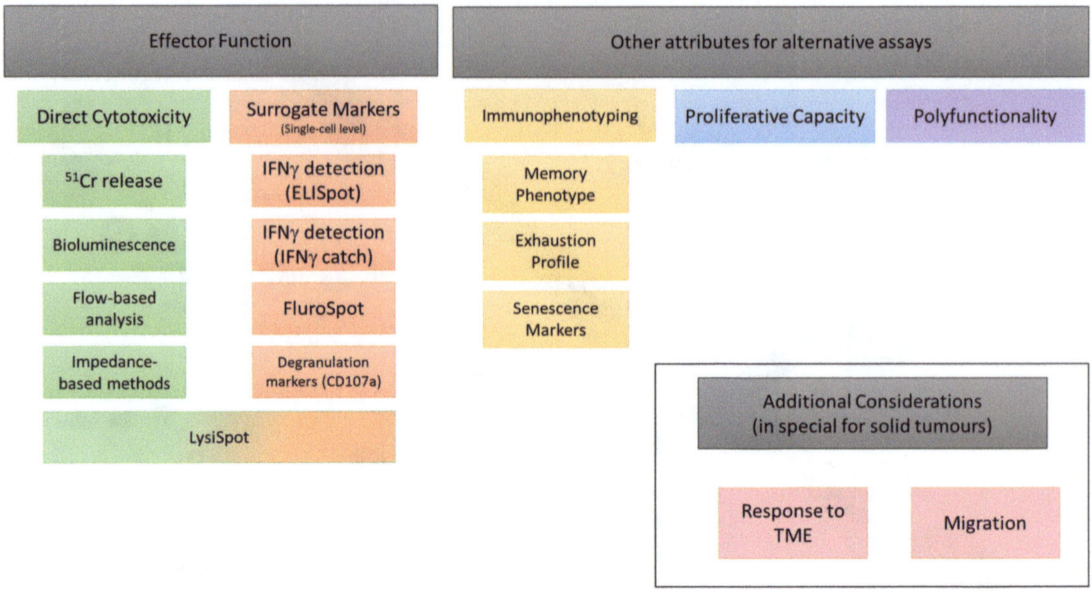

Fig. 8.3 Summary of most commonly used strategies for CAR T-cell products potency assessment. Most potency assays currently used for batch release are associated with assessment of CAR T-cell effector function, either via directly cytolytic activity of target-expressing cells or via the use of surrogate markers that reflect T-cell activation and function upon exposure to targets. Other attributes can also be relevant in the determination of CAR T-cell activity. Characterization of an immunophenotypic profile that correlates with product efficacy is being sought. Proliferative capacity and, most recently, polyfunctionality profile, have also been demonstrated to correlate with responses and can be considered for potency assessment. Special considerations for the development of potency assays for solid tumors include the effect of the complex tumor microenvironment (TME) and CAR T-cell migration capacity and ability to reach tumor site

It is essential to choose the most representative cell type(s) for the study. On occasion, primary patient-derived target cells can be used, but this adds complexity, as autologous target cell isolation and culture can be cumbersome, it can increase assay variability (and failure) between batches and can hinder inter-batch comparison [59]. Instead, CD19+ transduced cells or natively expressing CD19+ cell lines (e.g., Burkitt's lymphoma derived Raji cells) are easier to cultivate and are routinely used as targets for CD19 CAR T-cell cytotoxicity assays [31, 75], with the accepted limitation that they do not fully replicate the variable and complex metabolic and genetic profile of autologous tumor cells. Careful evaluation and selection of a suitable surrogate target cell line, including features like antigen expression and resistance to lysing activity are critical to ensure correlation with in vivo effect and tangible relevance [19, 33].

Chromium (^{51}Cr)-based cytotoxicity assays represent the gold standard for characterization of CAR T-cell cytolytic activity due to their high sensitivity. Target cells are labeled with radioactive ^{51}Cr, which is released to the supernatant upon effector-mediated target cell lysis [10]. As an endpoint assay, ^{51}Cr release is usually measured on a single short time point (usually 4 h), due to the spontaneous release of ^{51}Cr from the cell over time impairing longer analysis. The need for target cell labellng in a radiation-restricted area, alongside the hazards and technical/equipment requirements associated with the use of radiation, as well as the lack of target lysis kinetic information obtained are the main drawbacks of this technique.

Cytotoxicity assays using alternative target labeling techniques (e.g., calcein, europium, bioluminescence) are now emerging as more user-friendly approaches, although sensitivity must be evaluated [49, 89]. For bioluminescence analysis, target cells are transduced with a luciferase reporter gene. As the added bioluminescent substrate (Luciferin) is only processed by live cells, direct quantification of live target cells, and thus quantification of cytotoxic activity is measured as a decrease in bioluminescent signal over time [44].

Cytotoxic assays based on quantification of cytosolic enzymes that are naturally present in the cell and whose enzymatic activity can be measured upon release from damaged cells (e.g. lactate dehydrogenase [21]) could be used for cytotoxicity measurement. The main limitation of this approach is that these enzymes are present in both effector and target cells. This impairs the ability to discern the relative contributions from individual cell populations to the final enzymatic read-out, leading to poor assay sensitivity.

Impedance-based assays allow label-free, real-time monitoring of specific effector-induced cytolysis, measured by the detachment of target cells from a treated surface [95]. This technique was first validated for assessment of NK cell-induced cytotoxicity [35]. Briefly, nonadherent effector cells are incubated together with adherent target cells following which cytolytic action leads to target cell detachment, loss of impedance, and the restoration of electric current flow, which correlates with cytolytic activity. These automated platforms (e.g., xCELLigence systems) permit real-time monitoring of target cell populations over extended periods, combining high-sensitivity analysis with minimal cell manipulation. Furthermore, these platforms are not limited to adherent tumor cell targets. Antibody-coated plates can be used to immobilize nonadherent cell targets such that impedance assays can be used in CAR T-cell potency assessment [11].

Flow cytometry-based cytotoxicity assays can be used to study cytolysis in heterogenous cell populations. Target and effector cells can be recognized in terms of size, granularity, and specific staining while evaluating target cell death using standard DNA intercalating agents (e.g., propidium iodide or 7-AAD). Detailed product characterization and target cell phenotyping for antigen expression and density can be conducted in parallel and the resulting profile(s) can be correlated with differential susceptibility to cytolysis [39, 97].

An alternative approach is "fluorometric assessment of T lymphocyte antigen-specific lysis" (FATAL), which is purported to be a sensitive and reliable alternative to the ^{51}Cr assay [77]. Target cells are loaded with fluorescent dyes and cytotoxic activity detected by flow cytometry. This assay has the potential advantage of lower dye leakage, allowing longer incubation times in comparison to the ^{51}Cr assay. The VITAL assay, based on the same principle, adds a further potential advantage, permitting differential labeling of distinct target cell populations and measurement of cytolysis against a range of targets simultaneously [38].

Overall, flow-based assays have much utility in the potency space, characterizing the dynamic relationship between target and effector cells [50, 64]. Current limitations include the need for individual sample data acquisition, increasing the time required for analysis. High-throughput multiparametric assays that allow workflow automatization and timely cytotoxicity evaluation are key to scalability and validity [9, 55].

8.6.3 Indirect Assays

Indirect assays aim to measure the by-product(s) of effector cell activation upon exposure to target cells and can be particularly useful where product availability is limited. Indirect assays measure cytokines and chemotoxins (e.g., IFN-γ, granzyme B) secreted upon effector cell activation [76, 78]. Both FDA-approved *Tisagenlecleucel* and *Axicaptagene Ciloleucel* products utilize IFN-γ secretion in response to CD19 expressing targets as part of a potency assessment for product release. Interestingly, Novartis reported that IFN-γ secretion varied greatly from batch to batch, complicating the correlation between limited potency assessment and clinical effect [30, 47].

IFN-γ detection via ELISA reflects cytokine release from the whole incubated cell population (not restricted to CAR T-cells), which can lead to an overestimation of cytokine secretion by CAR T-cells [20]. For a more specific read-out, flow cytometry assays can detect cytokines intracellularly and allow investigators to differentiate cytokine secretion between different cell types. However, a few drawbacks of this approach are the need for prolonged intracellular staining protocols and the requirement of blocking cellular secretory pathways. This assay, therefore, reflects cytokine production rather than cytokine release. Another method, the IFN-γ catch assay, utilizes a capture reagent that combines a pan-leukocyte CD45 binder and an IFN-γ binder, immobilizing the cytokine as released by each cell to its surface [22]. This overcomes the limitations outlined for ELISA and flow cytometric methods.

Some centers use indirect methods based on the correlation between T-cell degranulation and killing activity. Upon interaction with target cells, markers of T-cell activation and degranulation (e.g., CD107a) are expressed on the CAR T-cell surface and can be detected via flow cytometry. This technique is also compatible with extend phenotyping with the use of additional markers [2].

More recently, efforts have concentrated in the measurement of effector-released cytokines at a single-cell level via enzyme-linked immunospot assay (ELISPot). ELISPot requires only a low cell number for analysis, but the disadvantage is that it does not allow further immunophenotyping of the product and it is limited to the detection of only one or two enzymes [43]. Alternative approaches based on fluorophores (such as the FluoroSpot assay) could allow the accurate detection of multiple cytokines per cell [42]. Further refinement of this approach has led to the LysisPot platform that uses target cell lines expressing β-galactosidase, a nonsecreted enzyme that is released from the cells upon lysis. This method allows characterization at a single cell level of both the direct cytotoxic activity of the CAR T-cell product and cytokine (IFN-γ) release. Of note, this assay has demonstrated that not all cytokine-producing cells have cytolytic activity [4].

8.6.4 Immunophenotyping

Detailed immunophenotyping can inform CAR T-cell potency assessment, provided a correlation between specific phenotypes and product efficacy can be made [56]. Cell exhaustion and senescence are related to loss of function and disease relapse, such that expression of the phenotypic marker programmed cell death protein 1 (PD-1) (and others) could predict for functionality [32, 37]. Further, CD45RA and CD62L expression are used as markers of T-cell memory, which appears to correlate with product efficacy [56]. Immunophenotyping assays are quick and simple and allow analysis at single cell level, but results should be evaluated with caution, as these are surrogate markers of CAR T-cell functionality and results may vary significantly from patient to patient. To date, no precise immunophenotypic profile has been determined as a direct predictor of CAR T-cell function in a validated, quantitative assay.

8.6.5 Target-Induced Proliferation

Proliferation capacity upon target antigen recognition has been demonstrated to predict efficacy of CAR T-cell therapies in vivo [58, 70]. This feature could also be used as an alternative potency assay in vitro, using fluorescent markers such as carboxyfluorescein succinimidyl ester (CFSE). To date, the correlation between proliferation upon target antigen recognition in vitro and in vivo potency is still pending [14]. Cytotoxic activity is the main MoA of CAR T-cell therapies to reduce in tumor burden and as such these assays tend to be preferred for batch release assessment.

8.6.6 Polyfunctionality

Novel, high-throughput single-cell analysis platforms have the potential to revolutionize the field of CAR T-cell potency assessment. Several studies have positively correlated the presence of polyfunctional cells (cells that co-secrete multiple cytokines), with potent and durable immunity against certain infections [17, 48] and tumors [8]. Recently, highly polyfunctional CD19 CAR T-cell products were demonstrated to be associated with clinical responses in non-Hodgkin's lymphoma (NHL) patients [72]. High-throughput platforms such as the IsoPlexis system, use barcode chip assays [6] that can accurately and simultaneously measure up to 16 cytokine/chemokines secreted by thousands of CAR T-cells at a single-cell level [96]. Implementation of such high-throughput assays in potency testing may permit a more thorough characterization of CAR T-cells at single cell level and improve prediction of clinical response.

8.7 Challenges and Potential Improvements for CAR T-Cell Potency Assays

In the future, assessment of potency will encompass in vitro assays designed to study cell behavior and activity in an environment that more closely mirrors what is found in vivo. Antigen-stress tests will assist in the investigation of maintained cytotoxic activity after several rounds of exposure to target, mirroring the chronic cell activation observed in vivo and providing a model to investigate mechanisms associated with CAR T-cell failure. Target cells expressing a continuum of antigen densities can also be used to investigate the correlation between antigen density and product cytotoxicity. Soluble factors or cytokines such us TGFβ are present in vivo and can influence the biological activity of CAR T-cells [19]. TGFβ challenge assays may help to quantitate this impact on CAR T-cell function.

Due to the inherent variability of autologous CAR T-cell product, assays selected for potency assessment for final product release should have appropriate acceptance criteria that consider inter-batch variability and should be defined prior to the commencement of pivotal clinical trials [25, 28] in order to accurately define potent versus non-potent products. In the solid tumor CAR T-cell space, potency assays may have additional requirements beyond those outlined here, such as measures of CAR T-cell migration capacity to remote and immunologically hostile tumor sites [45, 61].

8.8 Future Challenges and Directions for CAR T-Cell Product Release Testing

In recent years, the field of cell therapy has developed at unprecedented speed. New CAR designs, new manufacturing technologies, and new approaches to address the current limitations of CAR T-cell therapies continuously emerge, and researchers and regulatory agencies are faced with the challenges of developing new assays and guidelines to address additional unknowns and risks [46].

Although the evaluation of cytotoxicity against CD19-expressing cells is a relatively well-described measure of CD19 CAR T potency, the panorama can be complex in the case of not as well-characterized targets and more complex and heterogenous tumors. Particularly, in the field of solid tumors, more advanced in vitro antitumor efficacy assays are likely to be required, taking into account the differential expression levels of target antigens for definition of activation thresholds, the impact of prolonged antigen exposure and the effects of the immunomodulatory tumor microenvironment. Recent approaches to overcome these challenges include the establishment of cell libraries expressing different amounts of surface antigens using CRISPR/Cas9 *knock out*, reexpression, FACS sorting and single-cell cloning [52], and the development of ex vivo tumor-derived culture systems that can account for the environment-derived immunomodulation [80]. Conversely, in vitro assays to evaluate homing and tumor infiltration are challenging. The use and characterization of animal models [1], as well as the emergence of methods that combine the use of human tumor slices and real-time imaging [23] are likely to provide unvaluable insights into some of the key quality attributes associated with in vivo efficacy of CAR T-cell therapy in solid tumors.

On the other hand, one of the biggest developments in the CAR T-cell field is the move away from viral vectors and toward alternative gene delivery methods. Older methods such as the Transposon/Transposase platform relies upon DNA plasmids and mRNA transposase electro-

or lipo-transfected into T-cells [60]. Several groups have shown the feasibility of generating CAR T-cells using the Sleeping Beauty system and minicircle vectors [51, 60]. Safety concerns with this technology include residual DNA plasmids and transposase (activity) in the final cell product and the potential risks of insertional mutagenesis and transposon remobilization. Release assays for this type of product would require an additional set of safety analyses and risk assessments to investigate the additional risk(s) posed to product recipients through use of this manufacture methodology. As an alternative to Transposon/Transposase technology, genome-editing tools such as transcription activator-like effector nucleases (TALENs) and clustered regularly interspaced short palindromic repeats (CRISPR)/CRISPR (Cas) tools allow specific modification of target genes, via disruption, correction, or replacement and have unlimited potential to improve CAR T-cell therapies [3]. In an attempt to minimize risks associated with insertional mutagenesis, targeting genes into genomic safe harbors is now possible [65].

Genome-editing tools have found favor in the development of third party or "universal" CAR T-cell therapeutics. TCRαβ/CD3 disruption has been demonstrated as a feasible approach to develop CAR T-cells products from mismatched donors, minimizing the risks of graft-versus-host disease [59, 65]. Several groups are combining TCRαβ/CD3 knockdown with additional strategies to prevent allogenic CAR T-cell rejection, which is another formidable challenge in the universal CAR T-cell space. Universal CAR therapies may overcome some of the limitations associated with autologous products such as poor-quality patient starting material, logistic and manufacturing challenges, disease progression prior to product availability, and batch-to-batch product variability. Alternative approaches using an endoplasmic reticulum retention signal to prevent CD3 surface expression have also been reported and these have the potential advantage of overcoming some of the limitations described above with genome-editing tools [68].

Genome-edited products require extensive characterization to demonstrate safety. Off-target effects are a major concern when using the CRISPR/Cas9 system, as these can lead to unintended mutagenesis and increase the risks of malignant cellular transformation [3, 12]. Although in silico methods are used to screen for potential off-target sites, they cannot precisely predict mutations that occur in vivo. EMA indicates that for genetically modified cells derived using genome-editing tools, in vitro assays for editing efficiency and off-target editing should be conducted [26]. However, development of sensitive and specific assays to detect off-target editing remains a challenge. Approaches like the T7 endonuclease 1 (T7E1) assay, deep sequencing and Chromatin Immunoprecipitation Sequencing (ChIP-seq) have been employed for detection of off-target editing. In silico prediction tools have also been developed [98], but many shortcomings are associated with these techniques. Indeed, off-target mutations with a frequency below 0.5% remain mostly undetected by current genome-wide analyses. Furthermore, targeting more than one gene for editing confers additional risk, as complex, multiplex gene editing can potentially lead to translocations induced by simultaneous double-stranded breaks at multiple loci. These have been reported to occur with a frequency as high as 7% in T-cells and have been detected by different methods, including cytogenetic analysis, qPCR, and droplet digital PCR [5, 66, 92]. Base-editing is a next-generation approach to CRISPR-Cas, which may overcome some of the risks described.

An additional risk posed by allogeneic CAR T-cells is the presence of alloreactive cells in the final product. This confers a risk of GvHD and should be addressed by efficient cell selection/depletion strategies, along with stringent purity criteria for product release. Alternative allogeneic cell sources such as NK cells or γδ T-cells might avoid the development of GvHD although challenges with rejection and persistence remain [68]. Characterization of these cell products and development of assays for identity, purity, potency, and so on will mirror, but will not be the same as those required for T-cell products.

In conclusion, as new developments increase the efficacy, applicability, and accessibility of CAR T-cell therapies, adoption of this technology for more widespread use in cancer therapy should become a reality. It is crucial for the field to develop a solid understanding of individual products and their biological activity so that critical quality attributes can be defined to ensure efficiency, consistency, and safety. In the coming years, as new data emerges from preclinical research and early clinical trials, researchers and regulatory agencies worldwide will face the challenge to keep pace with clinical development. There will be a need to generate new and harmonized guidelines to ensure patient safety and product quality to cover the diversity of emerging novel CAR T-cell therapies.

References

1. Ahmadbeigi N, Alatab S, Vasei M, Ranjbar A, Aghayan S, Khorsand A, Moradzadeh K, Darvishyan Z, Jamali M, Muhammadnejad S (2021) Characterization of a xenograft model for anti-CD19 CAR T cell studies. Clin Transl Oncol 23:2181. https://doi.org/10.1007/s12094-021-02626-5
2. Aktas E, Kucuksezer UC, Bilgic S, Erten G, Deniz G (2009) Relationship between CD107a expression and cytotoxic activity. Cell Immunol 254:149–154. https://doi.org/10.1016/j.cellimm.2008.08.007
3. Ashmore-Harris C, Fruhwirth GO (2020) The clinical potential of gene editing as a tool to engineer cell-based therapeutics. Clin Transl Med 9:15. https://doi.org/10.1186/s40169-020-0268-z
4. Bordignon V, Cordiali-Fei P, Rinaldi M, Signori E, Cottarelli A, Zonfrillo M, Ensoli F, Rasi G, Fuggetta M (2012) Evaluation of antigen specific recognition and cell mediated cytotoxicity by a modified LysisPot assay in a rat colon carcinoma model. J Exp Clin Cancer Res 31:9. https://doi.org/10.1186/1756-9966-31-9
5. Bothmer A, Gareau KW, Abdulkerim HS, Buquicchio F, Cohen L, Viswanathan R, Zuris JA, Marco E, Fernandez CA, Myer VE, Cotta-Ramusino C (2020) Detection and modulation of DNA translocations during multi-gene genome editing in T cells. CRISPR J 3:177–187. https://doi.org/10.1089/crispr.2019.0074
6. Bowman N, Liu D, Paczkowski P, Chen J, Rossi J, Mackay S, Bot A, Zhou J (2020) Advanced cell mapping visualizations for single cell functional proteomics enabling patient stratification. Proteomics 20:1900270. https://doi.org/10.1002/pmic.201900270
7. Bravery CA, Carmen J, Fong T, Oprea W, Hoogendoorn KH, Woda J, Burger SR, Rowley JA, Bonyhadi ML, Van't Hof W (2013) Potency assay development for cellular therapy products: an ISCT∗ review of the requirements and experiences in the industry. Cytotherapy 15:9–19.e9. https://doi.org/10.1016/j.jcyt.2012.10.008
8. Brummelman J, Mazza EMC, Alvisi G, Colombo FS, Grilli A, Mikulak J, Mavilio D, Alloisio M, Ferrari F, Lopci E, Novellis P, Veronesi G, Lugli E (2018) High-dimensional single cell analysis identifies stem-like cytotoxic CD8+ T cells infiltrating human tumors. J Exp Med 215:2520–2535. https://doi.org/10.1084/jem.20180684
9. Brummelman J, Haftmann C, Núñez NG, Alvisi G, Mazza EMC, Becher B, Lugli E (2019) Development, application and computational analysis of high-dimensional fluorescent antibody panels for single-cell flow cytometry. Nat Protoc 14:1946–1969. https://doi.org/10.1038/s41596-019-0166-2
10. Brunner KT, Mauel J, Cerottini JC, Chapuis B (1968) Quantitative assay of the lytic action of immune lymphoid cells on 51-Cr-labelled allogeneic target cells in vitro; inhibition by isoantibody and by drugs. Immunology 14:181–196
11. Cerignoli F, Abassi YA, Lamarche BJ, Guenther G, Santa Ana D, Guimet D, Zhang W, Zhang J, Xi B (2018) In vitro immunotherapy potency assays using real-time cell analysis. PLoS One 13:e0193498. https://doi.org/10.1371/journal.pone.0193498
12. Cheng Y, Tsai SQ (2018) Illuminating the genome-wide activity of genome editors for safe and effective therapeutics. Genome Biol 19:226. https://doi.org/10.1186/s13059-018-1610-2
13. Chmielewski M, Abken H (2020) TRUCKS, the fourth-generation CAR T cells: current developments and clinical translation. Adv Cell Gene Ther 3:e84. https://doi.org/10.1002/acg2.84
14. Clay TM, Hobeika AC, Mosca PJ, Lyerly HK, Morse MA (2001) Assays for monitoring cellular immune responses to active immunotherapy of cancer. Clin Cancer Res 75:1127–1135
15. Council of Europe (2018) European Pharmacopoeia
16. Dai X, Mei Y, Cai D, Han W (2019) Standardizing CAR-T therapy: getting it scaled up. Biotechnol Adv 37:239–245. https://doi.org/10.1016/j.biotechadv.2018.12.002
17. Darrah PA, Patel DT, De Luca PM, Lindsay RWB, Davey DF, Flynn BJ, Hoff ST, Andersen P, Reed SG, Morris SL, Roederer M, Seder RA (2007) Multifunctional TH1 cells define a correlate of vaccine-mediated protection against Leishmania major. Nat Med 13:843–850. https://doi.org/10.1038/nm1592
18. Dautzenberg IJC, Rabelink MJWE, Hoeben RC (2021) The stability of envelope-pseudotyped lentiviral vectors. Gene Ther 28:89–104. https://doi.org/10.1038/s41434-020-00193-y
19. de Charette M, Marabelle A, Houot R (2016) Turning tumour cells into antigen presenting cells: the next step to improve cancer immunotherapy? Eur J Cancer 68:134–147. https://doi.org/10.1016/j.ejca.2016.09.010
20. de Wolf C, van de Bovenkamp M, Hoefnagel M (2018) Regulatory perspective on in vitro potency assays for human T cells used in anti-tumor immunotherapy.

Cytotherapy 20:601–622. https://doi.org/10.1016/j.jcyt.2018.01.011
21. Decker T, Lohmann-Matthes ML (1988) A quick and simple method for the quantitation of lactate dehydrogenase release in measurements of cellular cytotoxicity and tumor necrosis factor (TNF) activity. J Immunol Methods 115:61–69. https://doi.org/10.1016/0022-1759(88)90310-9
22. Desombere I, Meuleman P, Rigole H, Willems A, Irsch J, Leroux-Roels G (2004) The interferon gamma secretion assay: a reliable tool to study interferon gamma production at the single cell level. J Immunol Methods 286:167–185. https://doi.org/10.1016/j.jim.2004.01.001
23. Donnadieu E, Dupré L, Pinho LG, Cotta-de-Almeida V (2020) Surmounting the obstacles that impede effective CAR T cell trafficking to solid tumors. J Leukoc Biol 108:1067–1079. https://doi.org/10.1002/JLB.1MR0520-746R
24. EudraLex (2018) The rules governing medicinal products in the European Union. Volume 4 – Good manufacturing practice. European Comission. https://health.ec.europa.eu/medicinal-products/eudralex/eudralex-volume-4_en
25. European Medicines Agency, Committee for Advanced Therapies (2016) Guideline on potency testing of cell based immunotherapy medicinal products for the treatment of cancer. EMA/CHMP/BWP/271475/2006 rev.1
26. European Medicines Agency, Committee for Advanced Therapies (2021) Guideline on quality, non-clinical and clinical aspects of medicinal products containing genetically modified cells. EMA/CAT/GTWP/671639/2008 Rev 1
27. Finak G, Langweiler M, Jaimes M, Malek M, Taghiyar J, Korin Y, Raddassi K, Devine L, Obermoser G, Pekalski ML, Pontikos N, Diaz A, Heck S, Villanova F, Terrazzini N, Kern F, Qian Y, Stanton R, Wang K, Brandes A, Ramey J, Aghaeepour N, Mosmann T, Scheuermann RH, Reed E, Palucka K, Pascual V, Blomberg BB, Nestle F, Nussenblatt RB, Brinkman RR, Gottardo R, Maecker H, McCoy JP (2016) Standardizing flow cytometry immunophenotyping analysis from the human immunophenotyping consortium. Sci Rep 6:20686. https://doi.org/10.1038/srep20686
28. Food and Drug Administration (2008) Guidance for FDA reviewers and sponsors: content and review of chemistry, manufacturing, and control (CMC) information for human somatic cell therapy investigational new drug applications (INDs). FDA-2008-D-0206. U.S. Department of Health and Human Services, Food and Drug Administration, Center for Biologics Evaluation and Research. https://www.fda.gov/regulatory--information/search-fda-guidance-documents/content-and-review-chemistry-manufacturing-and-control-cmc-information-human-somatic-celltherapy
29. Food and Drug Administration (2011) Guidance for industry: potency tests for cellular and gene therapy products. U.S. Department of Health and Human Services, Food and Drug Administration, Center for Biologics Evaluation and Research. https://www.fda.gov/regulatory--information/search-fda-guidancedocuments/potency-tests-cellular-and-gene-therapy-products
30. Food and Drug Administration (2017) BLA 125646, tisagenlecleucel, Novartis Pharmaceuticals Corporation. U.S. Department of Health and Human Services, Food and Drug Administration, US Oncologic Drugs Advisory Committee Meeting. https://www.fda.gov/media/106081/download
31. Fousek K, Watanabe J, Joseph SK, George A, An X, Byrd TT, Morris JS, Luong A, Martínez-Paniagua MA, Sanber K, Navai SA, Gad AZ, Salsman VS, Mathew PR, Kim HN, Wagner DL, Brunetti L, Jang A, Baker ML, Varadarajan N, Hegde M, Kim Y-M, Heisterkamp N, Abdel-Azim H, Ahmed N (2021) CAR T-cells that target acute B-lineage leukemia irrespective of CD19 expression. Leukemia 35:75–89. https://doi.org/10.1038/s41375-020-0792-2
32. Fraietta JA, Lacey SF, Wilcox NS, Bedoya F, Chen F, Orlando E, Brogdon JL, Hwang W-T, Frey N, Young RM, Pequignot E, Ambrose DE, Levine BL, Bitter H, Porter DL, Xu J, June CH, Melenhorst JJ (2016) Biomarkers of response to anti-CD19 chimeric antigen receptor (CAR) T-cell therapy in patients with chronic lymphocytic leukemia. Blood 128:57–57. https://doi.org/10.1182/blood.V128.22.57.57
33. Friedl J, Stift A, Paolini P, Roth E, Steger GG, Mader R, Jakesz R, Gnant MFX (2000) Tumor antigen pulsed dendritic cells enhance the cytolytic activity of tumor infiltrating lymphocytes in human hepatocellular cancer. Cancer Biother Radiopharm 15:477–486. https://doi.org/10.1089/cbr.2000.15.477
34. Gebo JET, Lau AF (2020) Sterility testing for cellular therapies: what is the role of the clinical microbiology laboratory? J Clin Microbiol 58:e01492-19. https://doi.org/10.1128/JCM.01492-19
35. Glamann J, Hansen AJ (2006) Dynamic detection of natural killer cell-mediated cytotoxicity and cell adhesion by electrical impedance measurements. Assay Drug Dev Technol 4:555–563. https://doi.org/10.1089/adt.2006.4.555
36. Graham C, Jozwik A, Pepper A, Benjamin R (2018) Allogeneic CAR-T cells: more than ease of access? Cell 7:155. https://doi.org/10.3390/cells7100155
37. Gros A, Parkhurst MR, Tran E, Pasetto A, Robbins PF, Ilyas S, Prickett TD, Gartner JJ, Crystal JS, Roberts IM, Trebska-McGowan K, Wunderlich JR, Yang JC, Rosenberg SA (2016) Prospective identification of neoantigen-specific lymphocytes in the peripheral blood of melanoma patients. Nat Med 22:433–438. https://doi.org/10.1038/nm.4051
38. Hermans IF, Silk JD, Yang J, Palmowski MJ, Gileadi U, McCarthy C, Salio M, Ronchese F, Cerundolo V (2004) The VITAL assay: a versatile fluorometric technique for assessing CTL- and NKT-mediated cytotoxicity against multiple targets in vitro and in vivo. J Immunol Methods 285:25–40. https://doi.org/10.1016/j.jim.2003.10.017

39. Höppner M, Luhm J, Schlenke P, Koritke P, Frohn C (2002) A flow-cytometry based cytotoxicity assay using stained effector cells in combination with native target cells. J Immunol Methods 267:157–163. https://doi.org/10.1016/S0022-1759(02)00167-9
40. Hu Y, Huang J (2020) The chimeric antigen receptor detection toolkit. Front Immunol 11:1770. https://doi.org/10.3389/fimmu.2020.01770
41. Huang R, Li X, He Y, Zhu W, Gao L, Liu Y, Gao L, Wen Q, Zhong JF, Zhang C, Zhang X (2020) Recent advances in CAR-T cell engineering. J Hematol Oncol 13:86. https://doi.org/10.1186/s13045-020-00910-5
42. Jahnmatz P, Bengtsson T, Zuber B, Färnert A, Ahlborg N (2016) An antigen-specific, four-color, B-cell FluoroSpot assay utilizing tagged antigens for detection. J Immunol Methods 433:23–30. https://doi.org/10.1016/j.jim.2016.02.020
43. Janetzki S, Price L, Schroeder H, Britten CM, Welters MJP, Hoos A (2015) Guidelines for the automated evaluation of Elispot assays. Nat Protoc 10:1098–1115. https://doi.org/10.1038/nprot.2015.068
44. Karimi MA, Lee E, Bachmann MH, Salicioni AM, Behrens EM, Kambayashi T, Baldwin CL (2014) Measuring cytotoxicity by bioluminescence imaging outperforms the standard chromium-51 release assay. PLoS One 9:e89357. https://doi.org/10.1371/journal.pone.0089357
45. Kiesgen S, Chicaybam L, Chintala NK, Adusumilli PS (2018) Chimeric antigen receptor (CAR) T-cell therapy for thoracic malignancies. J Thorac Oncol 13:16–26. https://doi.org/10.1016/j.jtho.2017.10.001
46. Kiesgen S, Messinger JC, Chintala NK, Tano Z, Adusumilli PS (2021) Comparative analysis of assays to measure CAR T-cell-mediated cytotoxicity. Nat Protoc 16:1331–1342. https://doi.org/10.1038/s41596-020-00467-0
47. Kite Pharma Incorporated (2017) Summary basis for regulatory action for BLA 125643 (YESCARTA™). U.S. Food and Drug Administration, Review Committee. https://www.fda.gov/media/108788/download
48. Lewinsohn DA, Lewinsohn DM, Scriba TJ (2017) Polyfunctional CD4+ T cells as targets for tuberculosis vaccination. Front Immunol 8:1262. https://doi.org/10.3389/fimmu.2017.01262
49. Lichtenfels R, Biddison WE, Schulz H, Vogt AB, Martin R (1994) CARE-LASS (calcein-release-assay), an improved fluorescence-based test system to measure cytotoxic T lymphocyte activity. J Immunol Methods 172:227–239. https://doi.org/10.1016/0022-1759(94)90110-4
50. Liu L, Chahroudi A, Silvestri G, Wernett ME, Kaiser WJ, Safrit JT, Komoriya A, Altman JD, Packard BZ, Feinberg MB (2002) Visualization and quantification of T cell-mediated cytotoxicity using cell-permeable fluorogenic caspase substrates. Nat Med 8:185–189. https://doi.org/10.1038/nm0202-185
51. Magnani CF, Tettamanti S, Alberti G, Pisani I, Biondi A, Serafini M, Gaipa G (2020) Transposon-based CAR T cells in acute leukemias: where are we going? Cell 9:1337. https://doi.org/10.3390/cells9061337
52. Majzner RG, Rietberg SP, Sotillo E, Dong R, Vachharajani VT, Labanieh L, Myklebust JH, Kadapakkam M, Weber EW, Tousley AM, Richards RM, Heitzeneder S, Nguyen SM, Wiebking V, Theruvath J, Lynn RC, Xu P, Dunn AR, Vale RD, Mackall CL (2020) Tuning the antigen density requirement for CAR T-cell activity. Cancer Discov 10:702–723. https://doi.org/10.1158/2159-8290.CD-19-0945
53. Marks P (2019) The FDA's regulatory framework for chimeric antigen receptor-T cell therapies. Clin Transl Sci 12:428–430. https://doi.org/10.1111/cts.12666
54. Marofi F, Motavalli R, Safonov VA, Thangavelu L, Yumashev AV, Alexander M, Shomali N, Chartrand MS, Pathak Y, Jarahian M, Izadi S, Hassanzadeh A, Shirafkan N, Tahmasebi S, Khiavi FM (2021) CAR T cells in solid tumors: challenges and opportunities. Stem Cell Res Ther 12:81. https://doi.org/10.1186/s13287-020-02128-1
55. Martinez EM, Klebanoff SD, Secrest S, Romain G, Haile ST, Emtage PCR, Gilbert AE (2018) High-throughput flow cytometric method for the simultaneous measurement of CAR-T cell characterization and cytotoxicity against solid tumor cell lines. SLAS Discov Adv Sci Drug Discov 23:603–612. https://doi.org/10.1177/2472555218768745
56. McLellan AD, Ali Hosseini Rad SM (2019) Chimeric antigen receptor T cell persistence and memory cell formation. Immunol Cell Biol 97:664–674. https://doi.org/10.1111/imcb.12254
57. Milone MC, O'Doherty U (2018) Clinical use of lentiviral vectors. Leukemia 32:1529–1541. https://doi.org/10.1038/s41375-018-0106-0
58. Milone MC, Fish JD, Carpenito C, Carroll RG, Binder GK, Teachey D, Samanta M, Lakhal M, Gloss B, Danet-Desnoyers G, Campana D, Riley JL, Grupp SA, June CH (2009) Chimeric receptors containing CD137 signal transduction domains mediate enhanced survival of T cells and increased antileukemic efficacy in vivo. Mol Ther 17:1453–1464. https://doi.org/10.1038/mt.2009.83
59. Mitra A, Mishra L, Li S (2013) Technologies for deriving primary tumor cells for use in personalized cancer therapy. Trends Biotechnol 31:347–354. https://doi.org/10.1016/j.tibtech.2013.03.006
60. Monjezi R, Miskey C, Gogishvili T, Schleef M, Schmeer M, Einsele H, Ivics Z, Hudecek M (2017) Enhanced CAR T-cell engineering using non-viral Sleeping Beauty transposition from minicircle vectors. Leukemia 31:186–194. https://doi.org/10.1038/leu.2016.180
61. Morello A, Sadelain M, Adusumilli PS (2016) Mesothelin-targeted CARs: driving T cells to solid tumors. Cancer Discov 6:133–146. https://doi.org/10.1158/2159-8290.CD-15-0583
62. Mullard A (2021) FDA approves fourth CAR-T cell therapy. Nat Rev Drug Discov 20:166. https://doi.org/10.1038/d41573-021-00031-9
63. On behalf of the EuroFlow Consortium (EU-FP6, LSHB-CT-2006-018708), Kalina T, Flores-Montero J, van der Velden VHJ, Martin-Ayuso M, Böttcher

S, Ritgen M, Almeida J, Lhermitte L, Asnafi V, Mendonça A, de Tute R, Cullen M, Sedek L, Vidriales MB, Pérez JJ, te Marvelde JG, Mejstrikova E, Hrusak O, Szczepański T, van Dongen JJM, Orfao A (2012) EuroFlow standardization of flow cytometer instrument settings and immunophenotyping protocols. Leukemia 26:1986–2010. https://doi.org/10.1038/leu.2012.122
64. Packard BZ, Komoriya A (2008) Intracellular protease activation in apoptosis and cell-mediated cytotoxicity characterized by cell-permeable fluorogenic protease substrates. Cell Res 18:238–247. https://doi.org/10.1038/cr.2008.17
65. Papapetrou EP, Schambach A (2016) Gene insertion into genomic safe harbors for human gene therapy. Mol Ther 24:678–684. https://doi.org/10.1038/mt.2016.38
66. Qasim W, Zhan H, Samarasinghe S, Adams S, Amrolia P, Stafford S, Butler K, Rivat C, Wright G, Somana K, Ghorashian S, Pinner D, Ahsan G, Gilmour K, Lucchini G, Inglott S, Mifsud W, Chiesa R, Peggs KS, Chan L, Farzaneh F, Thrasher AJ, Vora A, Pule M, Veys P (2017) Molecular remission of infant B-ALL after infusion of universal TALEN gene-edited CAR T cells. Sci Transl Med 9:eaaj2013. https://doi.org/10.1126/scitranslmed.aaj2013
67. Quintarelli C, Locatelli F, Caruana I, De Angelis B (2016) Overcoming challenges in CAR T-cell product CGMP release. Mol Ther 24:845–846. https://doi.org/10.1038/mt.2016.72
68. Rafiq S, Hackett CS, Brentjens RJ (2020) Engineering strategies to overcome the current roadblocks in CAR T cell therapy. Nat Rev Clin Oncol 17:147–167. https://doi.org/10.1038/s41571-019-0297-y
69. Richter JR (2021) Multiple myeloma: "if you don't stop to look around once in a while… you could miss it". Cancer J 27:183–184. https://doi.org/10.1097/PPO.0000000000000527
70. Robbins PF, Dudley ME, Wunderlich J, El-Gamil M, Li YF, Zhou J, Huang J, Powell DJ, Rosenberg SA (2004) Cutting edge: persistence of transferred lymphocyte clonotypes correlates with cancer regression in patients receiving cell transfer therapy. J Immunol 173:7125–7130. https://doi.org/10.4049/jimmunol.173.12.7125
71. Roddie C, O'Reilly M, Dias Alves Pinto J, Vispute K, Lowdell M (2019) Manufacturing chimeric antigen receptor T cells: issues and challenges. Cytotherapy 21:327–340. https://doi.org/10.1016/j.jcyt.2018.11.009
72. Rossi J, Paczkowski P, Shen Y-W, Morse K, Flynn B, Kaiser A, Ng C, Gallatin K, Cain T, Fan R, Mackay S, Heath JR, Rosenberg SA, Kochenderfer JN, Zhou J, Bot A (2018) Preinfusion polyfunctional anti-CD19 chimeric antigen receptor T cells are associated with clinical outcomes in NHL. Blood 132:804–814. https://doi.org/10.1182/blood-2018-01-828343
73. Santeramo I, Bagnati M, Harvey EJ, Hassan E, Surmacz-Cordle B, Marshall D, Di Cerbo V (2020) Vector copy distribution at a single-cell level enhances analytical characterization of gene-modified cell therapies. Mol Ther Methods Clin Dev 17:944–956. https://doi.org/10.1016/j.omtm.2020.04.016
74. Schaft N (2020) The landscape of CAR-T cell clinical trials against solid tumors—a comprehensive overview. Cancers 12:2567. https://doi.org/10.3390/cancers12092567
75. Schneider D, Xiong Y, Wu D, Nölle V, Schmitz S, Haso W, Kaiser A, Dropulic B, Orentas RJ (2017) A tandem CD19/CD20 CAR lentiviral vector drives on-target and off-target antigen modulation in leukemia cell lines. J Immunother Cancer 5:42. https://doi.org/10.1186/s40425-017-0246-1
76. Shafer-Weaver K, Sayers T, Strobl S, Derby E, Ulderich T, Baseler M, Malyguine A (2003) The Granzyme B ELISPOT assay: an alternative to the 51Cr-release assay for monitoring cell-mediated cytotoxicity. J Transl Med 1:14
77. Sheehy ME, McDermott AB, Furlan SN, Klenerman P, Nixon DF (2001) A novel technique for the fluorometric assessment of T lymphocyte antigen specific lysis. J Immunol Methods 249:99–110. https://doi.org/10.1016/S0022-1759(00)00329-X
78. Streeck H, Frahm N, Walker BD (2009) The role of IFN-γ Elispot assay in HIV vaccine research. Nat Protoc 4:461–469. https://doi.org/10.1038/nprot.2009.7
79. Stroncek DF, Jin P, Wang E, Jett B (2007) Potency analysis of cellular therapies: the emerging role of molecular assays. J Transl Med 5:24. https://doi.org/10.1186/1479-5876-5-24
80. Tano Z, Kiesgen S, Chintala N, Dozier J, Messinger J, Tan KS, Adusumilli P (2018) MA06.06 An ex-vivo patient-derived, immunocompetent (PDI) culture system to evaluate immunotherapeutic agents' antitumor efficacy. J Thorac Oncol 13:S376. https://doi.org/10.1016/j.jtho.2018.08.362
81. Tokarew N, Ogonek J, Endres S, von Bergwelt-Baildon M, Kobold S (2019) Teaching an old dog new tricks: next-generation CAR T cells. Br J Cancer 120:26–37. https://doi.org/10.1038/s41416-018-0325-1
82. Tokarew N, Ogonek J, Endres S et al (2019) Teaching an old dog new tricks: next-generation CAR T cells. Br J Cancer 120:26–37. https://doi.org/10.1038/s41416-018-0325-1
83. Tokuno O, Hayakawa A, Yanai T, Mori T, Ohnuma K, Tani A, Minami H, Sugimoto T (2015) Sterility testing of stem cell products by broad-range bacterial 16S ribosomal DNA polymerase chain reaction. Lab Med 46:34–41. https://doi.org/10.1309/LMKT4P9FFI2BBSIU
84. Tyagarajan S, Spencer T, Smith J (2020) Optimizing CAR-T cell manufacturing processes during pivotal clinical trials. Mol Ther Methods Clin Dev 16:136–144. https://doi.org/10.1016/j.omtm.2019.11.018
85. U.S. Department of Health and Human Services, Food and Drug Administration (2021) Code of Federal Regulations Title 21. Part 610 – General biological products standards. Rev. 2

86. U.S. Department of Health and Human Services, Food and Drug Administration, Center for Biologics Evaluation and Research (2020) Chemistry, manufacturing, and control (CMC) information for human gene therapy investigational new drug applications (INDs)
87. U.S. Department of Health and Human Services, Food and Drug Administration, Center for Biologics Evaluation and Research (2020) Testing of retroviral vector-based human gene therapy products for replication competent retrovirus during product manufacture and patient follow-up; Guidance for Industry
88. United States Pharmacopeia and National Formulary (2018) USP 41-NF 36 U.S
89. von Zons P, Crowley-Nowick P, Friberg D, Bell M, Koldovsky U, Whiteside TL (1997) Comparison of europium and chromium release assays: cytotoxicity in healthy individuals and patients with cervical carcinoma. Clin Diagn Lab Immunol 4:202–207. https://doi.org/10.1128/CDLI.4.2.202-207.1997
90. Wang X, Rivière I (2016) Clinical manufacturing of CAR T cells: foundation of a promising therapy. Mol Ther Oncolytics 3:16015. https://doi.org/10.1038/mto.2016.15
91. Wang D, Prager BC, Gimple RC, Aguilar B, Alizadeh D, Tang H, Lv D, Starr R, Brito A, Wu Q, Kim LJY, Qiu Z, Lin P, Lorenzini MH, Badie B, Forman SJ, Xie Q, Brown CE, Rich JN (2021) CRISPR screening of CAR T cells and cancer stem cells reveals critical dependencies for cell-based therapies. Cancer Discov 11:1192–1211. https://doi.org/10.1158/2159-8290.CD-20-1243
92. Webber BR, Lonetree C, Kluesner MG, Johnson MJ, Pomeroy EJ, Diers MD, Lahr WS, Draper GM, Slipek NJ, Smeester BA, Lovendahl KN, McElroy AN, Gordon WR, Osborn MJ, Moriarity BS (2019) Highly efficient multiplex human T cell engineering without double-strand breaks using Cas9 base editors. Nat Commun 10:5222. https://doi.org/10.1038/s41467-019-13007-6
93. Wei J, Han X, Bo J, Han W (2019) Target selection for CAR-T therapy. J Hematol Oncol 12:62. https://doi.org/10.1186/s13045-019-0758-x
94. Wei J, Guo Y, Wang Y, Wu Z, Bo J, Zhang B, Zhu J, Han W (2020) Clinical development of CAR T cell therapy in China: 2020 update. Cell Mol Immunol 18:792. https://doi.org/10.1038/s41423-020-00555-x
95. Xi B, Berahovich R, Zhou H, Xu S, Wei Y, Guan J, Harto H, Guan J, Wu L, Santa Ana D, Cerignoil F, Lamarche B, Abassi YA, Golubovskaya V (2019) A real-time potency assay for chimeric antigen receptor T cells targeting solid and hematological cancer cells. J Vis Exp (153):59033. https://doi.org/10.3791/59033
96. Xue Q, Bettini E, Paczkowski P, Ng C, Kaiser A, McConnell T, Kodrasi O, Quigley MF, Heath J, Fan R, Mackay S, Dudley ME, Kassim SH, Zhou J (2017) Single-cell multiplexed cytokine profiling of CD19 CAR-T cells reveals a diverse landscape of polyfunctional antigen-specific response. J Immunother Cancer 5:85. https://doi.org/10.1186/s40425-017-0293-7
97. Zaritskaya L, Shurin MR, Sayers TJ, Malyguine AM (2010) New flow cytometric assays for monitoring cell-mediated cytotoxicity. Expert Rev Vaccines 9:601–616. https://doi.org/10.1586/erv.10.49
98. Zhang X-H, Tee LY, Wang X-G, Huang Q-S, Yang S-H (2015) Off-target effects in CRISPR/Cas9-mediated genome engineering. Mol Ther Nucleic Acids 4:e264. https://doi.org/10.1038/mtna.2015.37
99. Zhao Y, Stepto H, Schneider CK (2017) Development of the first World Health Organization lentiviral vector standard: toward the production control and standardization of lentivirus-based gene therapy products. Hum Gene Ther Methods 28:205–214. https://doi.org/10.1089/hgtb.2017.078

Illustrative Potency Assay Examples from Approved Therapies

9

Sílvia Torrents, Marta Grau-Vorster, and Joaquim Vives

Abbreviations

3D	Three-dimensional
ALP	Alkaline phosphatase
API	Active pharmaceutical ingredient
ATMP	Advanced therapy medicinal products
CAR	Chimeric antigen receptor
CQA	Critical quality attribute
DLBCL	Diffuse large B-cell lymphoma
DP	Drug product
DS	Drug substance
ECFA	Ectopic cartilage formation assay
ELISA	Enzyme-linked immunosorbent assay
EMA	European Medicines Agency
FA	Fanconi anemia
FDA	Food and Drug Administration
GMP	Good manufacturing practice
HLA	Human leukocyte antigen
HSC	Hematopoietic stem cells
ICH	International Conference on Harmonization
iPSC	Induced pluripotent stem cells
ISCT	International Society for Cell and Gene Therapy
LESC	Limbal epithelial stem cells
MoA	Mechanisms of action
MHC	Major histocompatibility complex
MSC	Multipotent mesenchymal stromal cells
NSCLC	Non-small cell lung cancer
OTAT	Office of Tissues and Advanced Therapies
PBMC	Peripheral blood mononuclear cells
qPCR	Quantitative polymerase chain reaction
PFSB	Pharmaceutical and Food Safety Bureau
PMDA	Pharmaceuticals and Medical Devices Agency
PRIME	Priority Medicine
QC	Quality Control
TIL	Tumour infiltrating lymphocytes
TPP	Target product profile
US	United States
VCN	Vector copy number
VST	Virus-specific T cells

S. Torrents · M. Grau-Vorster
Banc de Sang i Teixits, Edifici Dr. Frederic Duran i Jordà, Barcelona, Spain

Transfusion Medicine group, Vall d'Hebron Research Institute, Universitat Autònoma de Barcelona, Barcelona, Spain

J. Vives (✉)
Banc de Sang i Teixits, Edifici Dr. Frederic Duran i Jordà, Barcelona, Spain

Musculoskeletal Tissue Engineering Group, Vall d'Hebron Research Institute (VHIR), Universitat Autònoma de Barcelona, Barcelona, Spain

Departament de Medicina, Universitat Autònoma de Barcelona, Barcelona, Spain
e-mail: jvives@bst.cat

J. S. Burns (ed.), *Potency Assays for Advanced Stem Cell Therapy Medicinal Products*, Advances in Experimental Medicine and Biology 1420, https://doi.org/10.1007/978-3-031-30040-0_9

9.1 Introduction

Advanced therapy medicinal Products (ATMP) including cell- and gene-based therapies are at the forefront of regenerative medicine, holding the potential to treat debilitating diseases and conditions [1–3]. After some timid initial positive results of clinical testing, a huge effort is currently underway to define specific target conditions that may be realistically treated with this new generation of medicines [4]. Clearly, this endeavour needs robust methods of production and suitable assays for determining critical quality attributes (CQA) of new medicines. From all CQA, those related to the potency of products are of extreme help to predict the pharmaceutical activity of such products after administration in the patients. However, this becomes challenging when actual mechanisms of action (MoA) are poorly understood, so both developers and regulators must agree on the choice of appropriate potency assays according to well-documented justification [5–9]. Importantly, such assays must be continuously improved, in a manner consistent with scientific and technological progress. Other limitations such as batch variability of starting materials, limited final product stability, and relatively small lot size (even in the context of allogeneic product banking) hinder the establishment of comprehensive product specifications in the potency tests [10].

9.2 Regulatory Framework

The field of cell and gene therapy is experiencing a rapid growth of approved cell-based ATMP medicines by the principal regulatory authorities worldwide [11]. This is remarkable given the complex procedures for production and quality control (QC) that reflect the variabilities of nature, plasticity of the drug substance (DS), and the relatively short lifespan of the fresh drug products (DP) upon batch release [10, 12–14]. Most new therapies approved in the past decade have been conditionally authorized for treating rare diseases [11, 15].

Important differences exist in the regulation, definition, scope, and approval of cell and gene therapy products by competent regulatory authorities in different parts of the world [16]. For instance, both the Food and Drug Administration (FDA) in the United States (US) of America and the European Medicines Agency (EMA) explicitly exclude gene-based prophylactic vaccines of infectious diseases from their definition of gene therapy, whereas the Japanese Pharmaceuticals and Medical Devices Agency (PMDA) does not [17, 18]. In addition, genetically modified oncolytic viral therapy falls within the definition of gene therapy in the European Union (EU) but not in the US. The EMA considers that hematopoietic stem cells (HSC) from cord blood do not fit the definition of ATMP, because these cells are not subject to substantial manipulation and are intended to be used for the same essential function in the recipient and the donor. In the US, however, HSC are classified in the category of cellular and gene therapies, which are regulated by the FDA's Office of Tissues and Advanced Therapies (OTAT). Another remarkable example of diverse criteria for classification in different countries is illustrated by the case of the melanoma treatment Imlygic® (talimogene laherparepvec), which was categorized as a gene therapy in Europe but not in the US. Alternatively, Swissmedic, the national authorization and supervisory authority for drugs and medical products in Switzerland, classified cell- and tissue-based products as transplant products, and specifically excluded cellular therapy intended for cosmetic use [15]. In all cases, proper definition of the CQA was necessary for better understanding of key parameters in the production bioprocess and specifications of the final DP that can potentially impact on patient's safety and success of the treatment. Therefore, major efforts are required from all stakeholders involved (including governments, public and private developers, and scientific societies) in order to define and standardize criteria in compliance with current quality standards and regulatory guidelines (Fig. 9.1).

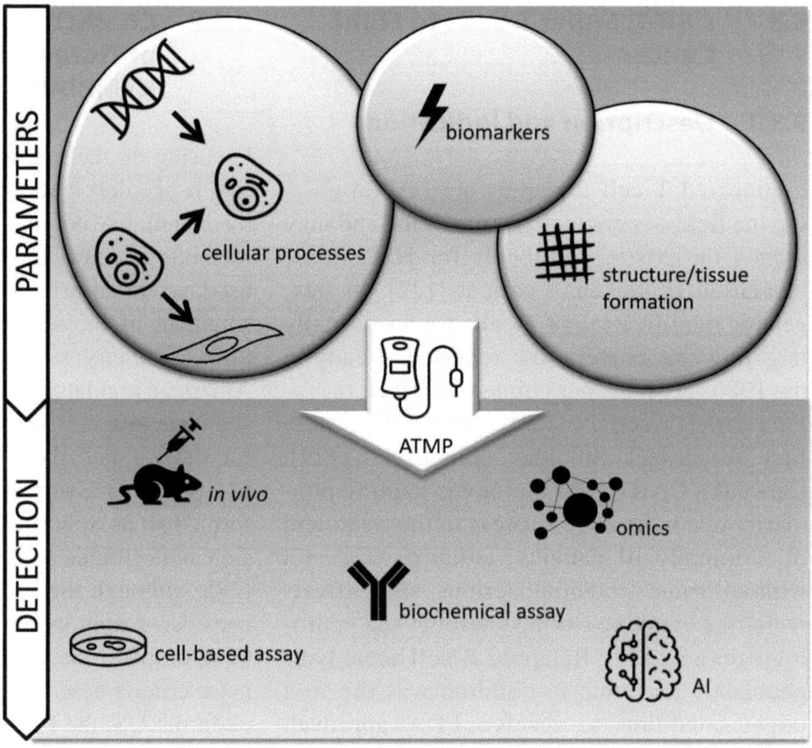

Fig. 9.1 The design of potency assays must reflect the mechanism of action. Potency assays can rely on the cellular processes and biomarkers of structures formed or remodeled in the course of action of the advanced therapy medicinal product (ATMP). Several methods are currently used to assess potency (i.e., in vitro, in vivo) and many others are currently being explored in the *omics* and artificial intelligence (AI) fields (e.g., using deep learning tools)

According to the International Conference on Harmonization (ICH) of Technical Requirements for Registration of Pharmaceuticals for Human Use, potency refers to the quantitative measure of biological activity based on the attribute of the product, which is linked to the relevant biological properties [19]. An assay aiming to demonstrate the biological activity of an ATMP should be based on the intended biological effect, which in turn should ideally correlate with the expected clinical response.

It is clear that a one-size-fits-all solution to address potency does not exist, and a case-by-case analysis is required in order to adopt existing assays or develop new designs. The FDA regulation for biological products allows some flexibility in establishing potency tests, understanding that it needs to be adapted to the unique characteristics of each product. In this sense, the FDA provides industry guidance for industry on potency assays for cellular and gene products. Similarly, the EMA has published a full guideline on potency testing of cell-based immunotherapy products for the treatment of cancer [20]. Of note, scientific advice from regulatory authorities is offered to developers to agree on appropriate tests for each specific ATMP [21]. Moreover, regulatory authorities publish reports with information on each one of the approved products, so the type of assays used for determining identity, purity, and potency are publicly available, although technical details are sometimes difficult to find, with the exception of those cases in which the developers publish their results in the scientific literature.

Hereunder, we describe three different ATMP, based on (1) an example pf genetically modified chimeric antigen receptor T-cells (CAR-T cell therapy), KYMRIAH® (tisagenlecleucel); (2) a tissue engineered product, Holoclar® (cornea-derived limbal epithelial stem cells); and (3) Remestemcel-L (bone marrow-derived multipotent stromal cells). We discuss further potency assay development selected from recent ATMP currently under clinical investigation.

9.3 CAR-T: Super T Cells to Fight Cancer

9.3.1 Description and Indication

Engineered T cell therapies are revolutionizing the field of personalized medicine and may impact on existing protocols for HSC transplantation in leukemia patients [22]. In fact, genetic modification of T cells for specifically targeting cancer were first reported already in the 1980s [23]. Proper chimeric antigen receptor (CAR) T-cells, as we know today, were first described in the mid-1990s [24]. Currently, CAR-T cell therapy is gaining popularity due to its huge success in the treatment of terminally ill patients, although it is not without some potential serious side effects including cytokine storm syndrome and neurologic toxicity [25]. Relapsed B-cell acute lymphoblastic leukemia in children was the first target condition to receive FDA approval. KYMRIAH® was designated as an orphan medicinal product for the treatment of B-lymphoblastic leukemia/lymphoma (EU/3/14/1266 on April 29, 2014) and for the treatment of diffuse large B-cell lymphoma (EU/3/16/1745 on October 14, 2016). In 2016, KYMRIAH® was granted eligibility to Priority Medicine (PRIME) scheme for the treatment of pediatric patients with relapsed or refractory B cell acute lymphoblastic leukemia. Its formal FDA approval as CAR-T cell therapy for this condition came in 2017, with subsequent approval in 2018, for diffuse large B-cell lymphoma and certain other types of lymphoma.

Another CAR-T cell therapy named YESCARTA (axicabtagene ciloleucel), manufactured by Kite Pharma, was also approved in 2017 for the treatment of adult patients with relapsed or refractory large B-cell lymphoma after two or more lines of systemic therapy, including diffuse large B-cell lymphoma (DLBCL) not otherwise specified, primary mediastinal large B-cell lymphoma, high grade B-cell lymphoma, and DLBCL arising from follicular lymphoma.

9.3.2 Characterization of Tisagenlecleucel Product KYMRIAH®

Focusing on the first approved CAR-T cell therapy, tisagenleucleucel, identity was demonstrated by quantitative polymerase chain reaction (qPCR) methods specific for the CAR gene sequence used, vector copy number (VCN) assay, and measurement of the surface expression of CAR by flow cytometry techniques [26]. European and American regulatory authorities published reports showing different product characterization tests for CAR-T cell therapy product release. In the EU, potency measurement was reported to be performed so as to ensure appropriate CAR expression and cytokine secretion upon T cell activation [27]. Although the proposed specifications were considered appropriate, the applicant was asked to re-evaluate the release tests and their acceptance criteria based on post-approval data.

In the US, the FDA reported that the potency of tisagenlecleucel was evaluated by measuring IFN-γ production in response to tumour antigen-bearing cells [26]. IFN-γ production was considered an indicator of T cell activation and a prerequisite for CAR-T cell activity. However, in the clinical trials, IFN-γ production varied greatly from batch to batch, making it difficult to correlate IFN-γ production in vitro to the tisagenlecleucel safety or efficacy. This clearly illustrated the difficulty of finding a suitable and robust potency assay while showing that this need not represent an insurmountable roadblock for progress toward clinical use if issues can be conveniently justified and reevaluated when new information is available.

Recent reports insist upon the need for joint efforts between product developers and regulatory authorities for better bioassays providing clinically relevant potency assessment and standardization [28, 29]. Moreover, tracking of the ATMP in patients is important and the fact that CAR-T cells are edited genetically offers a unique opportunity to study persistence, biodistribution, and phenotype of cells after infusion by tracking their unique characteristics [30].

9.4 Holoclar®: A Tissue Engineering Product to Regenerate Cornea

9.4.1 Description and Indication

The loss of corneal stem cells due to injury or disease results in impaired repair of the damaged cornea and an overgrowth of conjunctival epithelium that can subsequently lead to vision loss [31]. Specifically, limbal stem cell deficiency can be caused by physical or chemical ocular burns, and it is considered a rare disease by the EMA [32]. Holoclar® is the registered name for Chiesi Farmaceutici's therapeutic product based on autologous limbal epithelial stem cell (LESC)-based therapy [33]. Expanded to a clinical dose ex vivo on a cellular matrix LESC provide a treatment that allows the preservation of undifferentiated stem cells that can form an epithelial cell sheet suitable for transplantation and a sustained protection of the corneal tissue in the patient [31, 34].

9.4.2 Characterization of Limbal Epithelial Stem Cell Product Holocar®

The cell type of interest in Holoclar® is defined as a $p63^{bright}$ stem cell subset forming undifferentiated holoclones with high growth potential as the main functional component of the final product, since these are the cells expected to mediate long-term regenerative efficacy. Potency was therefore addressed by quantification of $p63^{bright}$ cells. Further differentiated cell populations present in the drug product (DP) were considered to be supportive, but functionally contributing only to short- or medium-term efficacy. Extensive earlier research had indicated that LESC cultures containing more than 3% $p63^{bright}$ cells led to successful corneal epithelial repair outcomes [35]. Thus, this served as a quantitative pharmacodynamic marker that was selected for characterization and control of the medicinal product instead of a using an extended panel composed of p63 and the newly described markers ABCB5, PAX6, and WNT/A. This decision was accepted by the regulatory authorities and both parties agreed to include the extended characterization as part of the confirmatory study post-marketing authorization. Importantly, the product was released also on the basis of macroscopic and microscopic appearance and results of an intermediate control testing established as an in-process controls (IPC) due to the short shelf life of the DP that impairs further testing before implantation in the patient. Microscopic and macroscopic assessment were conveniently validated and acceptable levels of precision, accuracy, and reliability were provided to the regulatory authorities by Chiesi Farmaceutici [33]. This example illustrates very well how potency testing can advantageously be kept simple and focused on what is certainly known about the attributes of the candidate medicine.

9.5 Remestemcel-L: MSC for the Management of GvHD

9.5.1 Description and Indication

Multipotent/mesenchymal stromal cells (MSC) represent a type of stem cells with multipotent differentiation potential mesodermal lineages that also display immunomodulatory properties with a notable capacity to exert various therapeutic effects via paracrine signaling [36, 37]. Although this cell type was first described in the 1970s [38], the understanding of their MoA in different conditions has become one of the main challenges for advancing the development of novel MSC-based therapies [39, 40]. Under different commercial names (e.g. Prochymal, Ryoncil, TEMCELL HS Inj.), products based on allogeneic expanded MSC have become available in different regions of the world for the treatment of graft versus host disease (GvHD) [11].

9.5.2 Characterization of Remestemcel-L Product PROCHYMAL® (or RYONCIL™)

Purity of MSC-based products is determined by the absence of hematopoietic cells, and their identity is characterized by broad though nonspecific MSC markers (e.g., CD105, CD73, CD90), as proposed by the International Society for Cell and Gene Therapy (ISCT) [41, 42]. It is believed that MSC share essential MoA mediating their immunomodulatory function regardless of the tissue source and/or in vitro expansion procedures [5]. Therefore, pro-angiogenic and immunomodulatory effects are commonly tested in the manufacture of MSC-based products [43]. Indeed, the ISCT proposes immune functional assays as a potency release criterion [44], since the minimal criteria for characterizing MSC seem to be insufficient indicators of therapeutic success [42]. In this context, a potency assay is certainly an indispensable tool to ensure that MSC-based products exert a differential therapeutic effect at a specific dosage [6, 45].

Three key parameters conforming to established specifications of Ryoncil included cell viability, the ability of MSC to inhibit IL-2Rα expression by activated T lymphocytes, and levels of tumour necrosis factor receptor 1 (TNFR1) expression associated with significant inhibition of the proliferation of peripheral blood mononuclear cells (PBMC). These parameters were the focus of investigations of the potency assays for QC and stability studies along the product development program [41, 46]. In addition to these criteria, it had been observed that low major histocompatibility complex (MHC) expression supports the immuno-privileged status of MSC to avoid rejection in the host, the manufacturer included in vitro evidence that MSC do not express human leukocyte antigen (HLA)-DR (i.e., MHC type II) molecules and express low levels of MHC type I molecules on the cell surface [47]. However, there is controversy regarding this point and other authors, including our group, have reported conflicting data [39, 48–51]. In any case, MSC seem to be indeed immuno-privileged through a mechanism named efferocytosis, which is a recently proposed hypothesis that is gaining relevance and is based on the short-living time of MSC in vivo, and it is independent of the expression of HLA markers [5].

9.6 Addressing Potency in Other Selected ATMP

9.6.1 Approved ATMP

As a result of continued improvements of both product quality and also strategic vision, some traditional blood-related products have been marketed as therapeutics in the recent years [11, 52]. This is the case of products such as Hemacord (from the NY Blood Center) consisting of allogeneic HSC from cord blood. In this regard, due to the wide-spread use of HSC, the FDA elaborated the "Guidance for Industry Biologics License Applications for Minimally Manipulated, Unrelated Allogeneic Placental/Umbilical Cord Blood Intended for Hematopoietic and Immunologic Reconstitution in Patients with Disorders Affecting the Hematopoietic System," including a requirement for recommended tests for identity, purity, and potency (Table 9.1). Interestingly, in this case, purity and potency bioassays are common [53]. Beyond marketing approval mechanisms for making ATMP available to patients include the hospital exemption clause in Europe, applicable to those products not intended to be commercialized or for situations when it is produced in a nonindustrial manner [54, 55]. Such has been the case for academic institutions that developed their own CAR-T cell therapies for specific uses not covered by other marketed products. Notably, the research led by Dr. Manel Juan (Hospital Clínic, Barcelona, Spain) resulted in a CAR-T treatment against $CD19^+$ for the treatment of B-cell malignancies [56].

Interestingly, potency testing used for product release was based on an in vitro cytotoxicity assay using the CAR-T manufactured cells cocultured with NALM6 cells (a B cell precursor leukemia cell line). After 4 h of coculture, the percentage of live CD19 cells was measured by flow cytometry with an established acceptance criteria for product release of less than 70% viable $CD19^+$ cells [56]. Evaluation of the clinical

Table 9.1 FDA guidance for blood-products characterization in terms of purity, potency, and identity

Characteristics	Testing	Sample	Expected results
Identity	HLA typing	Cord blood	Report
	Confirmatory HLA typing	Attached segment of HSC	Confirmatory results
	Blood group and Rh type	Cord blood	Report
Purity and potency	TNC	HSC (pre-cryopreservation)	$\geq 5.0 \times 10^8$ TNC/unit
	Viable nucleated cells	HSC (pre-cryopreservation)	≥85% viable nucleated cells
	Viable CD34+ cells (flow cytometry)	HSC (pre-cryopreservation)	$\geq 1.25 \times 10^6$ viable CD34+ cells/unit

Presentation of different tests associated to each type of sample and results of accepted criteria for each testing. *HLA* human leucocyte antigen, *HSC* hematopoietic stem cells, *TNC* total nucleated cells

outcomes in patients receiving CAR-T cell therapy was challenging due to differences in CAR and vector design, effector T-cell selection, CAR-T-cell production methods and choice of treatment cohorts. Therefore, it was all the more important to ensure relevant potency assays were harmonized among production facilities [28].

For MSC, great efforts have been made to improve the methods for assessing potency, from traditional potency assays (e.g., proliferation, in vitro differentiation assays, inhibition of the proliferation of stimulated PBMC) to the use of omics technologies [57]. These methods include anti-inflammatory and immunomodulatory potency assays such as the endothelial tube formation assay used for MultiStem® [58]. For an in vivo test of immunoregulatory effectiveness, an ovalbumin challenge model of acute asthma has been developed [59]. Angiogenic potency assays have been based on the secretion of "pro-angiogenic factors" e.g., vascular endothelial growth factor (VEGF), Interleukin 6 (IL-6), platelet-derived growth factor (PDGF), and C-X-X motif chemokine ligand 5 (CXCL5) among others [58, 60, 61]. Following observation that when treating inflammatory diseases with MSC conditioned-medium, patient serum Interleukin 10 (IL-10) levels increased, a potency assay based on MSC release of anti-inflammatory IL-10 was developed [62]. Further examples of secreted proteins may serve as potency assay candidates, emerged from the observation that local or systemic administration of MSC could ameliorate corneal chemical injury by MSC secretion of TNF-stimulated gene/protein 6 (TSG-6) in response to injury signals from the cornea [63]. MSC-derived elevated levels of anti-inflammatory soluble mediators, such as the heme-containing enzyme indoleamine-pyrrole-2,3-dioxygenase (IDO), prostaglandin E_2 (PGE2), transforming growth factor beta (TGF-β), nitric oxide, HLA-G5, and interleukins, have encouraged interest in MSC paracrine signaling [64]. More recently, innovative assays have been developed and validated based on the measurement of an expanded set of potency biomarkers types (e.g., gene expression analysis, telomere length, telomerase activity, cell size) [65, 66], proteomics, analysis of the secretome, and transcriptomics [67], complementing the study of surface markers by flow cytometry [68]. In the context of MSC-based therapy for the treatment of immunological of inflammatory disorders, the potency assays most commonly used are based on the determination of their in vitro immunomodulation capacity [6, 44].

9.6.2 ATMP Under Clinical Investigation

The number of marketed ATMP is gradually increased each year and new developments, especially those concerning products related to immunotherapy are likely to boost this list in the coming years [3, 11, 69]. Most probably for this reason, the specific guidelines from regulatory authorities and scientific societies made available for developers have focused on these products [6, 21, 44]. A couple of very interesting developments in this immunotherapy field are the tumour-infiltrating lymphocytes (TIL) or the virus-specific T cells (VST).

TIL therapy development was pioneered by Dr. Steven A. Rosenberg (National Cancer Institute, Bethesda, MD, USA) and IOVANCE Biotherapeutics has conducted several Phase II clinical trials to assess the efficacy and safety of autologous TIL for treatment of patients with metastatic melanoma, squamous cell carcinoma of the head and neck, non-small cell lung cancer (NSCLC) and cervical cancer [70]. Potency testing, consisted of subjecting cells to an IFN-γ release assay by restimulating TIL products with anti-CD3/CD28/CD137-coated beads for 18–24 h with subsequent harvesting of supernatants for assessment of IFN-γ secretion using an ELISA assay.

The second immunotherapy already undergoing clinical trials in several countries, predominantly in the US led by the Baylor College of Medicine (Houston, TX, USA), seeks to take advantage of virus-specific T cells. VST aims to treat different viral infections including cytomegalovirus, Epstein-Barr (EBV) virus, BK virus, adenovirus, and human herpesvirus 6 (HHV-6) [71]. Fanconi anemia (FA), a defective DNA repair syndrome, is associated with congenital abnormalities, cancer predisposition, and bone marrow failure during the first decade of a patient's life. An innovative gene-based strategy has shown promise for the correction of the pathogenic mutations present in the FANCA gene sequence, introducing the prospect of a low-toxicity therapeutic option for this life-threatening disorder. The team led by Dr. Juan Bueren (CIEMAT, Madrid, Spain) demonstrated that lentiviral-mediated hematopoietic gene therapy reproducibly conferred engraftment and a proliferation advantage of gene-corrected HSC in non-conditioned patients with FA subtype A [72]. Demonstrating the potency of the procedure for phenotypic correction of blood and bone marrow cells, the targeted hematopoietic progenitors and T lymphocytes acquired resistance to DNA cross-linking agents. To test the repopulating ability of $CD34^+$ cells edited with the therapeutic cassette, samples were transplanted into mice. Analysis of the hematopoietic organs confirmed a multilineage human hematopoietic engraftment. In addition to the in vivo data, qPCR analyses confirmed successful gene editing in the target HSC. Interestingly, corrected HSC showed repopulating properties in vivo that provided a selective advantage with respect to HSC carrying FANCA mutations [73].

9.7 The Case of Pluripotent Stem Cells

Since their discovery by Prof. Shinya Yamanaka (Kyoto University, Japan) [74], the development of induced pluripotent stem cells (iPSC) has offered an evolved understanding of mechanisms governing cell type-specific differentiation furthering the possibility of scalable manufacture of cellular therapies for regenerative medicine [75, 76]. Potency is understood as the capacity of iPSC to differentiate into clinically relevant cells having specific phenotypic and functional qualities that can be found in cells from any of the three germ layers [77]. Of particular potential benefit, this approach allows for donor selection on the basis of major transplant antigen systems, thus enabling MHC matching to improve engraftment and complement other immunoregulatory approaches to allow iPS treatment for the widest range of patients worldwide, reducing the likely risk of immunological rejection and the degree of immune suppression or tolerance required [78–80]. In this regard, a few initiatives are currently addressing the production of cell banks of clinical grade iPSC of specific haplotypes of high frequency in the population as active pharmaceutical ingredients (API) according to current good manufacturing practice (cGMP) regulations [81].

9.8 Final Remarks

Current examples of potency testing among approved ATMP have confirmed the complexity of finding a suitable potency assay, yet they have shown that relatively straightforward strategies can be followed by developers to comply with regulatory requirements. It is encouraging that our understanding of MoA has been able to advance rapidly together with technologies enabling novel bioassay designs, with potency assays stimulating collaborative progress.

References

1. Naji A, Eitoku M, Favier B, Deschaseaux F, Rouas-Freiss N, Suganuma N (2019) Biological functions of mesenchymal stem cells and clinical implications. Cell Mol Life Sci 76:3323
2. Fischbach MA, Bluestone JA, Lim WA (2013) Cell-based therapeutics: the next pillar of medicine. Sci Transl Med 5(179):179ps7
3. Boráň T, Menezes-Ferreira M, Reischl I, Celis P, Ferry N, Gänsbacher B et al (2017) Clinical development and commercialization of advanced therapy medicinal products in the European Union: how are the product pipeline and regulatory framework evolving? Hum Gene Ther Clin Dev 28(3):126–135
4. Galipeau J (2013) The mesenchymal stromal cells dilemma–does a negative phase III trial of random donor mesenchymal stromal cells in steroid-resistant graft-versus-host disease represent a death knell or a bump in the road? Cytotherapy 15(1):2–8
5. Galipeau J, Sensébé L (2018) Mesenchymal stromal cells: clinical challenges and therapeutic opportunities. Cell Stem Cell 22(6):824–833
6. de Wolf C, van de Bovenkamp M, Hoefnagel M (2017) Regulatory perspective on in vitro potency assays for human mesenchymal stromal cells used in immunotherapy. Cytotherapy 19(7):784–797
7. de Wolf C, van de Bovenkamp M, Hoefnagel M (2018) Regulatory perspective on in vitro potency assays for human dendritic cells used in anti-tumor immunotherapy. Cytotherapy 20(11):1289–1308
8. de Wolf C, van de Bovenkamp M, Hoefnagel M (2018) Regulatory perspective on in vitro potency assays for human T cells used in anti-tumor immunotherapy. Cytotherapy 20(5):601–622
9. Hematti P (2016) Characterization of mesenchymal stromal cells: potency assay development. Transfusion 56(4):32S–35S
10. Schneider CK, Salmikangas P, Jilma B, Flamion B, Todorova LR, Paphitou A et al (2010) Challenges with advanced therapy medicinal products and how to meet them. Nat Rev Drug Discov 9(3):195–201
11. Cuende N, Rasko JEJ, Koh MBC, Dominici M, Ikonomou L (2018) Cell, tissue and gene products with marketing authorization in 2018 worldwide. Cytotherapy 20(11):1401–1413
12. Mirabel C, Puente-Massaguer E, Del Mazo-Barbara A, Reyes B, Morton P, Gòdia F et al (2018) Stability enhancement of clinical grade multipotent mesenchymal stromal cell-based products. J Transl Med 16(1):291
13. Veronesi E, Murgia A, Caselli A, Grisendi G, Piccinno MS, Rasini V et al (2014) Transportation conditions for prompt use of ex vivo expanded and freshly harvested clinical-grade bone marrow mesenchymal stromal/stem cells for bone regeneration. Tissue Eng Part C Methods 20(3):239–251
14. Gálvez-Martín P, Hmadcha A, Soria B, Calpena-Campmany AC, Clares-Naveros B (2014) Study of the stability of packaging and storage conditions of human mesenchymal stem cell for intra-arterial clinical application in patient with critical limb ischemia. Eur J Pharm Biopharm 86(3):459–468
15. Shukla V, Seoane-Vazquez E, Fawaz S, Brown L, Rodriguez-Monguio R (2019) The landscape of cellular and gene therapy products: authorization, discontinuations, and cost. Hum Gene Ther Clin Dev 30(3):102–113
16. Coppens DGM, De Bruin ML, Leufkens HGM, Hoekman J (2018) Global regulatory differences for gene- and cell-based therapies: consequences and implications for patient access and therapeutic innovation. Clin Pharmacol Ther 103(1):120–127
17. Nakayama Y, Aruga A (2015) Comparison of current regulatory status for gene-based vaccines in the U.S., Europe and Japan. Vaccines (Basel) 3(1):186–202
18. Kusakabe T (2015) Regulatory perspectives of Japan. Biologicals 43(5):422–424
19. Guideline IHT (1999) ICH topic Q6B. Note for guidance on specifications: test procedures and acceptance criteria for biotechnological/biological products. European Medicines Agency, London
20. Administration U.S. Department of Health and Human Services Food and Drug Administration (2011) Guidance for industry potency tests for cellular and gene therapy products, Rockville
21. Agency EM (2016) Guideline on potency testing of cell based immunotherapy medicinal products for the treatment of cancer
22. Jacoby E (2019) The role of allogeneic HSCT after CAR T cells for acute lymphoblastic leukemia. Bone Marrow Transplant 54(Suppl 2):810–814
23. Gross G, Waks T, Eshhar Z (1989) Expression of immunoglobulin-T-cell receptor chimeric molecules as functional receptors with antibody-type specificity. Proc Natl Acad Sci U S A 86(24):10024–10028
24. Eshhar Z, Waks T, Gross G, Schindler DG (1993) Specific activation and targeting of cytotoxic lymphocytes through chimeric single chains consisting of antibody-binding domains and the gamma or zeta subunits of the immunoglobulin and T-cell receptors. Proc Natl Acad Sci U S A 90(2):720–724
25. Bonifant CL, Jackson HJ, Brentjens RJ, Curran KJ (2016) Toxicity and management in CAR T-cell therapy. Mol Ther Oncolytics 3:16011
26. Administration FaD (2017) Tis agenlecleucel. Contract No.: BLA 125646

27. Agency EM (2018) Kymriah. London. Contract No.: EMA/485563/2018
28. Quintarelli C, Locatelli F, Caruana I, De Angelis B (2016) Overcoming challenges in CAR T-cell product CGMP release. Mol Ther 24(5):845–846
29. Wang L, Gong W, Wang S, Neuber B, Sellner L, Schubert ML et al (2019) Improvement of in vitro potency assays by a resting step for clinical-grade chimeric antigen receptor engineered T cells. Cytotherapy 21(5):566–578
30. Vives J, Casademont-Roca A, Martorell L, Nogués N (2020) Beyond chimerism analysis: methods for tracking a new generation of cell-based medicines. Bone Marrow Transplant 55:1229
31. Notara M, Daniels JT (2008) Biological principals and clinical potentials of limbal epithelial stem cells. Cell Tissue Res 331(1):135–143
32. Milazzo G, Ardigò D, Toschi M, Matuska S, Rama P, De Luca M et al (2016) Holoclar®: first of its kind in more ways than one. Cell Gene Ther Insights 2(2):15
33. Agency EM (2014) Assessment report: Holoclar. London. Contract No.: EMA/25273/2015
34. Pellegrini G, Rama P, Di Rocco A, Panaras A, De Luca M (2014) Concise review: hurdles in a successful example of limbal stem cell-based regenerative medicine. Stem Cells 32(1):26–34
35. Rama P, Matuska S, Paganoni G, Spinelli A, De Luca M, Pellegrini G (2010) Limbal stem-cell therapy and long-term corneal regeneration. N Engl J Med 363(2):147–155
36. Gomzikova MO, James V, Rizvanov AA (2019) Therapeutic application of mesenchymal stem cells derived extracellular vesicles for immunomodulation. Front Immunol 10:2663
37. Spees JL, Lee RH, Gregory CA (2016) Mechanisms of mesenchymal stem/stromal cell function. Stem Cell Res Ther 7(1):125
38. Friedenstein AJ, Gorskaja JF, Kulagina NN (1976) Fibroblast precursors in normal and irradiated mouse hematopoietic organs. Exp Hematol 4(5):267–274
39. Mendicino M, Bailey AM, Wonnacott K, Puri RK, Bauer SR (2014) MSC-based product characterization for clinical trials: an FDA perspective. Cell Stem Cell 14(2):141–145
40. López-Beas J, Guadix JA, Clares B, Soriano-Ruiz JL, Zugaza JL, Gálvez-Martín P (2020) An overview of international regulatory frameworks for mesenchymal stromal cell-based medicinal products: from laboratory to patient. Med Res Rev 40:1315
41. Daly A (2012) Remestemcel-L, the first cellular therapy product for the treatment of graft-versus-host disease. Drugs Today (Barc) 48(12):773–783
42. Dominici M, Le Blanc K, Mueller I, Slaper-Cortenbach I, Marini F, Krause D et al (2006) Minimal criteria for defining multipotent mesenchymal stromal cells. The International Society for Cellular Therapy position statement. Cytotherapy 8(4):315–317
43. Bartaula-Brevik S, Pedersen TO, Finne-Wistrand A, Bolstad AI, Mustafa K (2016) Angiogenic and immunomodulatory properties of endothelial and mesenchymal stem cells. Tissue Eng Part A 22(3–4):244–252
44. Galipeau J, Krampera M, Barrett J, Dazzi F, Deans RJ, DeBruijn J et al (2016) International Society for Cellular Therapy perspective on immune functional assays for mesenchymal stromal cells as potency release criterion for advanced phase clinical trials. Cytotherapy 18(2):151–159
45. Oliver-Vila I, Ramírez-Moncayo C, Grau-Vorster M, Marín-Gallén S, Caminal M, Vives J (2018) Optimisation of a potency assay for the assessment of immunomodulative potential of clinical grade multipotent mesenchymal stromal cells. Cytotechnology 70(1):31–44
46. Kebriaei P, Isola L, Bahceci E, Holland K, Rowley S, McGuirk J et al (2009) Adult human mesenchymal stem cells added to corticosteroid therapy for the treatment of acute graft-versus-host disease. Biol Blood Marrow Transplant 15(7):804–811
47. Administration TG (2015) Australian Public Assessment Report for Remestemcel-L, ex vivo adult human mesenchymal stem cells. Wooden, Australia. Contract No.: Prochymal
48. Codinach M, Blanco M, Ortega I, Lloret M, Reales L, Coca MI et al (2016) Design and validation of a consistent and reproducible manufacture process for the production of clinical-grade bone marrow-derived multipotent mesenchymal stromal cells. Cytotherapy 18(9):1197–1208
49. Dighe PA, Viswanathan P, Mruthunjaya AK, Seetharam RN (2013) Effect of bFGF on HLA-DR expression of human bone marrow-derived mesenchymal stem cells. J Stem Cells 8(1):43–57
50. Grau-Vorster M, Rodríguez L, Torrents-Zapata S, Vivas D, Codinach M, Blanco M et al (2019) Levels of IL-17F and IL-33 correlate with HLA-DR activation in clinical-grade human bone marrow-derived multipotent mesenchymal stromal cell expansion cultures. Cytotherapy 21(1):32–40
51. Grau-Vorster M, Laitinen A, Nystedt J, Vives J (2019) HLA-DR expression in clinical-grade bone marrow-derived multipotent mesenchymal stromal cells: a two-site study. Stem Cell Res Ther 10(1):164
52. Vives J, Oliver-Vila I, Pla A (2015) Quality compliance in the shift from cell transplantation to cell therapy in non-pharma environments. Cytotherapy 17(8):1009–1014
53. Administration FaD (2014) Guidance for industry biologics license applications for minimally manipulated, unrelated allogeneic placental/umbilical cord blood intended for hematopoietic and immunologic reconstitution in patients with disorders affecting the hematopoietic system, Rockville
54. Cuende N, Boniface C, Bravery C, Forte M, Giordano R, Hildebrandt M et al (2014) The puzzling situation of hospital exemption for advanced therapy medicinal products in Europe and stakeholders' concerns. Cytotherapy 16(12):1597–1600
55. Roura S, Gálvez-Montón C, Mirabel C, Vives J, Bayes-Genis A (2017) Mesenchymal stem cells for cardiac repair: are the actors ready for the clinical scenario? Stem Cell Res Ther 8(1):238

56. Castella M, Caballero-Baños M, Ortiz-Maldonado V, González-Navarro EA, Suñé G, Antoñana-Vidósola A, Boronat A, Marzal B, Millán L, Martín-Antonio B, Cid J, Lozano M, García E, Tabera J, Trias E, Perpiña U, Canals JM, Baumann T, Benítez-Ribas D, Campo E, Yagüe J, Urbano-Ispizua Á, Rives S, Delgado J, Juan M (2020) Point-of-care CAR T-cell production (ARI-0001) using a closed semi-automatic bioreactor: experience from an academic phase I clinical trial. Front Immunol 11(482)
57. Samsonraj RM, Rai B, Sathiyanathan P, Puan KJ, Rötzschke O, Hui JH et al (2015) Establishing criteria for human mesenchymal stem cell potency. Stem Cells 33(6):1878–1891
58. Lehman N, Cutrone R, Raber A, Perry R, Van't Hof W, Deans R et al (2012) Development of a surrogate angiogenic potency assay for clinical-grade stem cell production. Cytotherapy 14(8):994–1004
59. Bonfield TL, Nolan Koloze MT, Lennon DP, Caplan AI (2010) Defining human mesenchymal stem cell efficacy in vivo. J Inflamm (Lond) 7:51
60. Bloom DD, Centanni JM, Bhatia N, Emler CA, Drier D, Leverson GE et al (2015) A reproducible immunopotency assay to measure mesenchymal stromal cell-mediated T-cell suppression. Cytotherapy 17(2):140–151
61. Tao H, Han Z, Han ZC, Li Z (2016) Proangiogenic features of mesenchymal stem cells and their therapeutic applications. Stem Cells Int 2016:1314709
62. Jiao J, Milwid JM, Yarmush ML, Parekkadan B (2011) A mesenchymal stem cell potency assay. Methods Mol Biol 677:221–231
63. Roddy GW, Oh JY, Lee RH, Bartosh TJ, Ylostalo J, Coble K et al (2011) Action at a distance: systemically administered adult stem/progenitor cells (MSCs) reduce inflammatory damage to the cornea without engraftment and primarily by secretion of TNF-α stimulated gene/protein 6. Stem Cells 29(10):1572–1579
64. Guadix JA, López-Beas J, Clares B, Soriano-Ruiz JL, Zugaza JL, Gálvez-Martín P (2019) Principal criteria for evaluating the quality, safety and efficacy of hMSC-based products in clinical practice: current approaches and challenges. Pharmaceutics 11(11)
65. Oliver-Vila I, Coca MI, Grau-Vorster M, Pujals-Fonts N, Caminal M, Casamayor-Genescà A et al (2016) Evaluation of a cell-banking strategy for the production of clinical grade mesenchymal stromal cells from Wharton's jelly. Cytotherapy 18(1):25–35
66. Murgia A, Veronesi E, Candini O, Caselli A, D'souza N, Rasini V et al (2016) Potency biomarker signature genes from multiparametric osteogenesis assays: will cGMP human bone marrow mesenchymal stromal cells make bone? PLoS One 11(10):e0163629
67. Wagner W, Wein F, Seckinger A, Frankhauser M, Wirkner U, Krause U et al (2005) Comparative characteristics of mesenchymal stem cells from human bone marrow, adipose tissue, and umbilical cord blood. Exp Hematol 33(11):1402–1416
68. Sousa BR, Parreira RC, Fonseca EA, Amaya MJ, Tonelli FM, Lacerda SM et al (2014) Human adult stem cells from diverse origins: an overview from multiparametric immunophenotyping to clinical applications. Cytometry A 85(1):43–77
69. Salmikangas P, Schuessler-Lenz M, Ruiz S, Celis P, Reischl I, Menezes-Ferreira M et al (2015) Marketing regulatory oversight of advanced therapy medicinal products (ATMPs) in Europe: the EMA/CAT perspective. Adv Exp Med Biol 871:103–130
70. Rosenberg SA (2011) Cell transfer immunotherapy for metastatic solid cancer–what clinicians need to know. Nat Rev Clin Oncol 8(10):577–585
71. Tzannou I, Papadopoulou A, Naik S, Leung K, Martinez CA, Ramos CA et al (2017) Off-the-shelf virus-specific T cells to treat BK virus, human herpesvirus 6, cytomegalovirus, Epstein-Barr virus, and adenovirus infections after allogeneic hematopoietic stem-cell transplantation. J Clin Oncol 35(31):3547–3557
72. Río P, Navarro S, Wang W, Sánchez-Domínguez R, Pujol RM, Segovia JC et al (2019) Successful engraftment of gene-corrected hematopoietic stem cells in non-conditioned patients with Fanconi anemia. Nat Med 25(9):1396–1401
73. Diez B, Genovese P, Roman-Rodriguez FJ, Alvarez L, Schiroli G, Ugalde L et al (2017) Therapeutic gene editing in CD34. EMBO Mol Med 9(11):1574–1588
74. Okita K, Ichisaka T, Yamanaka S (2007) Generation of germline-competent induced pluripotent stem cells. Nature 448(7151):313–317
75. Chang DJ, Lee N, Park IH, Choi C, Jeon I, Kwon J et al (2013) Therapeutic potential of human induced pluripotent stem cells in experimental stroke. Cell Transplant 22(8):1427–1440
76. Cyranoski D (2013) Stem cells cruise to clinic. Nature 494(7438):413
77. Baghbaderani BA, Tian X, Neo BH, Burkall A, Dimezzo T, Sierra G et al (2015) cGMP-manufactured human induced pluripotent stem cells are available for pre-clinical and clinical applications. Stem Cell Rep 5(4):647–659
78. Barry J, Hyllner J, Stacey G, Taylor CJ, Turner M (2015) Setting up a Haplobank: issues and solutions. Curr Stem Cell Rep 1(2):110–117
79. Murata K, Ikegawa M, Minatoya K, Masumoto H (2020) Strategies for immune regulation in iPS cell-based cardiac regenerative medicine. Inflamm Regen 40:36
80. Morizane A, Kikuchi T, Hayashi T, Mizuma H, Takara S, Doi H et al (2017) MHC matching improves engraftment of iPSC-derived neurons in non-human primates. Nat Commun 8(1):385
81. Alvarez-Palomo B, Vives J, Casaroli-Marano RPP, Gomez SG, Rodriguez Gómez L, Edel MJ et al (2019) Adapting cord blood collection and banking standard operating procedures for HLA-homozygous induced pluripotent stem cells production and banking for clinical application. J Clin Med 8(4)

From the Integrity of Potency Assays to Safe Clinical Intervention: Legal Perspectives

10

Waltter Roslin and Juli Mansnérus

10.1 Introduction

Stem cells are increasingly researched and applied within medicine, as their potential holds considerable promise. However, intensified research on their potential therapeutic applications ushers the development of medicinal products based on this technology, which also brings forth the need to consider how the novel products function within the pre-established legal framework intended to safeguard the research industry, and most importantly the patients. In regenerative medicine, potency of a stem cell-based product is a critical quality attribute [20]. Therefore, implementation of relevant potency assays is of paramount importance for assessment of quality, integrity, and consistency of an advanced therapy medicinal product (ATMP). They are also critical for clinical development as a tool for prediction of ATMP's clinical efficacy and effective dose. Finally, potency assays help ensure integrity, stability, and quality in manufacturing ATMPs. Establishment of potency for biological products is appreciably complicated (see for instance [15]) and in the case of stem cell-based ATMPs, even harder due to their highly complex molecular nature and modes of action involving many pathways that may not be entirely known or understood. A risk-based approach, relying on the most recent scientific publications on modes of action, coupled with robust development data is needed for creating a potency assay strategy for product development and authorization purposes.

This chapter outlines the EU regulatory framework for stem cell-based ATMPs, illustrating some possible ways to meet the regulatory challenges of potency assays associated with ATMP development. It also provides an overview of the ATMP Regulation from the perspective of stem cell products, focusing on their definition, requirements, and application within the framework of the law, while introducing the relevant legislations regarding the planned development and research of ATMPs. The main focus will be within EU legislation, paying attention to the publications of the European Medicines Agency, to establish and present the current interpretation of the EU legislation in place.[1] While the whole lifecycle of a medicinal product needs to be acknowledged, this overview will emphasize the premarket authorization stage, given the current lack of many stem-cell-based products beyond the authorization stage.

W. Roslin · J. Mansnérus (✉)
University of Helsinki, Helsinki, Finland
e-mail: juli.mansnerus@helsinki.fi

[1] As the pharmaceutical industry operates also on a global level, insight can be gained from observing foreign practices. For further reading, see: Takashima et al. [24]; Ghinea et al. [11]; for a US-EU overview of policies see: Iglesias-Lopez et al. [15]; however, this review intends to give a short introduction into the EU regulatory scheme, and is thus limited in scope.

J. S. Burns (ed.), *Potency Assays for Advanced Stem Cell Therapy Medicinal Products*, Advances in Experimental Medicine and Biology 1420, https://doi.org/10.1007/978-3-031-30040-0_10

10.2 The Evolving Regulatory Concept of Potency

The potency assay establishes a quantitative measure of biological activity that seeks to measure the ability of the medicinal product to trigger a specific response in a disease-relevant biological system [2, 6, 7, 12]. Initially, the ICH Q6B [12] guideline defined potency as "*[t]he measure of the biological activity using a suitably quantitative biological assay (also called potency assay or bioassay), based on the attribute of the product which is linked to the relevant biological properties.*" It should be noted that the notion of potency has evolved in course of recent decades and it has been applied to various different types of pharmaceuticals ranging from plant-based medicinal products to chemical ones, then later to more complex biological products now including ATMPs. Earlier, with limited knowledge of the active substance's biochemical structure, potency verification was actually the only way to ensure the therapeutic product would meet its requirements when applied in the clinic. Also in earlier times, reference was made to the concept of "biological activity measurement," while now in modern-day medicine the biochemical structure of and mode of action of a synthetically produced molecule is well characterized. Therefore, in case of synthesized chemical medicinal products, measuring the content (the actual strength) by a dosage in most cases is an adequate approach [21, pp. 5–10].

A new consideration when applying potency measurement to highly complex biological products, such as ATMPs, is the additional complexity and involvement of heterogeneous molecular properties resulting in modes of action that may not be entirely understood, since in some cases they may entail a number of different pathways. Consequently, the "old-school" mass and content-focused approach may no longer be an adequate measure to determine the biological activity, or potency of a particular ATMP. For this reason, as part of quality assessment and quality control, specific potency assays are needed to detect the actual ATMP functional activity [21, pp. 5–10]. In the case of some other types of biologicals, the modes of action are often at least partly specified, facilitating design and implementation of the chosen potency assay approach. In this respect, developers of stem cell-based ATMPs can encounter huge challenges due to variability of their starting materials and the highly complicated biological properties of their productsproduct development process [5].

The potency of these stem cells can be specified by means of diverse functional assays besides the evaluation of various molecular markers.[2] Usually, knowledge on a specific product is gradually compiled during the experimental product development process. Sometimes the process starts with identification of a simple indicator that is later refined toward more specific, relevant, and reliable quantification of its biological activity.[3] In the case of stem cell-based ATMPs, potency assessment is typically a protracted ongoing and evolving development process that, if successful, results in the determinant tests being applied to a particular product. They are used not only for monitoring stability but also for creating a link between quality criteria and efficacy [21, pp. 5–10]. When an experimental product is being developed and scaled up to a commercial one, potency assays establish comparability and consistency between ATMP batches in industry-scale manufacture. As a necessary cornerstone for robust ATMP development and quality assurance, potency assays play a key role in product characterization and authorization for its release.

10.3 The Emergence of the Regulatory Landscape for Stem Cell-Based ATMPs and Their Potency Assays

The rapid emergence of human tissue engineering technologies in the late 1990s raised some worries about insufficient regulatory governance in the field of regenerative medicine and it was

[2]These molecular markers could include diagnosis of transcriptional, epigenetic, and metabolic states of stem cells.

[3]A simple indicator could be, for instance, a specific cell surface marker.

evident that more harmonized EU legislation was needed [16, p. 172]. The urging need for creating a favorable regulatory atmosphere to support and facilitate the development of a strong internal market for ATMPs has persisted. Retrospectively, it has turned out to be a difficult task. Even today, developers of potency assays for stem cell-based ATMPs do not only encounter intrinsic[4] or operational challenges,[5] but they also frequently encounter legal and regulatory roadblocks during the development process.

In 2007, Regulation No. 1394/2007 of the European Parliament and of the Council (the ATMP Regulation) was issued as a consequence of heavy industry lobbying, resulting in introduction of industry-level current good manufacturing practice (cGMP) requirements to cover small-scale, niche ATMP products [16]. That made the market access process quite burdensome for small- and medium-sized enterprises and academia operating in the field. Yet, more recently, some adaptations and flexibilities have been introduced to the applicable cGMP guidelines that have been made more specific to ATMPs. The most recent adaptations to the ATMP-specific cGMP requirements, together with risk-proportionate adaptations to clinical trials, represent positive developments that may accelerate market entry of ATMPs, since several changes in applicable guidelines and standards aim to facilitate the development and manufacturing of ATMPs in the future [17, pp. 128–133].

[4]Further analysis of these intrinsic challenges is left outside the scope of this chapter. Still, it can be briefly mentioned that it is very difficult to characterize the complete mode of action of a stem cell-based ATMP. The mode of action can also be associated with a number of different factors that are not clearly indicated. Some of these factors may also take place in different stages of the in vivo response to the therapy. It can be an impossible task to develop an assay that reflects complete mode of action each and every element of a complex stem cell-based ATMP to qualify all steps of the organism's response to the therapy. Furthermore, stem cell-based ATMPs often comprise of may active cell types involving potential biological activity. There may also be different kind of synergies and interferences that depend on the composition of the product.

[5]Despite operational challenges are left outside the scope of this chapter, it can be mentioned that such challenges include, for instance, short shell lives of the products requiring potency assays that can be read fast, limited amount of starting materials resulting in small batch-sizes, and also any sample taken for purposes of quality assurance reduces the quantity of product available to the patient.

10.4 Overview of the Current EU Regulatory Framework for Stem Cell-Based Therapies

Stem cell-based products often fall under the definition of ATMPs consisting of products that generally encompass recombinant nucleic acids or engineered cells and/or tissues [13, p. 2]. These are covered by the EU-wide ATMP Regulation intended to harmonize the development, safety, and introduction of medicinal products within the Member States, promoting free movement of the products available to every State. The overall scheme of medicinal products for human use is governed by Directive 2001/83/EC and Regulation 726/2004/EC, that underpin the general authorization procedure, established the European Medicines Agency and harmonize definitions of medicinal products for all Member States. Table 10.1 provides an overview of relevant EU legislation covering the ATMP field.

10.4.1 ATMP Regulation Covering the Market Access, Supervision and Pharmacovigilance of Advanced Therapies

The scope of the Regulation is defined as the authorization, supervision, and pharmacovigilance of ATMPs. Clearly, the rules are intended to be *lex specialis*, and while the Regulation is to be read in conjunction with both Directive 2001/83/EC and Regulation 726/2004/EC, it holds the priority when the subject matter consists of ATMPs [13, p. 2]. The Regulation consists of eight chapters that concern the harmonization of definitions, specific requirements set for ATMPs for market authorization, the process both prior and

Table 10.1 Overview of the ATMP relevant EU legislation

EU legislation	Topic	Comment
Directive 2001/83/EC	The Community code relating to medicinal products for human use	Harmonizes definitions, sets rules for monitoring and the procedure for market authorization
Regulation (EC) No. 726/2004	Laying down Community procedures for the authorization and supervision of medicinal products for human and veterinary use and establishing a European Medicines Agency (EMA)	Builds upon Directive 2001/83 and sets up EMA
Regulation (EC) No. 141/2000	Orphan medicinal products	Defines the specific rules regarding orphan medicinal products and establishes the Committee for Orphan Medicinal Products
Regulation (EC) No. 1901/2006	Medicinal products for pediatric use and amending Regulation (EEC) No. 1768/92, Directive 2001/20/EC, Directive 2001/83/EC, and Regulation (EC) No. 726/2004	Defines the specific rules regarding medicinal products for children and establishes the Paediatric Committee. **N.B.** Articles 20, 49, and 51 are amended in Regulation (EC) No. 1902/2006
Regulation (EC) No. 1394/2007	Advanced therapy medicinal products and amending Directive 2001/83/EC and Regulation (EC) No. 726/2004	Defines the specific rules regarding ATMPs and amending the preexisting legislation to better encompass ATMPs. Establishes the Committee for Advanced Therapies
Directive 2001/20/EC	The approximation of the laws, regulations, and administrative provisions of the Member States relating to the implementation of good clinical practice in the conduct of clinical trials on medicinal products for human use (Clinical Trials Directive)	Predecessor to Regulation (EU) No. 536/2014, seeking to harmonize administrative procedure governing clinical trials within EU Member States
Regulation (EU) No. 536/2014	Clinical trials on medicinal products for human use and repealing Directive 2001/20/EC	Not yet in force, will be set to replace Clinical Trials Directive once the EU portal and database by the European Medicines Agency is fully functional (estimate end of 2021)
Directive 2004/23/EC	Setting standards of quality and safety for the donation, procurement, testing, processing, preservation, storage, and distribution of human tissues and cells	Harmonizes the procedures regarding the handling of human tissues and cells

post authorization, establishing the Committee for Advanced Therapies (CAT), and provision of incentives for technological development.

The current definition for ATMPs allows reference to any of the three already introduced categories, a gene therapy medicinal product (GTMP) as defined in Part IV of Annex I to Directive 2001/83/EC, a somatic cell therapy medicinal product (CTMP) as defined in Part IV of Annex I to Directive 2001/83/EC, or a tissue engineered product (TEP) as defined in point (b). Prior to the introduction of the ATMP Regulation, GTMPs and CTMPs had already been introduced with Directive 2003/63/EC amending Directive 2001/83/EC that had introduced the ATMPs in 2003; however, TEPs were considered an unregulated class of medicinal products, with varying practices between Member States regarding their authorization. This divergence had resulted in fragmentation of practice that threatened the free movement of TEPs, which required harmonization; hence, this was achieved via the ATMP Regulation [16, p. 24]. TEPs were added under the umbrella of ATMPs, as "*products that contains or consists of engineered cells or tissues, and is presented as having properties for, or is used in or administered to human beings with a view to regenerating, repairing or replacing a human tissue.*" The final subcategory are combined ATMPs, products that combine medical devices as an integrated part of the medicine. A product's classification can require profound scientific analysis. For instance, when considering cell therapy medicinal products and TEPs, both aspects require that the manipulation of a living material should be considered engineered. Article

2 1.c. defines ATMPs as "engineered" products that contain or consist of cells or tissues that have been subject to substantial manipulation, so that "biological characteristics, physiological functions or structural properties relevant for the intended regeneration, repair or replacement are achieved" and/or "the cells or tissues are not intended to be used for the same essential function or functions in the recipient as in the donor." [16, p. 24]. However, the distinction between CTMP and TEP is that instead of administering the product with a view to treating, preventing, or diagnosing a disease through the pharmacological, immunological, or metabolic actions of its cells or tissues, the product is used to regenerate, repair, or replace human tissue. What matters in the categorization is the intended action and effect of the product. Naturally, problems can arise when the distinction between the products is unclear; for instance, a product that exerts a pharmacological action in order to regenerate, repair, or replace a human tissue. For these cases, premises have been established in order to categorize a specific product: a product which may fall within the definition of a TEP and CTMP should be considered a TEP according to ATMP Regulation, although the final classification should be considered on a case-by-case basis [13, p. 8].

There is also an exception intended for specific, more experimental products excluding them from the jurisdiction of the ATMP Regulation, known as the "hospital exception." According to Article 28, if a product "*which is prepared on a non-routine basis according to specific quality standards, and used within the same Member State in a hospital under the exclusive professional responsibility of a medical practitioner, in order to comply with an individual medical prescription for a custom-made product for an individual patient*" then it will be up for the Member States to authorize the use these products (Directive 2001/83/EC Art 3(7)). The exception was adopted to ensure that non-profit entities would not be barred from developing ATMPs, due to the lack of financial and human resources required to comply with EU pharmaceutical legislation. This has remained one of the most controversial articles, as its rather broad interpretation, it could enable the circumvention of the requirements set by the ATMP Regulation. Additionally, Committee for Advanced Therapies (CAT) and EMA have recently published warnings toward unregulated cell-based treatments available to patients, as their impact can be ineffective or unsafe [9]. It is here where the easier affirmation of potency becomes increasingly relevant to shift the paradigm toward proven rather than unproven products.[6]

When considering the perspective of market authorization, the ATMP Regulation introduced the centralized procedure as the mandatory procedure for authorization. In the case of regular medicinal products, the applicant can in general choose between the national paths operated by the Member States' national authority or decide upon the centralized authorization procedure, coordinated by EMA. The ATMP requirement for a centralized procedure is due to the evaluation requiring a very specific technological knowledge that might not be readily available within Member States, and to ensure a high level of scientific evaluation of these medical products within the EU (ATMP Regulation, Preface (9)(10)). With the evaluation of the product, the CAT needs to be consulted prior to the decision being taken, giving them the pivotal role of providing scientific assessment of ATMPs. Based on this consultation, the decision is then taken by the Committee for Medicinal Products for Human Use (ATMP Regulation, Article 8). However, with clinical trials for ATMPs, the application needs to be submitted to the competent national authority, depending on where the clinical trials are to be conducted. This means that while the market authorization procedure must be centralized, clinical trials are still Member State specific.

As alluded in the previous paragraph, the ATMP Regulation established the CAT to provide an opinion on the quality, safety, and efficacy of an ATMP for the final approval before marketing authorization by the Committee for

[6]For further reading, see: Master et al. [18]; Smith et al. [23].

Medicinal Products for Human Use (ATMP Regulation, Article 22). However, the CAT also bears the role of providing advice and assistance in determining whether a product falls under the definition of an ATMP, or in providing general scientific assistance regarding ATMPs. Thus, overall, the CAT is the committee within EMA responsible for classifying; assessing the quality, safety, and efficacy of ATMPs; and following scientific progress within its mandate [13, p. 2]. Currently, the CAT is engaged with finalizing a guideline on quality, nonclinical, and clinical requirements for applications for clinical trials for ATMPs [1].

Notwithstanding the EMA's harmonization attempts, it appears that it is rather difficult to set quality criteria for stem cells, due to the heterogeneity of a cell preparation and also due to the presence of cells at various stages of induced differentiation. The quality control of a product is important for both patient safety and efficacy of the product, meeting a number of concerns regarding viral safety, characterization of cell population purity and differentiation status, potency testing, and process validation to control for consistency and potential risk of tumorigenicity [10].

The ATMP Regulation has several inconsistencies, both in the application of the hospital exception as well as the lack of harmonized classifications. Inconsistencies in the application of the hospital exemption is conducive to creating uncertainty among national competent authorities and developers of ATMPs, as it does not promote the harmonization of practices. Inconsistencies in the implementation of the ATMP Regulation, in particular the lack of harmonized ATMP classifications, constitute a barrier to the development of ATMPs across the EU, as national competent authorities cannot resort to classification procedure when they face difficulties with the classification of ATMPs.

Simultaneously, cGMP compliance-related costs have been reported to constitute a major bottleneck for the translation of research into advanced therapies. It has been argued that rigorous technical requirements (which are not negative as such) risk becoming disproportionately costly for SMEs and, consequently, impeding innovation [16]. Recent ATMP-specific adaptations to cGMP requirements have been welcomed by developers of ATMPs, as the specific characteristics of ATMPs are now better taken into consideration. These flexibilities are anticipated to decrease the costs related to compliance with cGMP guidelines. It appears that along with the EMA, some regulatory authorities in the Member States seem to have adopted a pragmatic approach already, allowing for a risk-based assessment of manufacturing procedures. The most recent adaptations to the ATMP-specific cGMP requirements together with the risk-proportionate adaptations to clinical trials represent positive developments that may facilitate the market entry of ATMPs.

Among other things, a number of changes in applicable guidelines and standards are likely to facilitate the development and manufacturing of ATMPs in the foreseeable future: ATMP-specific cGMP standards, Q & A document on the risk-based approach to non-substantially manipulated ATMPs, guidelines on GLP for ATMPs (all above published 2017), guidelines on good clinical practice (GCP) for ATMPs led by the European Commission (published 2019), guidelines on investigational ATMPs (consultation closed), scientific guidelines on ATMPs (a number of guidelines have been adopted lately or are being revised), scientific considerations on gene editing technologies (under preparation), as well as guidelines on safety, efficacy, and risk management plans (RMPs) for ATMPs and the re-revision of the EMA's procedural guidance on the evaluation of ATMPs (both revised in 2018). The introduction of the supplementary cGMP requirements for ATMPs is not only necessary to facilitate the market entry of new ATMPs but also serves to protect public health.[7]

[7]Interestingly, some individual action plans have been made within the EU to facilitate the introduction of new developments, see: Cuende et al. [3].

10.4.2 Allogeneic or Autologous: Does the Origin of the Source Materials Affect the Process?

In order to acquire the genetic material required for stem cell-based ATMPs, the donation, procurement, and testing of those cells or tissues shall be made in accordance with Directive 2004/23/EC (ATMP Regulation, Article 3). The supervision of which is left for the applicable Member States' national authority. There is no specific regulatory impact on whether the source of the cells is allogeneic or autologous; however, if the tissues and cells are removed and transplanted from the same individual, within the same surgical procedure, these are excluded from the jurisdiction of Directive 2004/23/EC according to Article 8. From the ATMP perspective, the substantial manipulation of stem cells is a mandatory feature in its determination; hence, when considering either allogeneic or autologous sources of cells, the extent that they are manipulated influences the applicable legislative framework.

The majority of ATMPs that progress to authorization or at least to the stage of clinical trials are manufactured from autologous mononuclear cells. The starting material is procured by hospital- or blood bank-operated apheresis facilities, which creates a peculiar situation in which a product starts under Directive 2004/23/EC, before falling under the ATMP Regulation, and where the hospitals' role transforms into a service provider for industry. By becoming a service provider, there should be further definition of the respective responsibilities and liabilities for both parties involved [19, p. 463].

10.4.3 EU Clinical Trials Regulation Streamlining the Application Procedure

While market access in the EU is granted via the centralized procedure, the approval of clinical trials for ATMPs is within the mandate of each Member States in which the sponsor plans to conduct the clinical trial. Member States have individual regulatory oversight for application, administrative procedures, execution, and surveillance of such studies. However, general guidance and framework has been implemented by the European Commission to approximate the rules and requirements for conducting clinical trials on medicinal products for human use, to ensure easier access within the European market. In 2001, the Clinical Trials Directive 2001/20/EC was issued by the European Commission to create such a framework through uniform good clinical practices. The directive also tackled aspects regarding the protection of clinical trial subjects as well as the formal and legal framework for the management of clinical trials, referring to the commencement, conduct, amendment, and suspension of a clinical trial [22, p. 87]. Naturally, as ATMPs are often the result of a complex manufacturing process, donor suitability, and quality control of procuring the genetic material is integral to the development of ATMPs. These requirements for quality and safety standards for the donation, procurement, and testing of human tissues and cells are specified in Directives 2004/23/EC and 2006/17/EC and Directive 2002/98/EC for human blood cells [22, pp. 87–98].

Overall, the CAT and the Committee for Medicinal Products for Human Use (CHMP) at the EMA have issued several guidelines addressing the scientific requirements for ATMPs. However, most of these guidance documents have described the set of quality, nonclinical, and clinical data needed at the level of the marketing authorization application of an ATMP. This often differs from the available data during the clinical trial development [22, pp. 91–92]. Nonetheless, with the Draft CAT guideline EMA/CAT/852602/2018, there is a specific plan to tackle ATMP-related clinical trials. This is most likely related to the enforcement of the new Clinical Trials Regulation.

The new Clinical Trials Regulation (Regulation 536/2014) entered into force on June 16, 2014; however, due to technical difficulties, an integral part for its functioning. The new Clinical Trials Information System (CTIS) was unable to be completed within the intended timeframe. It then followed that the old directive was applicable until January 31, 2022 when the CTIS became fully functional and operational.

From the perspective of ATMP developers, the Clinical Trial Regulation allows for a streamlined application procedure via a single-entry point EU portal and a more harmonized review of clinical trial applications with a view to faster approval times. Member States do retain the authorization and oversight of clinical trials with added supervision from the EMA through the management of CTIS. The EMA has also paid attention to the differing requirements across the EU Member States. In particular, the integration of assessment in clinical trial authorizations poses a challenge in the context of multicenter clinical trials on ATMPs. The timelines of such assessment should be aligned with those of a clinical trial authorization. Regardless of the harmonization of the application process, it should be noted that ethical approvals of clinical trials remain within the competence of the Member States. Consequently, the endorsement of a trial depends on the ethical position adopted by the ethical boards of the Member States [17, p. 134].

10.4.4 "Soft Law" Encountering "Hard Science": Flexibilities Are Needed to Deal with Rapid Scientific Advancements in an Ethically Sensitive Field

Not all legislative tools are heavily binding sources like the mandatory ATMP Regulation, as the nature of the ATMPs is highly technical, the creation of all-encompassing legislation can be considered a difficult feat, as the potential limits of technological development are constantly being stretched. Therefore, paired with the binding legislative framework consisting of the regulations and directives, EMA, CAT, and the Commission have issued guiding documents and recommendations to better facilitate the application of the laws as well as react to the changing potential within technology. An overview of the EU guidance covering ATMPs is provided in Table 10.2.

These so-called "soft-law" guidelines could be also used to address the EU's limited mandate to harmonize the ethical aspects of ATMPs; that is, the origin of ethical neutrality and the highly technical approach taken with the ATMP Regulation where the most disputed ethical aspects have not been dealt with (namely, the commercialization of altruistically donated material of human origin) [17, p. 128]. Through common targets and recommendations, unified practice can be reached. However, with ethical issues left to be dealt with by the Member State, there are still approaches lacking harmony with regard to the availability of certain types of raw materials or products based on such materials. In addition, the current wording of Article 4 Directive 2001/83/EC is drafted so ambiguously that the Member States may deny access to products based on cells or tissues on many possible grounds [16, p. 173; 17, p. 128].

Where soft law has been utilized more, is in relation to redefining cGMP compliance, especially with small-scale production of autologous products. For such tailor-made production, the expensive industrial cGMP manufacture model is not well suited [16, pp. 173–174]. In order to mitigate these issues, the European Commission launched ATMP-specific cGMP guidelines also addressing some particularities of autologous products [4]. In general, cGMP compliance-related costs have been reported to constitute a major bottleneck for the translation of research into advanced therapies. It has been argued that rigorous technical requirements (which are not negative as such) risk becoming disproportionately costly for SMEs and, consequently, impeding innovation. Recent ATMP-specific adaptations to cGMP requirements have been welcomed by developers of ATMPs, as the specific characteristics of ATMPs are now better taken into consideration. These flexibilities are anticipated to decrease the costs related to compliance with cGMP guidelines. In addition, the introduction of the supplementary cGMP requirements for ATMPs is not only necessary to facilitate the market entry of new ATMPs but also to protect public health. It appears that along with the EMA, some regulatory authorities in the Member States seem to have adopted a pragmatic approach

Table 10.2 Overview of the EU guidance documents covering ATMPs

Document title	Document description
Guideline on human cell-based medicinal products (EMEA/CHMP/410869/2006)	A multidisciplinary guideline addressing development, manufacturing, and quality control as well as nonclinical and clinical development of cell-based medicinal products.
Guideline on potency testing of cell-based immunotherapy medicinal products for the treatment of cancer (CHMP/BWP/271475/06)	This guidance document covers viable cell products for cancer-immunotherapy from autologous or allogeneic origin.
Guideline on safety and efficacy follow-up and risk management of advanced therapy medicinal products (EMEA/149995/2008)	The Regulation defines ATMPs as gene therapy medicinal products, somatic cell therapy medicinal products, and tissue engineered products. This Guideline focuses on unique characteristics of ATMPs as further detailed in the Chap. 6 – Scientific Rationale. Its applicability is restricted to ATMPs.
Quality, nonclinical, and clinical aspects of medicinal products containing genetically modified cells (CHMP/GTWP/671639/2008)	This Guideline defines scientific principles and provides guidance for the development and evaluation of medicinal products containing genetically modified cells intended for use in humans and presented for marketing authorization. Its focus is on the quality, safety, and efficacy requirements of genetically modified cells developed as medicinal products.
Reflection paper on stem cell-based medicinal products (EMA/CAT/571134/2009)	The aim of this reflection paper is to cover specific aspects related to stem cell-based medicinal products for marketing authorization application.
Reflection paper on in vitro cultured chondrocyte containing products for cartilage repair of the knee (EMA/CAT/CPWP/568181/2009)	This reflection paper addresses specific points related to medicinal products containing in vitro cultured autologous chondrocytes intended for the repair of cartilage lesions of the knee.
Guideline on xenogeneic cell-based medicinal products (EMEA/CHMP/CPWP/83508/2009)	This Guideline addresses the scientific requirements for xenogeneic cell-based medicinal products for human use.
Reflection paper on clinical aspects related to tissue engineered products (EMA/CAT/573420/2009)	This reflection paper is intended to provide specific guidance on clinical testing for tissue engineered products as defined in Regulation (EC) No. 1394/2007. This also applies to cells or tissues combined with a medical device and considered a combined advanced therapy medicinal product (ATMP) according to Art. 2(d) of Regulation (EC) No. 1394/2007.
Advanced therapy medicines: exploring solutions to foster development and expand patient access in Europe (EMA/345874/2016)	This report summarizes the main ideas and solutions proposed during the meeting as well as responses sent ahead of the meeting via a questionnaire. (Facilitating research and development, optimizing regulatory process for ATMPs, moving from hospital exemption to marketing authorization, improving funding, investment, and patient access.)
Guidelines of 22.11.2017 Good Manufacturing Practice for advanced therapy medicinal products (C(2017) 7694)	These Guidelines develop the GMP requirements that should be applied in the manufacturing of ATMPs that have been granted a marketing authorization and of ATMPs used in a clinical trial setting. These Guidelines do not apply to medicinal products other than ATMPs.
Questions and answers on comparability considerations for advanced therapy medicinal products (ATMP) (EMA/CAT/499821/2019)	The presented Q & A aims to address specific issues pertaining to the demonstration of comparability at the level of quality aspects for ATMPs.
Guidelines on Good Clinical Practice specific to advanced therapy medicinal products (C(2019) 7140)	These Guidelines develop the GCP requirements that are specific to clinical trials conducted with ATMPs. These Guidelines are to be read in conjunction with the International Council for Harmonisation of Technical Requirements for Pharmaceuticals for Human Use (ICH) guidelines on good clinical practice, which are also applicable to ATMPs. To the extent that there is a difference in the requirements, the content of these Guidelines prevails.

(continued)

Table 10.2 (continued)

Document title	Document description
Draft guideline on quality, nonclinical, and clinical requirements for investigational advanced therapy medicinal products in clinical trials (EMA/CAT/852602/2018)	The guideline provides guidance on the structure and data requirements for a clinical trial application for exploratory and confirmatory trials with advanced therapy investigational medicinal products (ATMPs).
ICH guideline S12 on nonclinical biodistribution considerations for gene therapy products – Step 2b (EMA/CHMP/ICH/318372/2021)	The objective of this guideline is to provide harmonized recommendations for the conduct of nonclinical biodistribution (BD) studies in the development of gene therapy (GT) products. This document provides recommendations for the overall design of nonclinical BD assessments.

already, allowing for a risk-based assessment of manufacturing procedures. The most recent adaptations to the ATMP-specific cGMP requirements together with the risk-proportionate adaptations to clinical trials represent positive developments that may facilitate the market entry of ATMPs. Among other things, several changes in applicable guidelines and standards are hoped to facilitate the development and manufacturing of ATMPs in the near future [17, pp. 128–133].

10.4.5 EU "Soft Law" Levelling the Playing Field for Potency Assay Developers

Interestingly, even today some of the EMA's guidelines discussing quality aspects of ATMPs seem to refer back to the ICH Q6B1 guideline from 1999, that specified expectations for a biological product. Later, the first EU guideline on quality and manufacturing issues for cell-based medical products was enacted in 2001. The focus was on quality management of cell-based medicinal products. It addressed among other things the use of validated potency assays for measuring biological activity. Already then, potency assays were seen as one of the quality attributes and criteria for qualification, validation, and control strategy of cell-based products. The superseding current guideline on human cell-based medicinal products (EMEA/CHMP/410869/06), notes the particular difficulties faced when characterizing the biological function of a cell-based medicinal product. In terms of potency assessment, it includes the possibility to use several potency assay types. Additionally, it permits use of surrogate markers[8] for in vitro assays linked to the intended biological activity of the cell-based medicinal product. Yet, it should be noted that if the mode of action entails some specific metabolic activity, surrogate endpoints alone are unlikely to be sufficient for potency assessment. The guideline also ambitiously requires that potency assay specifications should as much as possible rely on the efficacious dose based on correlations between potency results and (non) clinical outcomes.[9] In addition, there is also another product-specific guideline on potency testing of cell-based immunotherapy medicinal products for the treatment of cancer, which outlines the possibility to measure potency by means of in vivo or in vitro tests. It further states that potency assays must be created based on defined biological effect as close as possible to the mode of action or clinical response [12]. In addition, it requires potency assays to be "*sensitive enough to detect clinically meaningful changes*" [21, pp. 5–10].[10] As for the Phase I clinical trials, the guideline also refers to "a suitable potency assay,"

[8] Such as cell surface markers, activation markers, or expression pattern of specific genes.

[9] It is stated that: "*The selection of the dose should be based on the findings obtained in the quality and the non-clinical development of the product and it should be linked with the potency of product.*"

[10] In practice, this necessitates characterization of the cells' phenotypic and functional properties, which will help to tailor the assays.

but no further clarification regarding "suitability" is available.

As mentioned above, there is a particular product-specific guideline on genetically modified cells, which outlines that in potency assessment, various different assays can be used in combination [8].[11] A public consultation was organized to gather experiences of the ATMP Regulation in 2013. At that time, commentators expressed the need for ATMP-specific adaptations to the quality requirements and the need for more detailed guidance that took into consideration particularities of ATMPs. As for the challenges regarding development and validation of ATMP-specific potency assays, it was highlighted that further EU-level guidance on potency testing was needed.

Currently, EMA seems to shifting toward risk-based approaches, providing more flexibility for the developers of ATMPs to specify and apply relevant product-specific pathways instead of issuing rigorous predefined requirements.[12] This approach encourages innovation as it among other things creates novel possibilities for developers of chemistry manufacturing and control approaches that are suitable or even tailored for a particular product, instead of strictly adhering to some general requirements that may not be relevant for the product under development. Hence, in order to gain regulatory approvals, there is a possibility to make product-specific adaptations to the development strategy, as far as these approaches rely on robust scientific understanding and supportive data (Fig. 10.1).[13]

10.5 Conclusions

The ATMP Regulation aims to harmonize the legislative landscape for ATMPs throughout the EU, with the EMA offering regulatory support to developers. The near future will reveal how widely the reformed risk-based approach in cGMP manufacture and clinical trials gains wider general acceptance among the national regulatory authorities and ATMP developers in Europe. In this context, potency assay strategy should be taken into consideration as early as possible in the ATMP development process. Potency assays are not only a critical quality attribute, but in context of the ATMP Regulation and its relevant EU "soft law" guidance, characterization and correlations studies of potency assays can be used to justify the complete quality control and release strategy of an ATMP. It has been reported that a number of cell-based medicinal products (CBMPs) gained access to the internal markets after some struggle fulfilling the potency assay-related requisites, which could have been addressed by conducting appropriate studies earlier in the ATMP development process. Likewise, difficulties with potency assays have been noted to lead to withdrawals during the regulatory assessment process [21, pp. 5–10]. The EMA's risk-based approach allows for a more flexible strategy that takes into account particularities of each product.

[11] These could for instance include the number of genetically modified cells, the gene copy number, the expression level of the transgene, and the product activity level, as shown to be efficacious in clinical studies.

[12] See for instance Mansnérus [17].

[13] Interestingly, it is noted by Pimpaneau et al. that some products have been granted approvals when utilizing the surrogate endpoints as a measure of potency. This shows how adaptive regulatory pathways are now emerging in Europe and can be justified by means of a robust scientific rationale and data. It is further noted that development of a potency assay strategy is a long-term-process, that depends on a number of factors ranging from the aetiology and the knowledge of the disease, availability of relevant scientific publications, the desired composition of the final product to result of characterizations as well as available modes of action studies. Altogether these elements have impact on the chosen regulatory strategy for development of potency assays for stem cell-based ATMPs. This process could start with (1) selection of a first potency test followed by (2) continued investigation of modes of action and product characterization; resulting in (3) proposal of orthogonal methods as knowledge is gained; thereafter (4) building correlations between tests and select the most relevant ones in order to refine the selection of the most relevant potency assays and specifications; and thereafter (5) building correlation with biological activity and clinical outcome in order to verify whether the potency assay can be used to analyze meaningful clinical differences between batches allowing identification of a linkage to the dose; and finally (6) create the final strategy using surrogates, taking into consideration complementarity to comprehensively cover and correlate well with the modes of action.

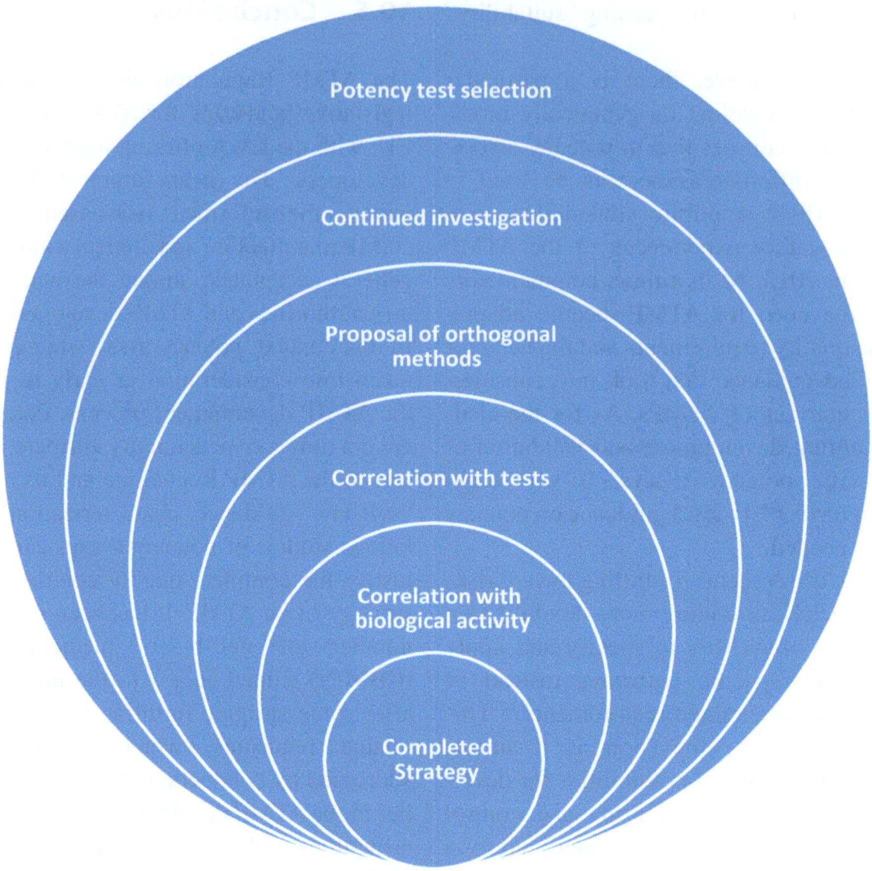

Fig. 10.1 Development strategy relying on robust scientific understanding

Furthermore, a careful assessment of benefit–risk balances should constitute a part of the development strategy early on. Successful implementation of risk-based approaches requires regular interaction with regulatory agencies as it is highly advised to exchange ideas about feasibility of the contemplated potency assay approach. In this context, access to the EMA's scientific advice provides a constructive opportunity to discuss particular challenges arising in development process of potency assays in relation to the characterization of the active substance. Sometimes early cooperation with regulatory authorities may also turn out to be helpful when setting the expectations when mixtures of cells constitute the actual substance of an ATMP. The insights of regulatory authorities can also be valuable when a complex product or mode of action is expected, inevitably affecting development of potency assays. Scientific advice also provides the opportunity to discuss correlations between tests and clinical outcomes. It can be helpful for selection of appropriate tests as well as specifications. Early interaction is likely to result in creation of additional data, which may turn out to be helpful when building a knowledge base by means of orthogonal methods, that in turn can be useful when proposing surrogate tests for the purpose of ATMP release.

Nonetheless, it remains to be seen whether these risk-based adaptations to our regularity framework are sufficient to nurture the ATMP field and ameliorate the availability and accessibility of valuable treatments. It is prudent to discuss challenging aspects with regulators and health technology assessment bodies at the early stages of ATMP development. The evolution of authorization and its related procedures is likely to occur under accelerated access pathways that need to be duly aligned with payment and reimbursement structures to ensure and facilitate patient access to new technologies.

References

1. CAT work plan 2021. Available from https://www.ema.europa.eu/en/documents/other/cat-work-plan-2021_en.pdf. Accessed 28 July 2021
2. CBER/FDA (2011) Guidance for industry potency tests for cellular and cene therapy products
3. Cuende N, Álvarez-Márquez AJ, Díaz-Aunión C, Castro P, Huet J, Pérez-Villares JM (2020) Promoting the ethical use of safe and effective cell-based products: the Andalusian plan on regenerative medicine. Cytotherapy 22:712–717
4. European Commission (2017) Guidelines of 22.11.2017 good manufacturing practice for advanced therapy medicinal products
5. European Medicines Agency (2008) Guideline on Human cell-based medicinal products (EMEA/CHMP/410869/06)
6. European Medicines Agency (2016) Guideline on potency testing of cell based immunotherapy medicinal products for the treatment of cancer. Available from: https://www.ema.europa.eu/en/documents/scientific-guideline/guideline-potency-testing-cell-based-immunotherapy-medicinal-products-treatment-cancer-revision-1_en.pdf
7. European Medicines Agency (2019) Guideline on quality, non-clinical and clinical requirements for investigational advanced therapy medicinal products in clinical trials. Available from: https://www.ema.europa.eu/en/documents/scientific-guideline/draft-guideline-quality-non-clinical-clinical-requirements-investigational-advanced-therapy_en.pdf. EMA/CHMP/BWP/271475/2006 rev.1
8. European Medicines Agency (2021a) Quality, non-clinical and clinical aspects of medicinal products containing genetically modified cells (CHMP/GTWP/671639/2008)
9. European Medicines Agency (2021b) EMA warns against using unproven cell-based therapies. Available from: https://www.ema.europa.eu/en/documents/public-statement/ema-warns-against-using-unproven-cell-based-therapies_en.pdf. Accessed 28 July 2021
10. Flory E et al (2015) Regulatory viewpoints on the development of advanced stem cell–based medicinal products in light of the first EU-approved stem cell product. Cell Gene Ther Insights
11. Ghinea N, Munsie M, Rudge C, Stewart C (2020) Australian regulation of autologous human cell and tissue products: implications for commercial stem cell clinics. Regen Med 15:1361–1369
12. ICH Q6B. Specifications: Test Procedures and Acceptance Criteria for Biotechnological/biological Products. ICH Harmonised Tripartite Guideline; 1999. Available from: https://database.ich.org/sites/default/files/Q6B%20Guideline.pdf. EMA/CAT/852602/2018
13. Iglesias-Lopez C et al (2019) Regulatory framework for advanced therapy medicinal products in Europe and United States. Front Pharmacol, 30 August 2019. Available from: https://www.frontiersin.org/articles/10.3389/fphar.2019.00921/full. Accessed 1 August 2021
14. Iglesias-Lopez C, Obach M, Vallano A, Agustí A (2021) Comparison of regulatory pathways for the approval of advanced therapies in the European Union and the United States. Cytotherapy 23:261–274
15. Kabir K, Moreino SS, Siam M (2019) The breakthrough of Biosimilars: a twist in the narrative of biological therapy. Biomol Ther 9(9):410
16. Mansnérus J (2016) Commercialisation of advanced therapies: a study of the EU regulation of advanced therapy medicinal products. University of Helsinki
17. Mansnérus J (2020) Over ten years since the adoption of the EU regulation on advanced therapy medicinal products – lessons learned thus far. In: Mansnérus J, Lahti R, Blick A (eds) Personalized medicine legal and ethical challenges, pp 128–133
18. Master Z, Matthews KRW, Abou-El-Enein M (2021) Unproven stem cell interventions: a global public health problem requiring global deliberation. Stem Cell Rep 16:1435–1445
19. McGrath E, Chabannon C (2019) Regulatory aspects of ATMP versus minimally manipulated immune cells. In: Carreras E, Dufour C, Mohty M et al (eds) The EBMT handbook: hematopoietic stem cell transplantation and cellular therapies [Internet], 7th edn. Springer, Cham; Chapter 62
20. National Academies of Sciences, Engineering, and Medicine, Health and Medicine Division, Board on Health Sciences Policy, Forum on Regenerative Medicine (2017) Navigating the manufacturing process and ensuring the quality of regenerative medicine therapies: proceedings of a workshop, National Academies Press (US), Washington (DC). 3, Identifying and measuring critical quality attributes. Available from: https://www.ncbi.nlm.nih.gov/books/NBK475684/. Accessed 27 July 2021
21. Pimpaneau V et al (2015) The challenges of potency assay development for cell-based medicinal products in Europe. Regul Rapp 12:5–10
22. Renner M et al (2015) Regulation of clinical trials with advanced therapy medicinal products in Germany. Adv Exp Med Biol 871:87–101
23. Smith C, Crowley A, Munsie M, DeMartino ES, Staff NP (2021) Academic physician specialists' views toward the unproven stem cell intervention industry: areas of common ground and divergence. Cytotherapy 23:348–356
24. Takashima K, Morrison M, Minari J (2021) Reflection on the enactment and impact of safety laws for regenerative medicine in Japan. Stem Cell Rep 16:1425–1434

References

11 The Evolving Landscape of Potency Assays

Jorge S. Burns

11.1 Getting Potency Assays Just Right

There is a "goldilocks" aspect to potency assays [88]. On the one hand, a comprehensive evaluation of the cell product with detailed quantitative measurement of the critical quality attribute/s (CQA) of the desired biological activity is required. On the other, the potency assay benefits from simplification and lean approaches that avoid unnecessary complication and enhance robustness, to provide a reproducible and scalable product. There is a need to balance insightful knowledge of complex biological healing processes with straightforward manufacture of an advanced therapeutic medicinal product (ATMP) that can be administered in a trustworthy cost-effective manner. Earlier chapters within this book have highlighted numerous challenges facing the potency assay conundrum; however, this chapter offers a forward-looking perspective regarding the many advances that are likely to facilitate potency assay development in the future.

J. S. Burns (✉)
Department of Environmental and Prevention Sciences, University of Ferrara, Ferrara, Italy
e-mail: js.burns@unife.it

11.2 Finessing the Potency of ATMP

Great delicacy and skill accompany the long path between observations that guide new ideas for therapeutic intervention and the eventual clinical practice [111]. Enhanced focus on the potency assay experiment is best complemented by process-wide quality-by-design principles, replacing any "process is the product" complacency with motivation that well-understood processes can be further optimized to make manufacturing processes more efficient. Application of Lean thinking strategies can streamline research productivity and improve a number of key performance indicators [58]. When cell–gene therapy products involve manufacture outside the point of care, leading to highly specialized hospital/academic-based facilities having to collaborate with novel infrastructures [11], further factors, such as transportation conditions, become relevant [174]. Of note, improved multipotent stromal cell (MSC) clinical trial vial-to-vein reporting and reduction of any variability in cell handling that affects potency will be important to generate data that will allow retrospective analysis to advance the development of ATMP [185]. It has been recommended that guidelines for long-term stability data for a range of ATMPs based on risk analysis would help har-

J. S. Burns (ed.), *Potency Assays for Advanced Stem Cell Therapy Medicinal Products*, Advances in Experimental Medicine and Biology 1420, https://doi.org/10.1007/978-3-031-30040-0_11

monize specifications and procedures, in particular potency assays, among diverse cell therapy centres [20]. MSC viability, metabolic fitness, route of administration and host disposition are all key factors that impinge upon clinical potency [46]. Therefore, ATMP require a comprehensive development strategy and incremental improvement in organizational and technological approaches. As a legal requirement of medical approval, the potency assay may be responsible for a significant late-stage delay in the availability of a clinical product. Fortunately, prompt focus on potency assay development at early stages in the product lifecycle is increasingly popular. Providing a carefully conducted adequate potency measure for each product serves to efficiently integrate the many aspects governing the quality of the whole manufacturing process. Potency assays inform whether the product dose can be expected to provide a desired clinical effect, providing reassurance of manufacturing process consistency and product reproducibility. Ideal potency assays, though unique to each product, have common specific main aims. The assay needs to reflect the mechanism of action (MoA). If biological pathways cannot be reproduced in their entirety, the assay should focus on the most relevant specific aspects of the MoA. Direct correlation of the potency of the product to a predictable clinical efficacy may be difficult to achieve, hence surrogate biomarkers or functional assays may be needed to substitute clinical data. Even minor changes in the potency of a cell product may still have a large or unpredictable impact on the clinical outcome. The ability for the potency assay to distinguish between a high-quality targeting product and a degraded product should be tested with multiple tests during development [143] using intentionally degraded samples. Theoretically, potency can be quantified absolutely, yet in practice this may not always be achievable if there is too much variability. Consequently, the potency assay may assume a more binary "on/off" quantification according to specific measurable thresholds. Good performance parameters of a potency assay are accuracy, sensitivity and specificity, although there is appreciation that the limits of performance may vary quite widely according to the assay type. It may be necessary to derive a relative potency, calibrating a response against a recognized reference standard, expressing the relative potency as a percentage drift of the sample from the reference standard, reporting the outcome with confidence intervals. Selection of a suitable reference standard becomes a crucial factor determining the success of such relative potency assays. It is often generated internally, e.g. a well-characterized development batch available at sufficient quantities that allow it to be supplemented with an alternative batch after appropriate comparability studies. Acceptable performance limits are to be defined, in some situations an assay detecting a change in potency of 5% may be considered good, yet in others a change of 30% may be expected. To what extent such quantities guide the decision process may be influenced by how the single potency assay fits into a wider product assessment matrix. Performance and specification limits will need to be scientifically justifiable to the regulatory authorities. Ease of reproducibility and consistency are important concerns, otherwise the potency assay will be of little value. In this regard, in vivo assays are usually less consistent than assays based on primary cells or cell lines, which in turn are less reproducible than physiochemical measures. Ethical considerations raise questions as to whether an animal-based assay with highly variable outcomes would be justifiable. Since a potency assay may be used repeatedly at different stages of ATMP manufacture, a simple pragmatic design is advantageous and worth considering at a very early stage of development to keep the number of process and assay steps to a minimum for cost-effective implementation. Risk assessment should accompany determination of how critical a step is for the manufacturing process, those deemed unnecessary can be discarded. Since in vivo assays often take a long time to perform, requiring animal sourcing and acclimatization, they present a high risk if the assay is invalidated and needs to be repeated. Alternative ex vivo assay platforms are available but although these can be sophisticated and provide high-content data, unique and expensive equipment may not be widely available and prove problematic for data transferability and reproducibility. Beyond cost-effectiveness, prudent regulatory authorities may be reluctant to accept data from highly innovative assays that have yet to be proven in the field and require further corroborative data.

Regulatory authorities appreciate that there are a number of challenges for ATMP potency assays—cell therapy can have multiple aspects to the MoA, the assay may detect a biomarker yet lack sensitivity, reference standards may be hard to define—it may be difficult to achieve a prompt fast cellular assay and cell-based assays typically show much higher variability than assays concerning a physiochemical drug. Scientific research that is carefully documented will not necessarily overcome the various challenges, but will be crucial for a constructive dialogue regarding the feasibility of the potency assay and its acceptance. Advisable approaches for building potency assays include a focus on the science, identifying and measuring the critical product attributes and their correlations to specific potency assays. Early interaction with the regulatory authorities is key and improved when information has already been gathered regarding published guidelines and jurisdiction-specific regulations, the latter a particularly nuanced consideration following Brexit [48]. Regulatory authorities contribute experience and expertise to help develop an acceptable potency assay and can provide helpful input regarding necessary scientific justifications. Detailed method reporting can provide empirical demonstration of due diligence that would be important when, for example, justifying substitution of more complex labour-intensive functional bioassays with simpler yet effective gene-expression surrogate assays. There may be more than one potency assay required according to the number of steps in the product MoA. A strategic focus on the most meaningful and relevant information will be needed. Assays will need to be carefully chosen, and there may need to be compromise between simplicity and true relevance of the information obtained, with considerations perhaps not at first apparent, such as sensitivity of the assay towards product degradation. Practical approaches will be needed when compiling what will most probably need to be a matrix of experiments, aiming to accommodate fast turnaround times when possible.

Multiple types of cell-based therapy, with functional differences between different MSC sources [29, 70], will each bring particular considerations. Autologous cell therapies introduce an inherent patient-specific variability that may be difficult to control. Hard to define reference standards can present a hurdle to allogeneic cell therapies. Stem cell therapies can incur challenges for defining the final cell population or measuring the critical biological activity responsible for their clinical effect. A patient cohort may behave quite differently and in an unexpected way if the assay was first exclusively established using healthy donor material. The interrelatedness of biological activity constituting a MoA and the potential involvement of numerous cell types can complicate potency assay development. If the cell therapy also involves a gene therapy component, this will need specific assessment, allowing each aspect of the MoA to be assessed separately. Inevitably, potency assays will require considerable investment of time and resources, the value of each assay needing independent assessment and refinement towards final performance and definition of acceptance criteria; only then will sound science be successful in meeting regulatory requirements for validation.

11.3 Potency Assays for Acellular Products

A more refined understanding of the natural functions of MSC has brought growing consensus that their MoA need not be limited to direct replenishment of stromal tissue cells. Alternative functions include the secretion of soluble mediators that can influence endogenous tissue and immune cells [5, 122] and does not exclude a role for cell death by apoptosis [119]. Since confirmation of their existence in 1967, there has been interest in a family of particles released from the cell. These have been termed exosomes, microvesicles, microparticles or ectosomes, the favoured nomenclature now being extracellular vesicles (EV) [9].

Beyond knee cartilage repair [177], MSC-secreted extracellular vesicles (EV) may exert a therapeutic effect in other contexts, e.g. osteogenesis [43], intervertebral disc repair [38], chronic kidney disease [19] and neurodegenerative pathologies [7]. Cell-derived-secreted products call for specific MSC-EV harmonization criteria, with quantifiable metrics to identify cellular origin and integrity of the vesicles [187]. As acknowledged by the ISCT Exosomes Scientific Committee, identification of MSC-EV attributes for potency assays with establishment of dose metrics and derivation of reference standards remains a current challenge [47], complemented by exploration of the optimal 2D or 3D culture environment for generating the desired MSC-EV properties [69, 83]. A systematic review of MSC-derived EV has highlighted the enormous global interest in MSC-EV, since they appear to be beneficial in many settings [149]. However, their precise mechanism of action remains poorly understood and details among scientific studies are often incomplete, failing to mention all the critical parameters, such as quantification of the particles per milligram of protein, per one million cells or micrograms of protein per one million cells. Such metrics are needed for greater consistency in reporting, that can be helpful for discerning functional differences in immunomodulation and regenerative potential among a range of secreted products from different stromal cell sources [120, 160]. MSC-EV dosage is an elusive variable, needing to take into consideration that EV diameters can range from 30 to 200 nm, with classification into diverse types reflecting specific size and origin. Genetic and protein information contained within plasma membrane-derived vesicles can serve as diagnostic tools for various diseases [4]. In a similar manner, careful characterization of other types of isolated EV can be helpful for determining critical quality attributes. Fortunately, the International Society for Extracellular Vesicles (ISEV) has issued guidelines for minimal information for studies of extracellular vesicles (MISEV) to consolidate nomenclature and improve scientific rigour [168]. These standards will be updated [186] to accompany advances in our understanding as growing interest in developing MSC-derived EV therapy evolves [169]. It is noteworthy that systematic analysis of miRNA profiles and proteomes of EV derived from three different human tissue sources, namely, umbilical cord multipotent stromal cells (hUC-MSCs), embryonic stem cells (hESCs) and induced pluripotent stem cells (hiPSCs) revealed cell-source specific qualities. The EV isolated from hES and hiPSC were associated with functional molecules regulating development, metabolism and aging, whereas the miRNA and proteomic molecular profile of hUC-MSC-derived EV suggested that they contributed more to immune regulation [13]. Potency assays for the effectiveness of EV may involve additional cell types, since their mechanism of action may involve modulation of intercellular communication from other interacting cell types such as macrophages [100, 135].

11.4 Cryopreservation and Scale-Up: Balancing Complexity and Product Quality

The advantages and disadvantages of cellular ATMP cryopreservation, an induced preservation via metabolic stasis, are a highly debated topic involving complex cryobiology principles [10]. A clear benefit of cryopreserved "off-the-shelf" products is that they can overcome need for cGMP cell expansion to obtain a clinically effective cell dose, a problematical time constraint when therapeutic situations need very prompt or highly synchronized patient application. Before considering MSC, there is much already learned from myeloablative chemotherapy, when autologous peripheral blood stem cells obtained by apheresis are often cryopreserved before subsequent use to reconstitute the bone marrow following chemotherapy. Haematopoietic progenitor cell (HPC) transplantation processes monitored by regulatory authorities such as the Joint Accreditation Committee ISCT – Europe (JACIE) and European Society for Blood and Marrow Transplantation (EBMT) require release of cryopreserved products for clinical use to be conducted by accredited cell therapy laboratories. Post-thaw enumeration of viable CD34+ cells is a mandatory standard assay, but criteria for testing functionality, such as colony forming unit (CFU) assays, are less uniform among laboratories. The precise freezing protocol can be deterministic, yet a lack of a "gold-standard" freeze profile highlights the importance of a critical assessment of post-thaw CD34+ cells. Testing post-thaw CD34+ cell viability per se was found to be an inadequate metric to predict functional activity and prompt engraftment in patients [110]. Nonetheless, post-thaw viable CD34+ status predicted haematopoietic engraftment more accurately than pre-freeze determinations or pre-freeze viability cell counts [87]. A review of research to improve cryopreservation of haematopoietic cells, varying cell concentration, freezing rate and storage temperature highlighted that at cryoprotectant composition of 2.5% dimethyl sulfoxide (DMSO) with trehalose was most successful at maintaining differentiation potential and cell viability. So particular cryopreservation procedures may influence the extent to which CD34+ viability may be more closely correlated with function [63]. Although DMSO represents a very effective cryoprotectant of choice [77], it can introduce mild to moderate side effects, thus alternative approaches are being devised [74, 129, 139].

Clinically applicable cryopreservation for other non-haematopoietic therapeutic cell types, in particular MSC, is not necessarily straightforward since existing protocols may need modification to accommodate specific requirements of different cell types [31]. Cryopreservation processes need to be xeno-free, nontoxic and immunocompatible, whilst achieving long-term storage at low temperatures with subsequent freeze–thawing phase changes yielding a high cell recovery that ideally conserves the cell functionality of the pre-storage state. Research laboratories have extensively explored a range of different methods [93]. Optimal manufacturing and standardization parameters have yet to be widely attained for different tissues, but notably, cryopreserved umbilical cord tissue fragments can be used to derive MSC with preserved proliferation rates and potency [113]. Furthermore, the rationale for a cryopreserved MSC product is supported by preclinical studies. Differentiation capabilities can be retained long-term, even after cryopreservation for over two decades [150], and MSC can be cryopreserved at a high cell density [6]. Although freshly thawed cells may be functionally impaired, a post-thaw acclimatization period in culture can restore functional properties [117, 130]. A cryopreservation step in the manufacturing process can bring convenience and increased consistency to cell-based potency bioassays [166]. A review of freshly cultured versus cryopreserved MSC in animal models of inflammation suggested that overall, their outcomes for in vivo efficacy and in vitro potency may be near-equivalent [35]. Nonetheless, there is a need to harmonize practices between different laboratories to improve the long-term stability studies and risk analyses for specific ATMP, establishing shelf-life and guaranteed efficacy upon infusion [20].

The transition from a single batch to mass-produced ATMP requires considerable accomplishments, especially when implementing automated manufacturing systems. Though costly to introduce, automated manufacturing should become cost-effective when products can be processed in parallel, with more consistent manufacture of clinical grade ATMP [54]. A quality-by-design (QbD) regulatory initiative stipulates that quality needs to be built into the manufacturing process, promoting the development of novel analytical and informatics tools that support this objective. Pico-droplet microfluidics combined with high resolution plate-imaging platforms can greatly improve GMP grade bio-manufacture of high-quality single-cell progenitor-derived clonal cell lines [133]. Automated capillary electrophoresis western blots, for example, can promptly monitor specific proteins quantitatively [191]. High-throughput droplet digital PCR (ddPCR) technology can provide absolute quantitation of DNA copy number [60]. Directly compared to quantitative PCR (qPCR), ddPCR was considered to provide more concise, reproducible and statistically significant results, and this could prove particularly informative for low abundant targets with relatively small expression differences of only twofold [165]. For cellular therapy using autologous genetically modified chimeric antigen receptor (CAR) T-cells, ddPCR robustly provided accurate quantitation of average vector copy number [96] and correlated well with flow cytometry-based methods [24]. Both qPCR and ddPCR of comparable precision could complement each other as methods to monitor the kinetics of CAR T-cells in treated patients [146], raising the relevance high-throughput technologies for clinically congruent potency assays. The innovation leading to greater precision of ddPCR is that samples are partitioned into thousands of emulsion-based nanolitre-sized droplets, each serving as an individual sample chamber for target detection and amplification. At the end of the amplification process, droplets are counted as positively containing target sequence or negative. On the basis of Poisson distribution, the fraction of positive droplets helps determine the absolute concentration of template in the original sample. Unlike real-time qPCR, there is no need to extrapolate from a standard curve or rely on a reference. Absolute quantification is relatively straightforward and can overcome changes in amplification efficiency to deliver highly accurate results. Ultimately, ddPCR presents scalable advantages that make it a robust platform for potency assays.

Given that MSC are physiologically mechanically responsive cells, it is to be expected that they are very sensitive to changes in their microenvironment and their cell culture conditions. Moreover, MSC are versatile and have been grown in a broad range of bioreactor options favouring large-scale cell culture and expansion; for example, monolayer cultures in multi-layer vessels, hollow fibre, stirred tank bioreactors, rotating wall vessels or vertical wheel bioreactors with microcarriers [138]. Micro-carrier-expanded MSC differed from monolayer flask-expanded cells, with regard to size, morphology, proliferation, viability, surface biomarkers, differentiation potential and secretome profile, all factors that can contribute to potency assay outcomes. Cells cultured on microcarriers have a higher contact angle and are subject to higher mechanical force raising cytoskeletal tension, leading to induction of more actin stress fibres. Scientific reports indicate that microcarrier culture in bioreactors promotes MSC differentiation towards an osteogenic rather than adipose phenotype. For skeletal cell therapy, such phenotypic bias may be advantageous, augmenting cell secretion abilities and potency biomarker expression [173]. How cells are harvested from microcarriers needs careful consideration [92]. The many parameters that influence cell yield need to be carefully modulated to optimize the provision of the desired cell product, be it the cell itself or factors secreted by the cell [152]. Promising ATMP scale-up bioprocessing conditions have been demonstrated [25], including microfluidic high-throughput on-chip assays suitable for scaled-up manufacture [145]. It will be increasingly important to further optimize procedures, integrating novel ways to monitor and control key parameters, engineering efficient and trusted means of sample collection for potency analysis, leading to subsequent provision of high-quality ATMP.

11.5 Highlighting COGS in the Wheel of CAR T-Cell Therapy

Recent high-profile cell products, e.g. CAR T-cell therapies Yescarta [110] and KYMRIAH®, with high impact as accelerated approval ATMP, cost less than the world's most expensive drug to date, Haemophilia B-directed gene therapy HEMGENIX® (≈€3.5 million dollars per treatment) superseding gene therapy ZOLGENSMA® (≈€1.9 million dollars per vial). Yet current significant costs of ≈€300,000 (Kymriah) and ≈€400,000/vial (Yescarta) raise concern for how the cell therapy sector evaluates the cost of goods sold (COGS) [128] and what future personal medicine financing and reimbursement models may be reached [79]. The cost of a range of existing ATMP across nations varies depending on the type of treatment needed, with site injections of orthopaedic conditions being significantly different than intravenous CAR T-cell treatments (Fig. 11.1). The complex and rapidly evolving CAR T-cell therapy sector, beset by challenges for widescale deployment (Fig. 11.2), highlights the remarkable evolution of cell therapy and need for newly tailored potency assays sensitive to relatively small changes in one element of manufacture that can lead to abrupt changes in the system as a whole.

New approaches that lower costs, improve manufacturing capacity and expedite provision making CAR T-cell therapy more widely available are being sought [97, 102]. Strategically designed CD19 CAR with a fast off-rate CD19 binding domain can improve T-cell persistence, reduce CAR T-cell immune toxicity and improve engraftment [137]. The autologous circular approach, where the patient's own cells are engineered, may be logistically cumbersome, but has so far been favoured over an allogeneic cell therapy, where cells are extracted from a healthy donor, engineered and expanded to treat multiple other patients. Allogeneic cells offer exciting prospects for centralized manufacture, large-scale production, wide availability and improved cell fitness for the desirable quality of CAR T-cell persistence [94]. However, a significant risk of graft-versus-host disease (GvHD) [144] and challenge from the host immune system can curtail their effectiveness. The main autologous CAR T-cell therapy process steps have been categorized as (i) provision of starting material, (ii) cell selection and activation, (iii) genetic modification, (iv) expansion, (v) cryopreservation/formulation and (vi) injection into patient. Best avoided are a lack of process understanding, impractical manual processes and cumbersome logistics. Autologous therapies developed in a centralized accredited GMP facility can involve transportation of apheresis, vector and CAR T-cell products that can subsequently impinge upon potency outcomes. Authorized point-of-care CAR T-cell production can enhance patient access to CAR T-cell products [22, 37], and may take advantage of closed semi-automated culture systems developed to help minimize cross-contamination. However, such devices are not necessarily optimal for large-scale strategies and are not yet equipped to provide the most insightful process metrics to inform on cell performance and potential therapeutic outcome [131]. Improved analytical technology will be necessary to provide the data that can overcome a lack of process understanding. The co-introduction of artificial intelligence (AI) can be particularly beneficial to manage the complexity of manufacture and adapt scheduling to integrate the manufacturing process and potency assays with overall therapeutic requirements [64, 157].

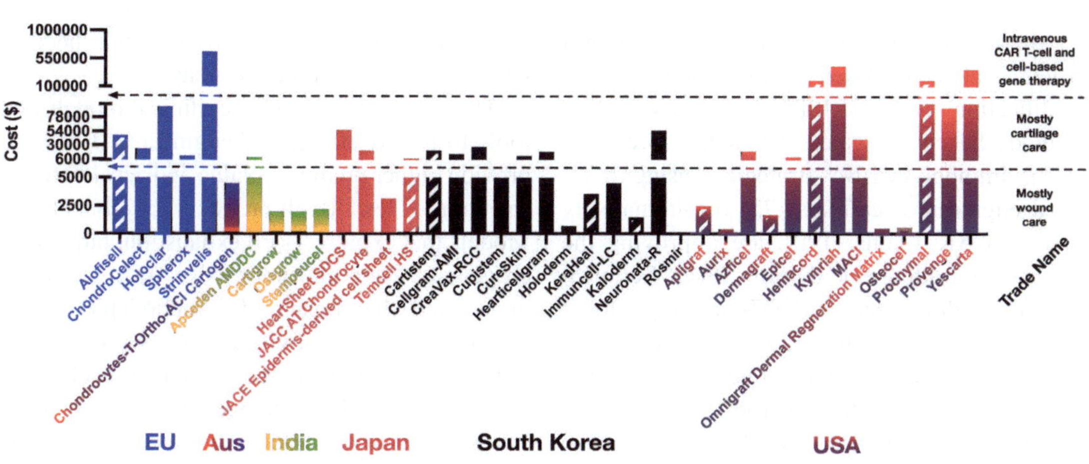

Fig. 11.1 Global pricing for a range of approved cell therapy products. The cost range can vary considerably according to the complexity of procedures, with attention to wound care and cartilage care generally costing less than intravenous treatments. Autologous therapies, solid bars; allogeneic therapies, hatched bars

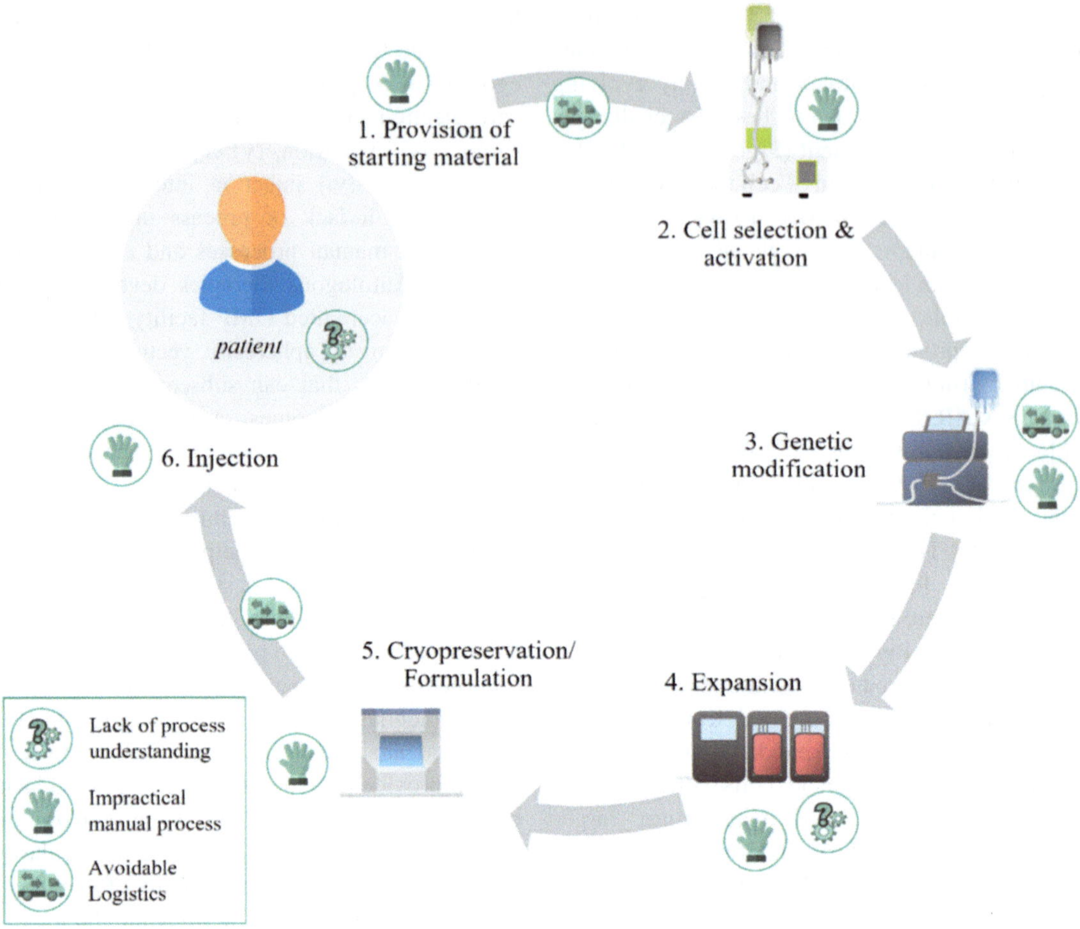

Fig. 11.2 Challenges for wide-scale deployment of autologous CAR T-cell therapy. (Reprinted from Hort et al., 2022 [64])

Alternative approaches to the costly retroviral or lentiviral vector-based products include T-cell modification via CRISPR-Cas gene-editing methods that can be scaled up with high precision [179]. Immuno-evasive strategies, using dedicated engineered scaffolds [72] or alternatively sourced CAR T-cells [190], are improving the prospects for allogeneic therapy. Phase I clinical trials using donor-derived CD7-targeting allogeneic CAR T-cells to treat CD7+ haematological malignancies have demonstrated encouraging safety and efficacy profiles [66]. A significant advantage of allogeneic CAR T-cells is that the single donor product can be tested simultaneously in different patients, a highly informative situation for determining the most consistent and significant critical quality attributes that need to be measured in the potency assay.

There are a number of pre-clinical models and tools to assess efficacy of engineered T-cells to provide higher predictive value and accelerate the selection of lead T-cell products for clinical use. Models that explore T-cell exhaustion, target heterogeneity, immunosuppressive microenvironments and the impact of lymphodepletion on engineered T-cell activity can be analysed by a number of analytical methods that each have advantages and disadvantages. It is by no means straightforward to extend the success of CAR T-cell therapy observed in haematological malignancies to solid tumours [52]. Gaps in models and tools make it difficult to derive potency tests that fully recapitulate the complex and dynamic clinical events modulating solid tumour microenvironments and how tumour biology can impact the antitumour efficacy of CAR T-cell therapy [98]. Nonetheless, extensive progress has been made, including examples that extend rodent models to more relevant larger mammals whilst combining CAR T-cell therapy with ways of modulating the tumour physical microenvironment to improve therapeutic effectiveness [194]. Both in vitro and in vivo pre-clinical models for CAR T-cell research have evolved with the needs of CAR T-cell research [151]. High-throughput assays provide increasingly sophisticated and efficient means of monitoring manufacture and clinical progress [112, 131]. Many aspects of CAR T-cell therapy remain to be optimized, with solid tumours presenting an additional complexity that increases upstream pre-clinical research costs. Progress in digitally controlled automated processes, coupled with recognition of the importance of international harmonization of product control and documentation, will ultimately enhance downstream product manufacturing efficiency. A forthright comprehensive and concerted approach to reduce CAR T-cell therapy COGS can make novel effective cell therapies more broadly available.

11.6 Potency Assays for Induced MSC

Applying the MSC acronym in its broadest sense, MSC research has grown to encompass a very broadly sourced cell type from different tissues and donors, subject to many different manufacturing processes and characterization methods, with numerous proposed functional mechanisms of potential therapeutic benefit in a broad range of disease contexts. The lack of a fully defined phenotype or truly unique specific biomarker that can characterize MSC and their recognized subpopulations with specific properties has been a challenging aspect to their use as ATMP, prompting reappraisal of their definition and reassessment of the approach to clinical investigation [75]. The use of highly cited minimal criteria to derive an MSC definition has been fundamental for steering debate and scientific consideration of their properties [39].

In seeking a stringently demonstrable quantitative functional attribute of the ATMP product with candidate markers correlated to bioactivity, the potency assay has provided motivation for excellent research, addressing many issues of potential confusion. An FDA survey of proposals submitted between 2006 and 2012 noted that less than half of the MSC-based product investigational new drug (IND) applications described marker-based bioactivity assays and most that did were submitted by commercial sponsors using MSC from allogeneic donors [105]. Commercial sponsors have an important role in driving progress for stringent MSC characterization and highly scalable allogeneic cells. Current data from over a thousand trials remains inconclusive as to whether the particular advantages and disadvantages of autologous or allogeneic MSC result in one providing better therapeutic benefit over the other [90].

Several possible restrictions limit the applicability of MSC, e.g. a relatively low number in source tissues, donor and cell source-dependent functional heterogeneity and limited proliferation with early senescence in culture. Donors with inherited syndromes such as Fanconi anaemia may be treated by haematopoietic stem cell transplantations and be asymptomatic yet still harbour MSC with defective functionality [55]. Such circumstances have prompted investigation into new alternative sources and ways of generating clinical grade MSC [175]. A novel approach for obtaining large populations of adult stem cells for use in regenerative medicine emerged from the discovery that it was possible to generate induced pluripotent stem cells (iPSC) from adult somatic cells via a reprogramming technique forcing expression of four transcription factor genes; *Oct3/4*, *Sox2*, *c-Myc* and *Klf4* [161]. Notable advantages of iPSC are that they can be generated from any tissue type and have unlimited proliferation capacity, providing the prospect for a sustainable source of stem cells for use in the clinic. An inherent drawback of iPSC is that their self-renewal and pluripotency qualities may result in instability and tumorigenicity, jeopardizing clinical use. However, outcomes from derivation of induced MSC (iMSC) allow optimism. Numerous methods for obtaining iMSC from iPSC have been established, often involving embryoid body formation as a preliminary step [42, 103]. The resulting iMSC cells have a blend of attractive traits. In particular, iMSC can be generated from readily accessible tissue sources and demonstrate greater proliferative capacity (>120 population doublings) than traditional bone marrow-sourced MSC. Moreover, iMSC have been found to be more genetically stable than iPSC, without the same tendency to form tumours [197]. Rather the epigenetic and chromatin remodelling changes that do occur during iMSC generation tend to provide a rejuvenated cell phenotype [86]. Thus, iMSC lack many of the epigenetic alterations incurred by aging MSC over the course of cell division that can promote premature replicative senescence and impair their functional capability. Significantly, this serves to largely circumvent the tissue and age-related heterogeneity associated with natively derived MSC [188]. Barcoding individual cells within preparations indicated that in contrast to iPSC or primary MSC, expanded cultures of iMSC tended to show clonal dominance, with a less pronounced heterogeneity in colony formation and in vitro differentiation potential than primary MSC [62]. This would be consistent with observations that iPSC can exhibit heterogeneity in the levels of telomerase and telomere length, with a strong influence of telomere length in iPSC re-programming driving a selection pressure for survival of cells with the longest telomeres [2]. Studies of iMSC from Werner Syndrome patient lineages indicated that iPSC-derived MSC were just as susceptible to telomere attrition and defective synthesis of lagging strand telomeres as primary MSC from the patients [26]. So, iMSC are likely to remain susceptible to a telomere length-based selection pressure restricting iMSC heterogeneity. The telomere length of iMSC may represent an important biomarker for potency assays. Telomere length was associated with the cardiomyocyte differentiation potential of murine iPSC [1] and human MSC, with enhanced telomerase activity have been shown to resist oxidative stress-induced genomic damage [172]. This accords with earlier studies demonstrating improved functional differentiation of telomerized MSC, with an enhanced ability to differentiate to bone in xenograft models [153]. A "rejuvenated" signature may be responsible for many of the favourable traits associated with iMSC [154]. Deriving the minimal criteria for defining iMSC needs to go beyond criteria used for MSC and will be critical for clinical applications [27].

11.7 Enhancing Potency Assays: Cell Priming, Nanotechnology and 3D Culture

High-dose infusion of MSC in clinical trials can result in promising outcomes and though it may be associated with transient fever, has been generally considered safe for a variety of clinical conditions [85], including acute myocardial infarction and ischemic heart failure [84]. Nonetheless, it would be desirable to reduce the number of required MSC required for effective therapy. In this regard, standard expansion protocols and potency assays may reduce risk of clinical failure, but do not necessarily promote the most effective clinical outcome. An improved fundamental understanding of MSC behaviour after infusion has led to the evolution of upstream "priming" approaches to boost the innate functions of the MSC for increased alignment with the therapeutic objectives [126].

The inaugural culture method for obtaining MSC was not without drawbacks, since in vitro surface adherent culture methods that isolate MSC according to fibroblast colony forming unit ability (CFU-F) introduced a phenotypic bias. Ultimately, this may hinder isolation and characterization of naïve tissue-resident MSC [118]. In particular, in vivo bone marrow analysis showed CD271+ MSC occupying hypoxic niches, whereas CD146+ MSC resided in a perivascular niche [171]. Reduced phenotypic heterogeneity with high therapeutic and secretory potency was a feature of CD146+ MSC [15]. Notably, CD271-selected MSC were less angiogenic than plastic-adherent MSC and better suited to cartilage repair [78]. Yet in serum-rich medium, MSC gradually lost CD271 expression during in vitro expansion [134]. Similarly, gingival MSC showed a passage-dependent loss of expression of the STRO-1 antigen [136], a reputed marker of osteogenic precursors [51]. Such insights have led to substantial investigation into many alternative ways of manufacturing MSC [116] and their derived products [198], based on the cell source [17, 30, 192], isolation [155, 167], culture microenvironment [33] and storage methods [113]. Extensively exploring how MSC respond to culture conditions allows better potency assay design, integral to rational strategies for priming MSC to enhance therapeutic potency [193]. In the bone marrow, MSC and haematopoietic cells form a unique low oxygen tension niche [104, 182]. Compatibly, in sites of ischemic injury, oxygen tension is also low and in addition, hepatocyte growth factor (HGF) is often activated. Given that MSC express the cMet/HGFR receptor and have an ability to migrate towards HGF, this is considered a key signal that recruits MSC to damaged hypoxic tissue [106]. Notably, MSC can secrete HGF among many paracrine factors that act on haemopoietic progenitor cells and T cells. The immunomodulatory impact of MSC is predominantly mediated by their response to microenvironmental inflammatory signals, in particular priming MSC with inflammatory cytokines interferon gamma (INFγ) and tumour necrosis factor alpha (TNFα) increases the expression of markers associated with MSC immunosuppressive function [73]. Beyond selection of MSC with favourable surface markers and biophysical attributes, strategies to maintain and enhance the immunomodulatory potency of MSC during ex vivo expansion include use of media formulations including heparin sulphate, a low-affinity co-receptor that can enhance chemokine activity and chemical or surface-mediate biophysical priming of culture MSC, that can have diverse immunoregulatory immunomodulatory profiles when aggregated [156]. Introducing MSC to hypoxic conditions [140] and INFγ [141] is consistent with the naïve MSC niche in vivo and represent stimuli that enhance the desired therapeutic effects of MSC and their extracellular vesicle (EV)-secreted products [162, 181]. To derive effective potency assays to accompany these cellular manufacturing approaches, it will be necessary to understand the underlying molecular mechanisms of action. Although micro-RNA (miRNA) molecules are often important mediators of EV actions, hypoxia and IFNγ priming of MSC had little effect on their overall miRNA profile [124]. In contrast, proteomic studies comparing MSC in normoxic and hypoxic conditions revealed numerous proteomic differences associated with altered glycolytic metabolism. Under hypoxic conditions they secreted more EV, that bore proteomic changes in immune system and extracellular matrix proteins consistent with subsequently greater anti-inflammatory and pro-regenerative effects than EV from normoxic conditions [16]. The application of such high-throughput "omic" tools are of great benefit for defining MoA and the development of potency assays [147].

Yet another powerful ally in potency assay development is emerging from convergence of biotechnology with nanotechnology, the application of nano-biomaterials (NBM) will necessarily exhort novel strategic scientific risk assessment to ensure their long-term safety [28, 49]. Cancer research luminaries have long appreciated a dominant influence of the extracellular matrix (ECM) microenvironment on the genome and cell fate [14, 184] with broad relevance for tissue homeostasis [12]. Nanoparticles can be designed to precisely target ECM components to influence cell behaviour [23] and remodel tumour microenvironments [163]. Pre-clinical experiments have demonstrated how nanomaterial–MSC interactions can activate mechanotransduction signals helpful for a number of therapeutic goals, including stimulation of osteogenic differentiation, enhancement of MSC secretomes and heightened immune regulation. Composed largely of organic collagen and inorganic nano-hydroxyapatite molecules, the hierarchical structure of bone has macroscale and nanoscale qualities. Indicative of the very significant physiological effect of nanoscale events, MSC responded to nanovibrations with osteogenic differentiation involving specific bioactive metabolites, with cytoskeletal contractility indicative of osteogenic potency [61]. Nanoscale engineered calcium phosphate materials that duplicate the microstructure of human bone can constitute bioscaffolds that support mimicry of native bone architecture and control cell fate [91], taking full advantage of rationally designed nanotopography to direct osteogenic differentiation of MSC [8, 21]. Controlled integration of MSC with regenerative biomaterials [81, 180] can be made all the more feasible with 3D printing technology [199]. For such strategies, human iMSC were found to be particularly responsive to matrix stiffness [53]. Bioengineered hydrogels show promise in vitro with regard to MSC encapsulation and direct modulation MSC secretions and longevity [115, 183], and they are being incorporated within clinical trials [108]. Furthermore, strategies such as individual cell encapsulation allow intravenous administration plus significantly increased MSC residence time in vivo at the therapeutic site [125].

Nanomaterials and 3D culture technology introduce significant novel aspects to stem cell biology and potency assay measurement. For example, the allotrope of carbon, graphene introduces new properties amenable to the development of biosensors, including Lab-on-PCB systems [121], which can combine enhanced cell performance and measurement [3]. Both bone marrow and adipose tissue-sourced MSC exhibited a graphene-induced osteogenic response, yet distinct genetic profiles could be characterized [99], consistent with observations that AD-MSC osteogenic differentiation in vitro may lag that of BM-MSC [114]. Tracking the chronological process of differentiation would benefit from non-invasive monitoring rather than conventional end-point assays that disrupt of the sample. Molecular Beacon (MB) nanosensors encapsulated in biodegradable poly(lactic-*co*-glycolic acid) (PLGA) nanoparticles can be internalized with release of MB into the cytoplasm that hybridize with intracellular mRNA to provide semi-quantitative mRNA expression measurement [164]. Beyond use for measurement, precision-engineered nanoparticles can counteract inflammation-induced MSC dysfunction by capturing Ca^{2+} ions around mitochondria to disrupt a damaging Ca^{2+} ion overload, thus acting to "nanorepair" the MSC and restore their physiological function [196]. Combining function with measurement, multifunctional theranostic nanoparticles were loaded as MSC "cargo" composed of photoluminescent quantum dots (QDs) and a photosensitizer chlorin e6 (Ce6) that together generated reactive oxygen species. MSC-mediated migration to tumour microenvironments could deliver the QD-Ce6 nanoparticles for photodynamic therapy to then locally activate the intratumorally accumulated selective photosensitizer. The QD element acted as a complexed energy donor and carrier of Ce6 that could also serve as a diagnostic tool [34]. Genetically engineered MSC are also being investigated as armed MSC, amid a growing interest in cell therapy against challenging cancer types [50]. Click chemistry has been used to engineer an enhanced collagen-binding affinity of MSC-derived EV for superior retention and therapeutic efficacy [56]. The above examples illustrate the high versatility of cell–nanomaterial combinations that will require new dedicated potency assays.

In contrast to two-dimensional (2D) monolater culture, three-dimensional (3D) spheroid cell aggregates better mimic features of the naïve MSC niche, introducing distinctive physical and biochemical qualities [71] considered beneficial for enhanced therapeutic applications [80]. Hydrogels and novel biomaterials provide a wide range of substrates and cues, presenting extensive opportunities to further enhance the in vitro MSC microenvironment [41, 65]. 3D culture environments can evoke stress responses that serve to enhance the MSC therapeutic efficacy [40]. There are reports of 3D MSC cultures producing a higher concentration of EV [69] and generation of therapeutically potent 3D MSC-derived EV [107]. However, the extent to which specific differences between EV derived from 3D versus 2D conditions leads to an increased therapeutic potential remains to be resolved [83]. Bioimpedance platforms can be used for non-invasive, real-time, spatially sensitive monitoring of 3D cultures in hydrogel scaffolds [18]. Of note, cells grown as 3D spheroids could reach a well-sustained dynamic metabolic equilibrium [189]. Cells maintaining a metabolic plateau advantageously extend the sampling time during which one can obtain more consistent and reproducible potency biomarker measurement, improving potency assay reliability.

11.8 Regulations, Guidelines and Evolving Institutional Roles

The first ATMP cell therapy granted marketing authorization throughout Europe in February 2015, named Holoclar®, was comprised of ex vivo expanded autologous human corneal epithelial cells for treating limbal stem cell deficiency after a chemical eye burn. Its development evolved amid a regulatory transition from pharmaceutical rules originally tailored for control and manufacture of chemical molecules to new regulations encompassing the alternative perspective of living cells. Monitoring the maintenance of the therapeutic effect in the product was particularly challenging, yet resolved through detailed understanding of the functional significance of p63 isoforms and their applicability as a quantitative potency biomarker of competent limbal stem cells [123]. Careful derivation of detailed standard operating procedures permitted the biological and clinical work at different sites, with maintenance of identity, purity and potency guaranteed. Follow-up data, up to 10 years post-implantation, allowed a risk–benefit study of 130 patients, with outcomes indicating 70–80% success rates. Future improvement could be anticipated from more sophisticated cell culture methodologies, gene therapies and improved stem cell characterization [142]. The Holoclar® precedent has helped focus attention on how it might be possible to streamline stem cell research guidelines and stringent translational and regulatory requirements with greater harmonization at both local [57, 176] and international levels [67, 68, 82, 127, 195]. A recurring theme emphasized within the field of human corneal endothelial cell therapy is how characterization of the critical quality attributes represents a significant challenge, given hurdles arising from inter-donor variation, sample handling, cell isolation techniques, culture medium, risk of karyotypic aberration, tissue contamination, transportation and storage. Regulatory frameworks can provide specific guidance for "Good Practice" applicable throughout ATMP development (Fig. 11.3) to address these challenges, thus early engagement with the local or regional authority is recommended as soon as reliable pre-clinical outcomes are obtained [170]. ATMP involving autologous chondrocytes, researched globally over the past two decades in response to a growing number of joint cartilage morbidities, provide an excellent example to compare how authorization review processes in the European Union, the United States, Japan, Australia and Korea have each classified and defined the eight approved products, with continuously evolving oversight on current clinical trials [76].

It may be difficult to fully define a potency assay at early stages, because the characterization is likely to be influenced by manufacturing procedures that may well change during product development. It would be prudent to identify a number of biological activities and phenotypic properties critical for clinical benefit, so that choices may ultimately align with how the process may contribute to potency assay selection. All specifications will need to be based on data obtained from manufactured batches, so comprehensive characterization early in development will help ensure sufficient data is available for informed decisions concerning suitably qualified assays that generate reliable data. To help accommodate changes in manufacturing process, comparability studies may be performed to ensure that the product is not harmed by the novel process. Comparability studies include molecular characterization with assessment of purity, potency and stability. A successful comparability study need not necessarily establish identical quality attributes, rather confirm that the product remains highly similar without any functionally deleterious differences. In-process quality

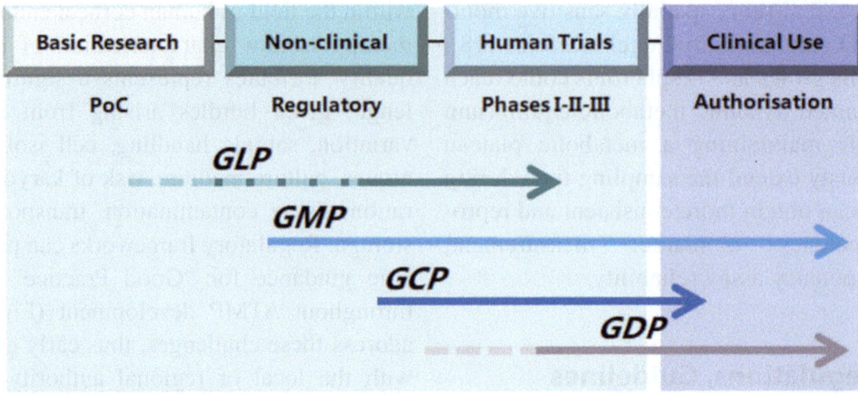

Fig. 11.3 Roadmap for compliance with pharmaceutical quality management standards in the development of cell-, gene- and tissue-based medicines in Europe. Cell-, gene- and tissue-based medicines (excluding vaccines) in Europe are regulated as advanced therapy medicinal products (ATMP) and their regulatory development has similarities to small-molecule drugs and biologicals, with the particularity of centralized authorization by the European Medicines Agency (EMA) and some considerations regarding the living nature of the drug substance (i.e. conditional batch release) or non-industrial manufacture (i.e. hospital exemption). Research stages are shaded in green. Abbreviations: GCP good clinical practice, GDP good distribution practice, GLP good laboratory practice, GMP good manufacturing practice, PoC proof of concept. (Reprinted from Lopez-Navas et al., 2022 [95])

controls are paramount to ensure cell efficiency and safety during all stages of the manufacturing process, with all aspects including cryopreservation, handling, transport and administration in accordance with regulatory authority guidelines and legislation [45]. Quality compliance even in pre-clinical stages of more basic research would be very beneficial for ATMP development, facilitating efficient clinical translation, but there are few regulatory pressures to uphold quality management in the non-clinical setting [95], fortunately, as indicated below, several initiatives are underway to improve circumstances.

Regulation (EC) No. 1394/2007, applied by the European Medical Agency (EMA) from 30 December 2008, was the first to specifically address cell, gene and tissue engineering therapies. It legislated an independent review of ATMP to guarantee high standards of quality, efficacy and safety equivalent to those of other pharmaceuticals before a product is distributed to patients. Directive 2001/83/EC defined (i) gene therapy medicinal products (GTMP) and (ii) somatic cell therapy medicinal products (sCTMP), with subsequent addition of (iii) tissue-engineered products (TEP) and (iv) combined ATMP in Regulation (EC) No. 1394/2007. Notably, classification in the United States by the Federal Drug Administration (FDA) differed since it highlighted only two product types, gene therapy and cellular therapy. Nonetheless, overall approval procedures for ATMP in both regions have strong similarities with expectation that greater regulatory convergence will help ATMP development globally [68]. Providing a central route for EU authorization of ATMP, the expert Committee for Advanced Therapies (CAT) inaugurated in January 2009, helped establish classifications and certification procedures to support small and medium enterprises (SME) and companies in early phases of ATMP development, clarifying whether a product falls within the definition of an ATMP in the EU. From the very large number of applications for authorization, disproportionately few have been approved. This has not only reflected the acknowledged deep intricacy of the therapeutic goals, but called into question whether the length and complexity of the regulatory procedures were themselves having detrimental consequences [57]. An unfortunate misguided impression would be that the regulatory framework is principally a significant obstacle that needs to be overcome before widescale deployment of a new cell therapy. Likewise, the need to provide a well-validated in vitro potency assay constitutes much more than just an approvability issue in a license application [101]. Significant accomplishments of the potency assay are proper activity of the product, help with dose selection and a means of demonstrating comparability when non-clinical/clinical batch production is extended to a commercial scale.

Three regulatory initiatives, namely, orphan drug (OD) designation, the hospital exemption (HE) clause and compassionate use (CU), have sought to mitigate hurdles, incentivize ATMP development and enable early patient accessibility. The EMA's Committee for Orphan Medicinal Products (COMP) is responsible for recommending whether ATMPs apt for providing new medicines for rare untreatable diseases qualify for OD designation according to Orphan Medicines Regulation (EC) No. 141/2000. OD designation introduces incentives that compensate for small patient populations offering limited return on investment. This includes protocol assistance, reduced procedural fees and product market exclusivity for 10 years after marketing authorization [44]. Most of the currently approved ATMP target orphan diseases. The HE clause, within regulation (EC) No. 1394/2007, allows ATMP manufacture to be authorized by a national competent authority (NCA) of the Member State rather than otherwise mandatory centralized marketing authorization procedures. Thereby, patients may benefit from specific ATMP treatments when no other authorized solutions remain available [36]. Valuable clinical experience can be gained, but the type of clinical data generated under the HE scheme fails to match clinical trial data and is not appropriate for supporting future marketing authorization applications (MAA). Despite broad ethical risk–benefit assessment and informed consent from the patient before treatment [32, 159], an analysis of HE implementation across seven countries revealed differences in HE interpretation and implementation. Although some countries required full compliance with good manufacturing practice, as well as non-clinical and clinical evidence, others did not [59]. Restricted scale HE treatment may apparently introduce parallel ATMP systems, difficult to harmonize and merge without full regulatory overlap [132]. However, although the quality of an ATMP under the HE scheme should be in line with regulations designed for commercialization, it is not intended for commercial development. Authorized by the NCA, under the exclusive responsibility of a medical practitioner, HE ATMP comply with a non-routine individual medicine prescription for a custom-made product for a named patient. Alternatively, in some situations, patients with a life-threatening or seriously debilitating disease have no effective authorized therapies and may not be able to participate in clinical trials. Nonetheless, article 28 of the ATMP regulation grants such patients access to an unregistered product as part of a compassionate use or extended access programme. CU ATMP are coordinated and implemented by NCA that may consult with EMA for opinion on how to administer, distribute and use such medicines. In general, CU ATMP must be undergoing clinical trials or have entered the MAA process. HE and CU ATMP ethically reconcile a patient's need for access to novel ATMP, yet risk abuse of the regulations. Countering unethical unproven ATMP-based interventions, there are calls for a Europe-wide registry [32] and adoption of MSC minimal criteria extended to include haematocompatibility assessment [109] to improve transparency, reduce patient risk and increase efficacy, whilst at the same time facilitating company awareness of opportunities, thereby encouraging entry of such ATMP into the mainstream system.

The hospital exemption scheme has usefully highlighted the many limitations encountered by academic and hospital institutions. Although their research may lead ATMP discovery, the full industrial capacity to fulfil the developmental path to marketing authorization is often lacking. Consequently, the role of hospitals for ATMP development has evolved to a focus on procurement of starting material, selected manufacturing steps, investigation to support the potency assay, clinical application and participation in clinical trials. Successful collaboration between academia and industry will be needed to integrate sector strengths for streamlined ATMP development. A number of contract research organizations (CRO) are progres-

Table 11.1 Examples of contract research organizations (CRO) providing potency assay information and support

Absorption Systems (a Pharmaron company)	https://www.absorption.com
AliraHealth	https://alirahealth.com/
Applied StemCell	https://www.appliedstemcell.com/
BioAgilytix	https://www.bioagilytix.com/
Bioassay Sciences	https://bioassaysciences.com/
Bioprocess Online	https://www.bioprocessonline.com/
Cell & Gene	https://www.cellandgene.com/
Charles River Laboratories	https://www.criver.com/
Cytiva	https://www.cytivalifesciences.com/en/us
Eurofins	https://www.eurofins.com/
HemoGenix, Inc.	https://www.hemogenix.com
IBR Inc. Institute for Biopharmaceutical Research	https://www.ibr-inc.com/
Innovacell	https://innovacell.com/en/
KYMOS Group	https://kymos.com/
Labcorp Drug Development	https://drugdevelopment.labcorp.com/
Marin Biologic Laboratories	https://www.marinbio.com/
North East Biolab	https://www.nebiolab.com/
Pacific Biolabs	https://pacificbiolabs.com/
PharmaLex	https://www.pharmalex.com
Pharmaron	https://www.pharmaron.com/
Promega	https://ita.promega.com/
Quality Assistance S.A.	https://www.quality-assistance.com
Reaction Biology	https://www.reactionbiology.com/
Sartorius	https://www.sartorius.com/en
Somru Bioscience	https://somru.ca/
Stemcell	https://www.stemcell.com/

sively placing focus on potency assay development (Table 11.1) and will likely play an increasingly important supportive role in ATMP development.

EMA and other principal regulatory authorities established for the governance of ATMP are continuously monitoring and appraising progress, responding to challenges when scientific discoveries outpace regulatory control [148]. In 2016, EMA launched a PRIority MEdicines (PRIME) scheme for fast-track development of medicines targeting unmet medical need [178], accelerating patient access with provision of active support without need for large data sets and accelerated assessment procedures when the Committee for Medicinal Products for Human Use (CHMP) deem a product of major interest for public health and therapeutic innovation. A series of initiatives have been introduced worldwide to expedite the translation of scientific invention into a health benefit product. Among the latest, a Strengthening Training of Academia in Regulatory Science (STARS) consortium of 18 European regulatory agencies including EMA has been established to reach out to innovative research scientists and bridge the regulatory knowledge gap. Bidirectional information exchange will stimulate the modification of regulatory requirements to help ensure a high-quality clinical development dossier can improve the likelihood of success when stakeholders apply for marketing authorization [127, 158]. Revolutionizing progress, technological advances are synergizing with clinical knowledge to foster rational strategies for designing ATMP with more clearly defined targets, ideally placing emphasis on the MoA and suitable potency assays for product verification from the beginning of a project [89]. Engineering strategies can be tailored accordingly to modulate and control cell performance, with a remarkable array of bioengineering tools capable of extending the capability of potency assays to attain ATMP of new levels of therapeutic achievement.

References

1. Aguado T, Gutiérrez FJ, Aix E, Schneider RP, Giovinazzo G, Blasco MA, Flores I (2017) Telomere length defines the cardiomyocyte differentiation potency of mouse induced pluripotent stem cells. Stem Cells 35:362–373
2. Allsopp R (2012) Telomere length and iPSC re-programming: survival of the longest. Cell Res 22:614–615
3. Amărandi RM, Becheru DF, Vlăsceanu GM, Ioniţă M, Burns JS (2018) Advantages of graphene biosensors for human stem cell therapy potency assays. Biomed Res Int 2018:1676851
4. Antwi-Baffour SS (2015) Molecular characterisation of plasma membrane-derived vesicles. J Biomed Sci 22:68
5. Baglio SR, Pegtel DM, Baldini N (2012) Mesenchymal stem cell secreted vesicles provide novel opportunities in (stem) cell-free therapy. Front Physiol 3:359
6. Bahsoun S, Brown MJ, Coopman K, Akam EC (2022) Cryopreservation of human bone marrow derived mesenchymal stem cells at high concentration is feasible. Biopreserv Biobank. https://doi.org/10.1089/bio.2022.0017
7. Bang OY, Kim JE (2022) Stem cell-derived extracellular vesicle therapy for acute brain insults and neurodegenerative diseases. BMB Rep 55:20–29
8. Barlian A, Vanya K (2022) Nanotopography in directing osteogenic differentiation of mesenchymal stem cells: potency and future perspective. Future Sci OA 8:FSO765
9. Bazzan E, Tinè M, Casara A, Biondini D, Semenzato U, Cocconcelli E, Balestro E, Damin M, Radu CM, Turato G, Baraldo S, Simioni P, Spagnolo P, Saetta M, Cosio MG (2021) Critical review of the evolution of extracellular vesicles' knowledge: from 1946 to today. Int J Mol Sci 22:6417
10. Benson E, Betson F, Fuller BJ, Harding K, Kofanova O (2013) Translating cryobiology principles into trans-disciplinary storage guidelines for biorepositories and biobanks: a concept paper. Cryo Letters 34:277–312
11. Bersenev A, Gustafson MP, Hanley PJ (2022) ISCT survey on hospital practices to support externally manufactured investigational cell-gene therapy products. Cytotherapy 24:27–31
12. Bhat R, Bissell MJ (2014) Of plasticity and specificity: dialectics of the microenvironment and macroenvironment and the organ phenotype. Wiley Interdiscip Rev Dev Biol 3:147–163
13. Bi Y, Qiao X, Liu Q, Song S, Zhu K, Qiu X, Zhang X, Jia C, Wang H, Yang Z, Zhang Y, Ji G (2022) Systemic proteomics and miRNA profile analysis of exosomes derived from human pluripotent stem cells. Stem Cell Res Ther 13:449
14. Bissell MJ, Hines WC (2011) Why don't we get more cancer? A proposed role of the microenvironment in restraining cancer progression. Nat Med 17:320–329
15. Bowles AC, Kouroupis D, Willman MA, Perucca Orfei C, Agarwal A, Correa D (2020) Signature quality attributes of CD146+ mesenchymal stem/stromal cells correlate with high therapeutic and secretory potency. Stem Cells 38:1034–1049
16. Braga CL, da Silva LR, Santos RT, de Carvalho LRP, Mandacaru SC, de Oliveira Trugilho MR, Rocco PRM, Cruz FF, Silva PL (2022) Proteomics profile of mesenchymal stromal cells and extracellular vesicles in normoxic and hypoxic conditions. Cytotherapy 24:1211–1224. S1465-3249(22)00783
17. Calcat-I-Cervera S, Sanz-Nogués C, O'Brien T (2021) When origin matters: properties of mesenchymal stromal cells from different sources for clinical translation in kidney disease. Front Med (Lausanne) 8:728496
18. Canali C, Heiskanen A, Muhammad HB, Høyum P, Pettersen FJ, Hemmingsen M, Wolff A, Dufva M, Martinsen ØG, Emnéus J (2015) Bioimpedance monitoring of 3D cell culturing—complementary electrode configurations for enhanced spatial sensitivity. Biosens Bioelectron 63:72–79
19. Cao Q, Huang C, Chen XM, Pollock CA (2022) Mesenchymal stem cell-derived exosomes: toward cell-free therapeutic strategies in chronic kidney disease. Front Med (Lausanne) 9:816656
20. Capelli C, Frigerio S, Lisini D, Nava S, Gaipa G, Belotti D, Cabiati B, Budelli S, Lazzari L, Bagnarino J, Tanzi M, Comoli P, Perico N, Introna M, Golay J (2022) A comprehensive report of long-term stability data for a range ATMPs: a need to develop guidelines for safe and harmonized stability studies. Cytotherapy 24:544–556
21. Carthew J, Taylor JBJ, Garcia-Cruz MR, Kiaie N, Voelcker NH, Cadarso VJ, Frith JE (2022) The bumpy road to stem cell therapies: rational design of surface topographies to dictate stem cell mechanotransduction and fate. ACS Appl Mater Interfaces. https://doi.org/10.1021/acsami.1c22109
22. Castella M, Caballero-Baños M, Ortiz-Maldonado V, González-Navarro EA, Suñé G, Antoñana-Vidósola A, Boronat A, Marzal B, Millán L, Martín-Antonio B, Cid J, Lozano M, García E, Tabera J, Trias E, Perpiña U, Canals JM, Baumann T, Benítez-Ribas D, Campo E, Yagüe J, Urbano-Ispizua Á, Rives S, Delgado J, Juan M (2020) Point-of-care CAR T-cell production (ARI-0001) using a closed semi-automatic bioreactor: experience from an academic phase I clinical trial. Front Immunol 11:482
23. Chaudhuri O, Cooper-White J, Janmey PA, Mooney DJ, Shenoy VB (2020) Effects of extracellular matrix viscoelasticity on cellular behaviour. Nature 584:535–546
24. Cheng J, Mao X, Chen C, Long X, Chen L, Zhou J, Zhu L (2022) Monitoring CAR19 T cell population by flow cytometry and its consistency with ddPCR. Cytometry A 103:16–26

25. Cherian DS, Bhuvan T, Meagher L, Heng TSP (2020) Biological considerations in scaling up therapeutic cell manufacturing. Front Pharmacol 11:654
26. Cheung HH, Liu X, Canterel-Thouennon L, Li L, Edmonson C, Rennert OM (2014) Telomerase protects Werner syndrome lineage-specific stem cells from premature aging. Stem Cell Rep 2:534–546
27. Choudhery MS, Mahmood R, Harris DT, Ahmad FJ (2022) Minimum criteria for defining induced mesenchymal stem cells. Cell Biol Int 46:986–989
28. Chowdhury N (2010) Regulation of nanomedicines in the EU: distilling lessons from the pediatric and the advanced therapy medicinal products approaches. Nanomedicine (Lond) 5:135–142
29. Christy BA, Herzig MC, Delavan CP, Abaasah I, Cantu C, Salgado C, Lovelace S, Garcia L, Jensen K, Montgomery R, Cap AP, Bynum JA (2020) Use of multiple potency assays to evaluate human mesenchymal stromal cells. J Trauma Acute Care Surg 89:S109–S117
30. Costela-Ruiz VJ, Melguizo-Rodríguez L, Bellotti C, Illescas-Montes R, Stanco D, Arciola CR, Lucarelli E (2022) Different sources of mesenchymal stem cells for tissue regeneration: a guide to identifying the most favorable one in orthopedics and dentistry applications. Int J Mol Sci 23:6356
31. Cottle C, Porter AP, Lipat A, Turner-Lyles C, Nguyen J, Moll G, Chinnadurai R (2022) Impact of cryopreservation and freeze-thawing on therapeutic properties of mesenchymal stromal/stem cells and other common cellular therapeutics. Curr Stem Cell Rep 8:72–92
32. Cuende N, Ciccocioppo R, Forte M, Galipeau J, Ikonomou L, Levine BL, Srivastava A, Zettler PJ (2022) Patient access to and ethical considerations of the application of the European Union hospital exemption rule for advanced therapy medicinal products. Cytotherapy 24:686–690
33. Czapla J, Matuszczak S, Kulik K, Wiśniewska E, Pilny E, Jarosz-Biej M, Smolarczyk R, Sirek T, Zembala MO, Zembala M, Szala S, Cichoń T (2019) The effect of culture media on large-scale expansion and characteristic of adipose tissue-derived mesenchymal stromal cells. Stem Cell Res Ther 10:235
34. Dapkute D, Pleckaitis M, Bulotiene D, Daunoravicius D, Rotomskis R, Karabanovas V (2021) Hitchhiking nanoparticles: mesenchymal stem cell-mediated delivery of theranostic nanoparticles. ACS Appl Mater Interfaces 13:43937–43951
35. Dave C, Mei SHJ, McRae A, Hum C, Sullivan KJ, Champagne J, Ramsay T, McIntyre L (2022) Comparison of freshly cultured versus cryopreserved mesenchymal stem cells in animal models of inflammation: a pre-clinical systematic review. elife 11:e75053
36. de Wilde S, Veltrop-Duits L, Hoozemans-Strik M, Ras T, Blom-Veenman J, Guchelaar HJ, Zandvliet M, Meij P (2016) Hurdles in clinical implementation of academic advanced therapy medicinal products: a national evaluation. Cytotherapy 18:797–805
37. Delgado J, Roddie C, Schmitt M (2022) Point-of-care production of CAR-T cells. In: Kröger N, Gribben J, Chabannon C, Yakoub-Agha I, Einsele H (eds) The EBMT/EHA CAR-T cell handbook. Springer, Cham
38. DiStefano TJ, Vaso K, Danias G, Chionuma HN, Weiser JR, Iatridis JC (2022) Extracellular vesicles as an emerging treatment option for intervertebral disc degeneration: therapeutic potential, translational pathways, and regulatory considerations. Adv Healthc Mater 11:e2100596
39. Dominici M, Le Blanc K, Mueller I, Slaper-Cortenbach I, Marini F, Krause D, Deans R, Keating A, Prockop D, Horwitz E (2006) Minimal criteria for defining multipotent mesenchymal stromal cells. The International Society for Cellular Therapy position statement. Cytotherapy 8:315–317
40. Domnina A, Ivanova J, Alekseenko L, Kozhukharova I, Borodkina A, Pugovkina N, Smirnova I, Lyublinskaya O, Fridlyanskaya I, Nikolsky N (2020) Three-dimensional compaction switches stress response programs and enhances therapeutic efficacy of endometrial mesenchymal stem/stromal cells. Front Cell Dev Biol 8:473
41. Doron G, Temenoff JS (2021) Culture substrates for improved manufacture of mesenchymal stromal cell therapies. Adv Healthc Mater 10:e2100016
42. Dupuis V, Oltra E (2021) Methods to produce induced pluripotent stem cell-derived mesenchymal stem cells: mesenchymal stem cells from induced pluripotent stem cells. World J Stem Cells 13:1094–1111
43. Eichholz KF, Woods I, Riffault M, Johnson GP, Corrigan M, Lowry MC, Shen N, Labour MN, Wynne K, O'Driscoll L, Hoey DA (2020) Human bone marrow stem/stromal cell osteogenesis is regulated via mechanically activated osteocyte-derived extracellular vesicles. Stem Cells Transl Med 9:1431–1447
44. Farkas AM, Mariz S, Stoyanova-Beninska V, Celis P, Vamvakas S, Larsson K, Sepodes B (2017) Advanced therapy medicinal products for rare diseases: state of play of incentives supporting development in Europe. Front Med (Lausanne) 4:53
45. Fernández-Santos M, García-Arranz M, Andreu E, García-Hernández A, López-Parra M, Villarón E, Sepúlveda P, Fernández-Avilés F, García-Olmo D, Prosper F, Sánchez-Guijo F, Moraleda JM, Zapata A (2022) Optimization of mesenchymal stromal cell (MSC) manufacturing processes for a better therapeutic outcome. Front Immunol 13:918565
46. Galipeau J, Krampera M, Leblanc K, Nolta JA, Phinney DG, Shi Y, Tarte K, Viswanathan S, Martin I (2021) Mesenchymal stromal cell variables influencing clinical potency: the impact of viability, fitness, route of administration and host predisposition. Cytotherapy 23:368–372
47. Gimona M, Brizzi MF, Choo ABH, Dominici M, Davidson SM, Grillari J, Hermann DM, Hill AF, de Kleijn D, Lai RC (2021) Critical considerations

for the development of potency tests for therapeutic applications of mesenchymal stromal cell-derived small extracellular vesicles. Cytotherapy 23:373–380
48. Ginty P (2020) Advanced therapy regulation in the UK: what might the future hold post-Brexit. Cell Gene Ther Insights 6:1171–1178
49. Giubilato E, Cazzagon V, Amorim MJB, Blosi M, Bouillard J, Bouwmeester H, Costa AL, Fadeel B, Fernandes TF, Fito C, Hauser M, Marcomini A, Nowack B, Pizzol L, Powell L, Prina-Mello A, Sarimveis H, Scott-Fordsmand JJ, Semenzin E, Stahlmecke B, Stone V, Vignes A, Wilkins T, Zabeo A, Tran L, Hristozov D (2020) Risk management framework for nano-biomaterials used in medical devices and advanced therapy medicinal products. Materials (Basel) 13:E4532
50. Golinelli G, Mastrolia I, Aramini B, Masciale V, Pinelli M, Pacchioni L, Casari G, Dall'Ora M, Soares MBP, Damasceno PKF, Silva DN, Dominici M, Grisendi G (2020) Arming mesenchymal stromal/stem cells against cancer: has the time come. Front Pharmacol 11:529921
51. Gronthos S, Graves SE, Ohta S, Simmons PJ (1994) The STRO-1+ fraction of adult human bone marrow contains the osteogenic precursors. Blood 84:4164–4173
52. Guedan S, Luu M, Ammar D, Barbao P, Bonini C, Bousso P, Buchholz CJ, Casucci M, De Angelis B, Donnadieu E, Espie D, Greco B, Groen R, Huppa JB, Kantari-Mimoun C, Laugel B, Mantock M, Markman JL, Morris E, Quintarelli C, Rade M, Reiche K, Rodriguez-Garcia A, Rodriguez-Madoz JR, Ruggiero E, Themeli M, Hudecek M, Marchiq I (2022) Time 2EVOLVE: predicting efficacy of engineered T-cells – how far is the bench from the bedside. J Immunother Cancer 10:e003487
53. Gultian KA, Gandhi R, Sarin K, Sladkova-Faure M, Zimmer M, de Peppo GM, Vega SL (2022) Human induced mesenchymal stem cells display increased sensitivity to matrix stiffness. Sci Rep 12:8483
54. Haeusner S, Herbst L, Bittorf P, Schwarz T, Henze C, Mauermann M, Ochs J, Schmitt R, Blache U, Wixmerten A, Miot S, Martin I, Pullig O (2021) From single batch to mass production-automated platform design concept for a phase II clinical trial tissue engineered cartilage product. Front Med (Lausanne) 8:712917
55. Haga CL, Boregowda SV, Booker CN, Krishnappa V, Strivelli J, Cappelli E, Phinney DG (2022) Mesenchymal stem/stromal cells from a transplanted, asymptomatic patient with Fanconi anemia exhibit an aging-like phenotype and dysregulated expression of genes implicated in hematopoiesis and myelodysplasia. Cytotherapy 25:362–368. S1465-3249(22)01023
56. Hao D, Lu L, Song H, Duan Y, Chen J, Carney R, Li JJ, Zhou P, Nolta J, Lam KS, Leach JK, Farmer DL, Panitch A, Wang A (2022) Engineered extracellular vesicles with high collagen-binding affinity present superior in situ retention and therapeutic efficacy in tissue repair. Theranostics 12:6021–6037
57. Hauskeller C (2017) Can harmonized regulation overcome intra-European differences? Insights from a European Phase III stem cell trial. Regen Med 12:599–609
58. Hildebrandt MG, Kidholm K, Pedersen JE, Naghavi-Behzad M, Knudsen T, Krag A, Ryg J, Gerke O, Lassen AT, Ellingsen T, Ditzel HJ, Andersen V, Langhoff A, Nielsen G, Masud T, Münster A-MB, Kyvik K, Brixen K (2022) How to increase value and reduce waste in research: initial experiences of applying Lean thinking and visual management in research leadership. BMJ Open 12:e058179
59. Hills A, Awigena-Cook J, Genenz K, Ostertag M, Butler S, Eggimann AV, Hubert A (2020) An assessment of the hospital exemption landscape across European Member States: regulatory frameworks, use and impact. Cytotherapy 22:772–779.e1
60. Hindson BJ, Ness KD, Masquelier DA et al (2011) High-throughput droplet digital PCR system for absolute quantitation of DNA copy number. Anal Chem 83:8604–8610
61. Hodgkinson T, Tsimbouri PM, Llopis-Hernandez V, Campsie P, Scurr D, Childs PG, Phillips D, Donnelly S, Wells JA, O'Brien FJ, Salmeron-Sanchez M, Burgess K, Alexander M, Vassalli M, Oreffo ROC, Reid S, France DJ, Dalby MJ (2021) The use of nanovibration to discover specific and potent bioactive metabolites that stimulate osteogenic differentiation in mesenchymal stem cells. Sci Adv 7:eabb7921
62. Hollmann J, Brecht J, Goetzke R, Franzen J, Selich A, Schmidt M, Eipel M, Ostrowska A, Hapala J, Fernandez-Rebollo E, Müller-Newen G, Rothe M, Eggermann T, Zenke M, Wagner W (2020) Genetic barcoding reveals clonal dominance in iPSC-derived mesenchymal stromal cells. Stem Cell Res Ther 11:105
63. Hornberger K, Yu G, McKenna D, Hubel A (2019) Cryopreservation of hematopoietic stem cells: emerging assays, cryoprotectant agents, and technology to improve outcomes. Transfus Med Hemother 46:188–196
64. Hort S, Herbst L, Bäckel N, Erkens F, Niessing B, Frye M, König N, Papantoniou I, Hudecek M, Jacobs JJL, Schmitt RH (2022) Toward rapid, widely available autologous CAR-T cell therapy – artificial intelligence and automation enabling the smart manufacturing hospital. Front Med (Lausanne) 9:913287
65. Hu X, Xia Z, Cai K (2022) Recent advances in 3D hydrogel culture systems for mesenchymal stem cell-based therapy and cell behavior regulation. J Mater Chem B 10:1486–1507
66. Hu Y, Zhou Y, Zhang M, Zhao H, Wei G, Ge W, Cui Q, Mu Q, Chen G, Han L, Guo T, Cui J, Jiang X, Zheng X, Yu S, Li X, Zhang X, Chen M, Li X, Gao M, Wang K, Zu C, Zhang H, He X, Wang Y, Wang D, Ren J, Huang H (2022) Genetically modified CD7-targeting allogeneic CAR-T cell therapy

with enhanced efficacy for relapsed/refractory CD7-positive hematological malignancies: a phase I clinical study. Cell Res 32:995–1007
67. Iglesias-López C, Agustí A, Obach M, Vallano A (2019) Regulatory framework for advanced therapy medicinal products in Europe and United States. Front Pharmacol 10:921
68. Iglesias-Lopez C, Obach M, Vallano A, Agustí A (2021) Comparison of regulatory pathways for the approval of advanced therapies in the European Union and the United States. Cytotherapy 23:261–274
69. Jalilian E, Massoumi H, Bigit B, Amin S, Katz EA, Guaiquil VH, Anwar KN, Hematti P, Rosenblatt MI, Djalilian AR (2022) Bone marrow mesenchymal stromal cells in a 3D system produce higher concentration of extracellular vesicles (EVs) with increased complexity and enhanced neuronal growth properties. Stem Cell Res Ther 13:425
70. Jansen BJ, Gilissen C, Roelofs H, Schaap-Oziemlak A, Veltman JA, Raymakers RA, Jansen JH, Kögler G, Figdor CG, Torensma R, Adema GJ (2010) Functional differences between mesenchymal stem cell populations are reflected by their transcriptome. Stem Cells Dev 19:481–490
71. Jauković A, Abadjieva D, Trivanović D, Stoyanova E, Kostadinova M, Pashova S, Kestendjieva S, Kukolj T, Jeseta M, Kistanova E, Mourdjeva M (2020) Specificity of 3D MSC spheroids microenvironment: impact on MSC behavior and properties. Stem Cell Rev Rep 16:853–875
72. Jo S, Das S, Williams A, Chretien AS, Pagliardini T, Le Roy A, Fernandez JP, Le Clerre D, Jahangiri B, Chion-Sotinel I, Rozlan S, Dessez E, Gouble A, Dusséaux M, Galetto R, Duclert A, Marcenaro E, Devillier R, Olive D, Duchateau P, Poirot L, Valton J (2022) Endowing universal CAR T-cell with immune-evasive properties using TALEN-gene editing. Nat Commun 13:3453
73. Kadle RL, Abdou SA, Villarreal-Ponce AP, Soares MA, Sultan DL, David JA, Massie J, Rifkin WJ, Rabbani P, Ceradini DJ (2018) Microenvironmental cues enhance mesenchymal stem cell-mediated immunomodulation and regulatory T-cell expansion. PLoS One 13:e0193178
74. Kaushal R, Jahan S, McGregor C, Pineault N (2022) Dimethyl sulfoxide-free cryopreservation solutions for hematopoietic stem cell grafts. Cytotherapy 24:272–281
75. Keating A (2012) Mesenchymal stromal cells: new directions. Cell Stem Cell 10:709–716
76. Kim J, Park J, Song S-Y, Kim E (2022) Advanced therapy medicinal products for autologous chondrocytes and comparison of regulatory systems in target countries. Regen Ther 20:126–137
77. Klbik I, Čechová K, Maťko I, Lakota J, Šauša O (2022) On crystallization of water confined in liposomes and cryoprotective action of DMSO. RSC Adv 12:2300–2309
78. Kohli N, Al-Delfi IRT, Snow M, Sakamoto T, Miyazaki T, Nakajima H, Uchida K, Johnson WEB (2019) CD271-selected mesenchymal stem cells from adipose tissue enhance cartilage repair and are less angiogenic than plastic adherent mesenchymal stem cells. Sci Rep 9:3194
79. Koleva-Kolarova R, Buchanan J, Vellekoop H, Huygens S, Versteegh M, Mölken MR, Szilberhorn L, Zelei T, Nagy B, Wordsworth S, Tsiachristas A, HEcoPerMed C (2022) Financing and reimbursement models for personalised medicine: a systematic review to identify current models and future options. Appl Health Econ Health Policy 20:501–524
80. Kouroupis D, Correa D (2021) Increased mesenchymal stem cell functionalization in three-dimensional manufacturing settings for enhanced therapeutic applications. Front Bioeng Biotechnol 9:621748
81. Kumawat VS, Bandyopadhyay-Ghosh S, Ghosh SB (2022) An overview of translational research in bone graft biomaterials. J Biomater Sci Polym Ed:1–44
82. Kurauchi R, Kasai H, Ito T (2020) Characteristics of medical products comprising human cells, genes, or tissues developed in Japan and the European Union compared via public assessment reports. Front Bioeng Biotechnol 8:606606
83. Kusuma GD, Li A, Zhu D, McDonald H, Inocencio IM, Chambers DC, Sinclair K, Fang H, Greening DW, Frith JE, Lim R (2022) Effect of 2D and 3D culture microenvironments on mesenchymal stem cell-derived extracellular vesicles potencies. Front Cell Dev Biol 10:819726
84. Lalu MM, Mazzarello S, Zlepnig J, Dong YYR, Montroy J, McIntyre L, Devereaux PJ, Stewart DJ, David Mazer C, Barron CC, McIsaac DI, Fergusson DA (2018) Safety and efficacy of adult stem cell therapy for acute myocardial infarction and ischemic heart failure (SafeCell Heart): a systematic review and meta-analysis. Stem Cells Transl Med 7:857–866
85. Lalu MM, McIntyre L, Pugliese C, Fergusson D, Winston BW, Marshall JC, Granton J, Stewart DJ, Canadian CCTG (2012) Safety of cell therapy with mesenchymal stromal cells (SafeCell): a systematic review and meta-analysis of clinical trials. PLoS One 7:e47559
86. Lee HR, Yang SJ, Choi HK, Kim JA, Oh IH (2021) The chromatin remodeling complex CHD1 regulates the primitive state of mesenchymal stromal cells to control their stem cell supporting activity. Stem Cells Dev 30:363–373
87. Lee S, Kim S, Kim H, Baek EJ, Jin H, Kim J, Kim HO (2008) Post-thaw viable CD34(+) cell count is a valuable predictor of haematopoietic stem cell engraftment in autologous peripheral blood stem cell transplantation. Vox Sang 94:146–152
88. Leijten J, Chai YC, Papantoniou I, Geris L, Schrooten J, Luyten FP (2015) Cell based advanced therapeutic medicinal products for bone repair: keep it simple. Adv Drug Deliv Rev 84:30–44

89. Levy O, Kuai R, Siren EMJ, Bhere D, Milton Y, Nissar N, De Biasio M, Heinelt M, Reeve B, Abdi R, Alturki M, Fallatah M, Almalik A, Alhasan AH, Shah K, Karp JM (2020) Shattering barriers toward clinically meaningful MSC therapies. Sci Adv 6:eaba6884
90. Li C, Zhao H, Cheng L, Wang B (2021) Allogeneic vs. autologous mesenchymal stem/stromal cells in their medication practice. Cell Biosci 11:187
91. Lin H, Sohn J, Shen H, Langhans MT, Tuan RS (2019) Bone marrow mesenchymal stem cells: aging and tissue engineering applications to enhance bone healing. Biomaterials 203:96–110
92. Lindskog U, Lundgren B, Billig D, Lindner E (1987) Alternatives for harvesting cells grown on microcarriers: effects on subsequent attachment and growth. Dev Biol Stand 66:307–313
93. Linkova DD, Rubtsova YP, Egorikhina MN (2022) Cryostorage of mesenchymal stem cells and biomedical cell-based products. Cell 11:2691
94. López-Cantillo G, Urueña C, Camacho BA, Ramírez-Segura C (2022) CAR-T cell performance: how to improve their persistence. Front Immunol 13:878209
95. Lopez-Navas L, Torrents S, Sánchez-Pernaute R, Vives J (2022) Compliance in non-clinical development of cell-, gene-, and tissue-based medicines: good practice for better therapies. Stem Cells Transl Med 11:805–813
96. Lu A, Liu H, Shi R, Cai Y, Ma J, Shao L, Rong V, Gkitsas N, Lei H, Highfill SL, Panch S, Stroncek DF, Jin P (2020) Application of droplet digital PCR for the detection of vector copy number in clinical CAR/TCR T cell products. J Transl Med 18:191
97. Lundh S, Maji S, Melenhorst JJ (2020) Next-generation CAR T cells to overcome current drawbacks. Int J Hematol 114:532–543
98. Luo Z, Yao X, Li M, Fang D, Fei Y, Cheng Z, Xu Y, Zhu B (2022) Modulating tumor physical microenvironment for fueling CAR-T cell therapy. Adv Drug Deliv Rev 185:114301
99. MacDonald AF, Trotter RD, Griffin CD, Bow AJ, Newby SD, King WJ, Amelse LL, Masi TJ, Bourdo SE, Dhar MS (2021) Genetic profiling of human bone marrow and adipose tissue-derived mesenchymal stem cells reveals differences in osteogenic signaling mediated by graphene. J Nanobiotechnol 19:285
100. Malvicini R, Santa-Cruz D, De Lazzari G, Tolomeo AM, Sanmartin C, Muraca M, Yannarelli G, Pacienza N (2022) Macrophage bioassay standardization to assess the anti-inflammatory activity of mesenchymal stromal cell-derived small extracellular vesicles. Cytotherapy 24:999–1012
101. Marks PW, Witten CM, Califf RM (2017) Clarifying stem-cell therapy's benefits and risks. N Engl J Med 376:1007–1009
102. Marple AH, Bonifant CL, Shah NN (2020) Improving CAR T-cells: the next generation. Semin Hematol 57:115–121
103. McGarvey SS, Ferreyros M, Kogut I, Bilousova G (2022) Differentiating induced pluripotent stem cells toward mesenchymal stem/stromal cells. Methods Mol Biol 2549:153–167
104. Méndez-Ferrer S, Michurina TV, Ferraro F, Mazloom AR, Macarthur BD, Lira SA, Scadden DT, Ma'ayan A, Enikolopov GN, Frenette PS (2010) Mesenchymal and haematopoietic stem cells form a unique bone marrow niche. Nature 466:829–834
105. Mendicino M, Bailey AM, Wonnacott K, Puri RK, Bauer SR (2014) MSC-based product characterization for clinical trials: an FDA perspective. Cell Stem Cell 14:141–145
106. Meng HF, Jin J, Wang H, Wang LS, Wu CT (2022) Recent advances in the therapeutic efficacy of hepatocyte growth factor gene-modified mesenchymal stem cells in multiple disease settings. J Cell Mol Med 26:4745–4755
107. Min Lim K, Kim S, Yeom J, Choi Y, Lee Y, An J, Gil M, Abdal Dayem A, Kim K, Kang GH, Kim A, Hong K, Kim K, Cho SG (2022) Advanced 3D dynamic culture system with transforming growth factor-β3 enhances production of potent extracellular vesicles with modified protein cargoes via upregulation of TGF-β signaling. J Adv Res. https://doi.org/10.1016/j.jare.2022.09.005. S2090-1232(22)00207
108. Mohapatra S, Mirza MA, Hilles AR, Zakir F, Gomes AC, Ansari MJ, Iqbal Z, Mahmood S (2021) Biomedical application, patent repository, clinical trial and regulatory updates on hydrogel: an extensive review. Gels 7:207
109. Moll G, Ankrum JA, Olson SD, Nolta JA (2022) Improved MSC minimal criteria to maximize patient safety: a call to embrace tissue factor and hemocompatibility assessment of MSC products. Stem Cells Transl Med 11:2–13
110. Morgenstern DA, Ahsan G, Brocklesby M, Ings S, Balsa C, Veys P, Brock P, Anderson J, Amrolia P, Goulden N, Cale CM, Watts MJ (2016) Post-thaw viability of cryopreserved peripheral blood stem cells (PBSC) does not guarantee functional activity: important implications for quality assurance of stem cell transplant programmes. Br J Haematol 174:942–951
111. Mousaei Ghasroldasht M, Seok J, Park HS, Liakath Ali FB, Al-Hendy A (2022) Stem cell therapy: from idea to clinical practice. Int J Mol Sci 23:2850
112. Mueller K, Saha K (2021) Single cell technologies to dissect heterogenous immune cell therapy products. Curr Opin Biomed Eng 20:100343
113. Muñoz-Domínguez N, Carreras-Sánchez I, López-Fernández A, Vives J (2022) Optimisation of processing methods to improve success in the derivation of human multipotent mesenchymal stromal cells from cryopreserved umbilical cord tissue fragments. Cryobiology 108:34–41
114. Ofiteru AM, Becheru DF, Gharbia S, Balta C, Herman H, Mladin B, Ionita M, Hermenean A, Burns JS (2020) Qualifying osteogenic potency

assay metrics for human multipotent stromal cells: TGF-β2 a telling eligible biomarker. Cell 9:E2559
115. Ogle ME, Doron G, Levy MJ, Temenoff JS (2020) Hydrogel culture surface stiffness modulates mesenchymal stromal cell secretome and alters senescence. Tissue Eng Part A 26:1259–1271
116. Oikonomopoulos A, van Deen WK, Manansala AR, Lacey PN, Tomakili TA, Ziman A, Hommes DW (2015) Optimization of human mesenchymal stem cell manufacturing: the effects of animal/xeno-free media. Sci Rep 5:16570
117. Oja S, Kaartinen T, Ahti M, Korhonen M, Laitinen A, Nystedt J (2019) The utilization of freezing steps in mesenchymal stromal cell (MSC) manufacturing: potential impact on quality and cell functionality attributes. Front Immunol 10:1627
118. Pal B, Das B (2017) In vitro culture of naïve human bone marrow mesenchymal stem cells: a stemness based approach. Front Cell Dev Biol 5:69
119. Pang SHM, D'Rozario J, Mendonca S, Bhuvan T, Payne NL, Zheng D, Hisana A, Wallis G, Barugahare A, Powell D, Rautela J, Huntington ND, Dewson G, Huang DCS, Gray DHD, Heng TSP (2021) Mesenchymal stromal cell apoptosis is required for their therapeutic function. Nat Commun 12:6495
120. Papait A, Ragni E, Cargnoni A, Vertua E, Romele P, Masserdotti A, Perucca Orfei C, Signoroni PB, Magatti M, Silini AR, De Girolamo L, Parolini O (2022) Comparison of EV-free fraction, EVs, and total secretome of amniotic mesenchymal stromal cells for their immunomodulatory potential: a translational perspective. Front Immunol 13:960909
121. Papamatthaiou S, Estrela P, Moschou D (2021) Printable graphene BioFETs for DNA quantification in Lab-on-PCB microsystems. Sci Rep 11:9815
122. Parekkadan B, Milwid JM (2010) Mesenchymal stem cells as therapeutics. Annu Rev Biomed Eng 12:87–117
123. Pellegrini G, Lambiase A, Macaluso C, Pocobelli A, Deng S, Cavallini G, Esteki R, Rama P (2016) From discovery to approval of an advanced therapy medicinal product-containing stem cells, in the EU. Regen Med 11:407–420
124. Peltzer J, Lund K, Goriot ME, Grosbot M, Lataillade JJ, Mauduit P, Banzet S (2020) Interferon-γ and hypoxia priming have limited effect on the miRNA landscape of human mesenchymal stromal cells-derived extracellular vesicles. Front Cell Dev Biol 8:581436
125. Peng H, Chelvarajan L, Donahue R, Gottipati A, Cahall CF, Davis KA, Tripathi H, Al-Darraji A, Elsawalhy E, Dobrozsi N, Srinivasan A, Levitan BM, Kong R, Gao E, Abdel-Latif A, Berron BJ (2021) Polymer cell surface coating enhances mesenchymal stem cell retention and cardiac protection. ACS Appl Bio Mater 4:1655–1667
126. Pittenger MF, Discher DE, Péault BM, Phinney DG, Hare JM, Caplan AI (2019) Mesenchymal stem cell perspective: cell biology to clinical progress. NPJ Regen Med 4:22
127. Pizevska M, Kaeda J, Fritsche E, Elazaly H, Reinke P, Amini L (2022) Advanced therapy medicinal products' translation in Europe: a developers' perspective. Front Med 9:757647
128. Pochopień M, Paterak E, Clay E, Janik J, Aballea S, Biernikiewicz M, Toumi M (2021) An overview of health technology assessments of gene therapies with the focus on cost-effectiveness models. J Mark Access Health Policy 9:2002006
129. Pollock K, Samsonraj RM, Dudakovic A, Thaler R, Stumbras A, McKenna DH, Dosa PI, van Wijnen AJ, Hubel A (2017) Improved post-thaw function and epigenetic changes in mesenchymal stromal cells cryopreserved using multicomponent osmolyte solutions. Stem Cells Dev 26:828–842
130. Pollock K, Sumstad D, Kadidlo D, McKenna DH, Hubel A (2015) Clinical mesenchymal stromal cell products undergo functional changes in response to freezing. Cytotherapy 17:38–45
131. Pratt CB, Alexander M (2022) Importance of CAR-T cell therapy monitoring using high-throughput assays. Drug Discov Today 27:103310
132. Priesner C, Hildebrandt M (2022) Advanced therapy medicinal products and the changing role of academia. Transfus Med Hemother 49(3):158–162
133. Pybus LP, Kalsi D, Matthews JT, Hawke E, Barber N, Richer R, Young A, Saunders FL (2022) Coupling picodroplet microfluidics with plate imaging for the rapid creation of biomanufacturing suitable cell lines with high probability and improved multi-step assurance of monoclonality. Biotechnol J 17:e2100357
134. Quirici N, Soligo D, Bossolasco P, Servida F, Lumini C, Deliliers GL (2002) Isolation of bone marrow mesenchymal stem cells by anti-nerve growth factor receptor antibodies. Exp Hematol 30:783–791
135. Rana N, Suliman S, Al-Sharabi N, Mustafa K (2022) Extracellular vesicles derived from primed mesenchymal stromal cells loaded on biphasic calcium phosphate biomaterial exhibit enhanced macrophage polarization. Cells 11(3):470
136. Ranga Rao S, Subbarayan R (2019) Passage-dependent expression of STRO-1 in human gingival mesenchymal stem cells. J Cell Biochem 120:2810–2815
137. Roddie C, Dias J, O'Reilly MA, Abbasian M, Cadinanos-Garai A, Vispute K, Bosshard-Carter L, Mitsikakou M, Mehra V, Roddy H, Hartley JA, Spanswick V, Lowe H, Popova B, Clifton-Hadley L, Wheeler G, Olejnik J, Bloor A, Irvine D, Wood L, Marzolini MAV, Domning S, Farzaneh F, Lowdell MW, Linch DC, Pule MA, Peggs KS (2021) Durable responses and low toxicity after fast off-rate CD19 chimeric antigen receptor-T therapy in adults with relapsed or refractory B-cell acute lymphoblastic leukemia. J Clin Oncol 39:3352–3363
138. Rodrigues CA, Fernandes TG, Diogo MM, da Silva CL, Cabral JM (2011) Stem cell cultivation in bioreactors. Biotechnol Adv 29:815–829
139. Rosell-Valle C, Antúnez C, Campos F, Gallot N, García-Arranz M, García-Olmo D, Gutierrez R,

Hernán R, Herrera C, Jiménez R, Leyva-Fernández L, Maldonado-Sanchez R, Muñoz-Fernández R, Nogueras S, Ortiz L, Piudo I, Ranchal I, Rodríguez-Acosta A, Segovia C, Fernández-Muñoz B (2021) Evaluation of the effectiveness of a new cryopreservation system based on a two-compartment vial for the cryopreservation of cell therapy products. Cytotherapy 23:740–753
140. Rosová I, Dao M, Capoccia B, Link D, Nolta JA (2008) Hypoxic preconditioning results in increased motility and improved therapeutic potential of human mesenchymal stem cells. Stem Cells 26:2173–2182
141. Ryan JM, Barry F, Murphy JM, Mahon BP (2007) Interferon-gamma does not break, but promotes the immunosuppressive capacity of adult human mesenchymal stem cells. Clin Exp Immunol 149:353–363
142. Sacchetti M, Rama P, Bruscolini A, Lambiase A (2018) Limbal stem cell transplantation: clinical results, limits, and perspectives. Stem Cells Int 2018:8086269
143. Samsonraj RM, Raghunath M, Nurcombe V, Hui JH, van Wijnen AJ, Cool SM (2017) Concise review: multifaceted characterization of human mesenchymal stem cells for use in regenerative medicine. Stem Cells Transl Med 6:2173–2185
144. Sanber K, Savani B, Jain T (2021) Graft-versus-host disease risk after chimeric antigen receptor T-cell therapy: the diametric opposition of T cells. Br J Haematol 195:660–668
145. Schneider RS, Vela AC, Williams EK, Martin KE, Lam WA, García AJ (2022) High-throughput on-chip human mesenchymal stromal cell potency prediction. Adv Healthc Mater 11:e2101995
146. Schubert ML, Berger C, Kunz A, Schmitt A, Badbaran A, Neuber B, Zeschke S, Wang L, Riecken K, Hückelhoven-Krauss A, Müller I, Müller-Tidow C, Dreger P, Kröger N, Ayuk FA, Schmitt M, Fehse B (2022) Comparison of single copy gene-based duplex quantitative PCR and digital droplet PCR for monitoring of expansion of CD19-directed CAR T cells in treated patients. Int J Oncol 60:48
147. Sebastião MJ, Serra M, Gomes-Alves P, Alves PM (2021) Stem cells characterization: OMICS reinforcing analytics. Curr Opin Biotechnol 71:175–181
148. Sepodes B, Mol P (2022) Editorial: insights in regulatory science 2021. Front Med (Lausanne) 9:1033558
149. Shekari F, Nazari A, Assar Kashani S, Hajizadeh-Saffar E, Lim R, Baharvand H (2021) Pre-clinical investigation of mesenchymal stromal cell-derived extracellular vesicles: a systematic review. Cytotherapy 23:277–284
150. Shen JL, Huang YZ, Xu SX, Zheng PH, Yin WJ, Cen J, Gong LZ (2012) Effectiveness of human mesenchymal stem cells derived from bone marrow cryopreserved for 23-25 years. Cryobiology 64:167–175
151. Si X, Xiao L, Brown CE, Wang D (2022) Preclinical evaluation of CAR T cell function: in vitro and in vivo models. Int J Mol Sci 23:3154
152. Silva Couto P, Rotondi MC, Bersenev A, Hewitt CJ, Nienow AW, Verter F, Rafiq QA (2020) Expansion of human mesenchymal stem/stromal cells (hMSCs) in bioreactors using microcarriers: lessons learnt and what the future holds. Biotechnol Adv 45:107636
153. Simonsen JL, Rosada C, Serakinci N, Justesen J, Stenderup K, Rattan SI, Jensen TG, Kassem M (2002) Telomerase expression extends the proliferative life-span and maintains the osteogenic potential of human bone marrow stromal cells. Nat Biotechnol 20:592–596
154. Spitzhorn LS, Megges M, Wruck W, Rahman MS, Otte J, Degistirici Ö, Meisel R, Sorg RV, Oreffo ROC, Adjaye J (2019) Human iPSC-derived MSCs (iMSCs) from aged individuals acquire a rejuvenation signature. Stem Cell Res Ther 10:100
155. Spohn G, Witte AS, Kretschmer A, Seifried E, Schäfer R (2021) More human BM-MSC with similar subpopulation composition and functional characteristics can be produced with a GMP-compatible fabric filter system compared to density gradient technique. Front Cell Dev Biol 9:638798
156. Srinivasan A, Sathiyanathan P, Yin L, Liu TM, Lam A, Ravikumar M, Smith RAA, Loh HP, Zhang Y, Ling L, Ng SK, Yang YS, Lezhava A, Hui J, Oh S, Cool SM (2022) Strategies to enhance immunomodulatory properties and reduce heterogeneity in mesenchymal stromal cells during ex vivo expansion. Cytotherapy 24:456–472
157. Srinivasan M, Thangaraj SR, Ramasubramanian K, Thangaraj PP, Ramasubramanian KV (2021) Exploring the current trends of artificial intelligence in stem cell therapy: a systematic review. Cureus 13:e20083
158. Starokozhko V, Kallio M, Kumlin Howell Å, Mäkinen Salmi A, Andrew-Nielsen G, Goldammer M, Burggraf M, Löbker W, Böhmer A, Agricola E, de Vries CS, Pasmooij AMG, Mol PGM, STARS C (2021) Strengthening regulatory science in academia: STARS, an EU initiative to bridge the translational gap. Drug Discov Today 26:283–288
159. Sugarman J, Barker R, Charo R (2019) A professional standard for informed consent for stem cell therapies. JAMA 322(17):1651–1652
160. Sung SE, Kang KK, Choi JH, Lee SJ, Kim K, Lim JH, Yang SY, Kim SK, Seo MS, Lee GW (2021) Comparisons of extracellular vesicles from human epidural fat-derived mesenchymal stem cells and fibroblast cells. Int J Mol Sci 22:2889
161. Takahashi K, Tanabe K, Ohnuki M, Narita M, Ichisaka T, Tomoda K, Yamanaka S (2007) Induction of pluripotent stem cells from adult human fibroblasts by defined factors. Cell 131:861–872
162. Takeuchi S, Tsuchiya A, Iwasawa T, Nojiri S, Watanabe T, Ogawa M, Yoshida T, Fujiki K, Koui Y, Kido T, Yoshioka Y, Fujita M, Kikuta J, Itoh T, Takamura M, Shirahige K, Ishii M, Ochiya T, Miyajima A, Terai S (2021) Small extracellular vesicles derived from interferon-γ pre-conditioned mes-

enchymal stromal cells effectively treat liver fibrosis. NPJ Regen Med 6:19
163. Tang L, Mei Y, Shen Y, He S, Xiao Q, Yin Y, Xu Y, Shao J, Wang W, Cai Z (2021) Nanoparticle-mediated targeted drug delivery to remodel tumor microenvironment for cancer therapy. Int J Nanomedicine 16:5811–5829
164. Tay LM, Wiraja C, Yeo DC, Wu Y, Yang Z, Chuah YJ, Lee EH, Kang Y, Xu C (2017) Noninvasive monitoring of three-dimensional chondrogenic constructs using molecular beacon nanosensors. Tissue Eng Part C Methods 23:12–20
165. Taylor SC, Laperriere G, Germain H (2017) Droplet Digital PCR versus qPCR for gene expression analysis with low abundant targets: from variable nonsense to publication quality data. Sci Rep 7:2409
166. TerWee JA, Chin CL, Watrin S, Tello RF, Rieder NJ, Lowell JD, Latham-Timmons D (2011) Increased consistency and efficiency in routine potency testing by bioassay with direct use of cryopreserved (ready-to-plate) cells. J Immunol Methods 370:65–74
167. Thamm K, Möbus K, Towers R, Baertschi S, Wetzel R, Wobus M, Segeletz S (2022) A chemically defined biomimetic surface for enhanced isolation efficiency of high-quality human mesenchymal stromal cells under xenogeneic/serum-free conditions. Cytotherapy 24:1049–1059
168. Théry C, Witwer KW, Aikawa E et al (2018) Minimal information for studies of extracellular vesicles 2018 (MISEV2018): a position statement of the International Society for Extracellular Vesicles and update of the MISEV2014 guidelines. J Extracell Vesicles 7:1535750
169. Tieu A, Lalu MM, Slobodian M, Gnyra C, Fergusson DA, Montroy J, Burger D, Stewart DJ, Allan DS (2020) An analysis of mesenchymal stem cell-derived extracellular vesicles for preclinical use. ACS Nano 14:9728–9743
170. Ting DSJ, Peh GSL, Adnan K, Mehta JS (2022) Translational and regulatory challenges of corneal endothelial cell therapy: a global perspective. Tissue Eng Part B Rev 28:52–62
171. Tormin A, Li O, Brune JC, Walsh S, Schütz B, Ehinger M, Ditzel N, Kassem M, Scheding S (2011) CD146 expression on primary nonhematopoietic bone marrow stem cells is correlated with in situ localization. Blood 117:5067–5077
172. Trachana V, Petrakis S, Fotiadis Z, Siska EK, Balis V, Gonos ES, Kaloyianni M, Koliakos G (2017) Human mesenchymal stem cells with enhanced telomerase activity acquire resistance against oxidative stress-induced genomic damage. Cytotherapy 19:808–820
173. Tsai AC, Jeske R, Chen X, Yuan X, Li Y (2020) Influence of microenvironment on mesenchymal stem cell therapeutic potency: from planar culture to microcarriers. Front Bioeng Biotechnol 8:640
174. Veronesi E, Murgia A, Caselli A, Grisendi G, Piccinno MS, Rasini V, Giordano R, Montemurro T, Bourin P, Sensebé L, Rojewski MT, Schrezenmeier H, Layrolle P, Ginebra MP, Panaitescu CB, Gómez-Barrena E, Catani F, Paolucci P, Burns JS, Dominici M (2014) Transportation conditions for prompt use of ex vivo expanded and freshly harvested clinical-grade bone marrow mesenchymal stromal/stem cells for bone regeneration. Tissue Eng Part C Methods 20:239–251
175. Vieira CP, McCarrel TM, Grant MB (2021) Novel methods to mobilize, isolate, and expand mesenchymal stem cells. Int J Mol Sci 22:5728
176. Vives J, Sòria MG, McGrath E, Magri M (2022) The quality management ecosystem in cell therapy in Catalonia (Spain): an opportunity for integrating standards and streamlining quality compliance. Cell 11:2112
177. Vonk LA, van Dooremalen SFJ, Liv N, Klumperman J, Coffer PJ, Saris DBF, Lorenowicz MJ (2018) Mesenchymal stromal/stem cell-derived extracellular vesicles promote human cartilage regeneration in vitro. Theranostics 8:906–920
178. Vreman RA, Heikkinen I, Schuurman A, Sapede C, Garcia JL, Hedberg N, Athanasiou D, Grueger J, Leufkens HGM, Goettsch WG (2019) Unmet medical need: an introduction to definitions and stakeholder perceptions. Value Health 22:1275–1282
179. Wagner DL, Koehl U, Chmielewski M, Scheid C, Stripecke R (2022) Review: sustainable clinical development of CAR-T cells – switching from viral transduction towards CRISPR-Cas gene editing. Front Immunol 13:865424
180. Wang J, Chen G, Chen ZM, Wang FP, Xia B (2022) Current strategies in biomaterial-based periosteum scaffolds to promote bone regeneration: a review. J Biomater Appl:8853282221135095
181. Watanabe Y, Fukuda T, Hayashi C, Nakao Y, Toyoda M, Kawakami K, Shinjo T, Iwashita M, Yamato H, Yotsumoto K, Taketomi T, Uchiumi T, Sanui T, Nishimura F (2022) Extracellular vesicles derived from GMSCs stimulated with TNF-α and IFN-α promote M2 macrophage polarization via enhanced CD73 and CD5L expression. Sci Rep 12:13344
182. Watt SM (2022) The long and winding road: homeostatic and disordered haematopoietic microenvironmental niches: a narrative review. Biomater Transl 3:31–54
183. Wechsler ME, Rao VV, Borelli AN, Anseth KS (2021) Engineering the MSC secretome: a hydrogel focused approach. Adv Healthc Mater 10:e2001948
184. Werb Z, Sympson CJ, Alexander CM, Thomasset N, Lund LR, MacAuley A, Ashkenas J, Bissell MJ (1996) Extracellular matrix remodeling and the regulation of epithelial-stromal interactions during differentiation and involution. Kidney Int Suppl 54:S68–S74
185. Wiese DM, Wood CA, Braid LR (2022) From vial to vein: crucial gaps in mesenchymal stromal cell clinical trial reporting. Front Cell Dev Biol 10:867426
186. Witwer KW, Goberdhan DC, O'Driscoll L, Théry C, Welsh JA, Blenkiron C, Buzás EI, Di Vizio D, Erdbrügger U, Falcón-Pérez JM, Fu QL, Hill AF,

Lenassi M, Lötvall J, Nieuwland R, Ochiya T, Rome S, Sahoo S, Zheng L (2021) Updating MISEV: evolving the minimal requirements for studies of extracellular vesicles. J Extracell Vesicles 10:e12182
187. Witwer KW, Van Balkom BWM, Bruno S, Choo A, Dominici M, Gimona M, Hill AF, De Kleijn D, Koh M, Lai RC, Mitsialis SA, Ortiz LA, Rohde E, Asada T, Toh WS, Weiss DJ, Zheng L, Giebel B, Lim SK (2019) Defining mesenchymal stromal cell (MSC)-derived small extracellular vesicles for therapeutic applications. J Extracell Vesicles 8:1609206
188. Wruck W, Graffmann N, Spitzhorn LS, Adjaye J (2021) Human induced pluripotent stem cell-derived mesenchymal stem cells acquire rejuvenation and reduced heterogeneity. Front Cell Dev Biol 9:717772
189. Wrzesinski K, Rogowska-Wrzesinska A, Kanlaya R, Borkowski K, Schwämmle V, Dai J, Joensen KE, Wojdyla K, Carvalho VB, Fey SJ (2014) The cultural divide: exponential growth in classical 2D and metabolic equilibrium in 3D environments. PLoS One 9:e106973
190. Wu X, Schmidt-Wolf IGH (2022) An alternative source for allogeneic CAR T cells with a high safety profile. Front Immunol 13:913123
191. Xu D, Marchionni K, Hu Y, Zhang W, Sosic Z (2017) Quantitative analysis of a biopharmaceutical protein in cell culture samples using automated capillary electrophoresis (CE) western blot. J Pharm Biomed Anal 145:10–15
192. Yen BL, Liu KJ, Sytwu HK, Yen ML (2022) Clinical implications of differential functional capacity between tissue-specific human mesenchymal stromal/stem cells. FEBS J. https://doi.org/10.1111/febs.16438
193. Yin JQ, Zhu J, Ankrum JA (2019) Manufacturing of primed mesenchymal stromal cells for therapy. Nat Biomed Eng 3:90–104
194. Yin Y, Boesteanu AC, Binder ZA, Xu C, Reid RA, Rodriguez JL, Cook DR, Thokala R, Blouch K, McGettigan-Croce B, Zhang L, Konradt C, Cogdill AP, Panjwani MK, Jiang S, Migliorini D, Dahmane N, Posey AD, June CH, Mason NJ, Lin Z, O'Rourke DM, Johnson LA (2018) Checkpoint blockade reverses anergy in IL-13Rα2 humanized scFv-based CAR T cells to treat murine and canine gliomas. Mol Ther Oncolytics 11:20–38
195. Yui H, Muto K, Yashiro Y, Watanabe S, Kiya Y, Kamisato A, Inoue Y, Yamagata Z (2022) Comparison of the 2021 International Society for Stem Cell Research (ISSCR) guidelines for "laboratory-based human stem cell research, embryo research, and related research activities" and the corresponding Japanese regulations. Regen Ther 21:46–51
196. Zhai Q, Chen X, Fei D, Guo X, He X, Zhao W, Shi S, Gooding JJ, Jin F, Jin Y, Li B (2022) Nanorepairers rescue inflammation-induced mitochondrial dysfunction in mesenchymal stem cells. Adv Sci (Weinh) 9:e2103839
197. Zhang J, Chen M, Liao J, Chang C, Liu Y, Padhiar AA, Zhou Y, Zhou G (2021) Induced pluripotent stem cell-derived mesenchymal stem cells hold lower heterogeneity and great promise in biological research and clinical applications. Front Cell Dev Biol 9:716907
198. Zhuang Y, Jiang S, Yuan C, Lin K (2022) The potential therapeutic role of extracellular vesicles in osteoarthritis. Front Bioeng Biotechnol 10:1022368
199. Zieliński PS, Gudeti PKR, Rikmanspoel T, Włodarczyk-Biegun MK (2023) 3D printing of bio-instructive materials: toward directing the cell. Bioact Mater 19:292–327

Glossary of Abbreviations

1002F dry photoresist 1002F resin (EPON™) combined with triarylsulfonium hexafluroantimonate salts. Relative to the photoresist SU-8, 1002F offers advantages for use as a substrate to microstructures and bioanalytical devices, such as improved cell attachment and lower autofluorescence.

2D Two dimensional. In the case of cell culture, 2D refers to cells grown on a monolayer culture flask or flat dishes. Despite limitations, 2D culture methodology is still used because it allows comparison to previous studies and serves as a convenient platform for cell analysis.

3D Three dimensional. For cell culture, 3D models serve to better imitate parental tissue architecture promoting proper interactions within the cell–cell and cell–extracellular microenvironments.

ABCB5 **A**TP-**b**inding **c**assette subfamily **B** member **5**, also known as P-glycoprotein. A plasma membrane-spanning protein encoded in humans band by the *ABCB5* gene on chromosome band 7p21.1. A limbal stem cell gene required for corneal development and repair.

AChE **A**cetyl**ch**olin**e**sterase. An enzyme that catalyzes the hydrolysis of the neurotransmitter acetylcholine into choline and acetic acid. Found at cholinergic synapses, its actions on acetylcholine result in termination of synaptic transmission.

ACI **A**utologous **C**hondrocyte **I**mplantation. A type of surgery used to treat damaged areas of cartilage in the knee joint. It involves harvesting healthy cartilage cells (chondrocytes) from a donor's own body, growing them in a laboratory, before then injecting them into the damaged area. ACI is a minimally invasive procedure that can repair injuries that have not responded to other treatments.

AD-MSC **A**dipose-**d**erived **m**esenchymal **s**tem **c**ells. Stem cells that are obtained from the adipose (fat) tissue of an individual considered applicable to a variety of medical treatments; approved in Japan as a therapy to enhance the wound healing and closure of chronic fistulas.

ADAMTS4 **A** **d**isintegrin **a**nd **m**etalloproteinase with **t**hrombo**s**pondin motifs **4**. An enzyme that in humans is encoded by the *ADAMTS4* gene on chromosome band 1q23.3 and member of a the large ADAMTS family of zinc-dependent proteases.

ADAMTS5 **A** **d**isintegrin **a**nd **m**etalloproteinase with **t**hrombo**s**pondin motifs **5**. An enzyme that in humans is encoded by the *ADAMTS5* gene on chromosome band 21q21.3 with the preprotein proteolytically processed to generate a mature enzyme containing two C-terminal TS motifs functioning as an aggrecanase that cleaves aggrecan, a major proteoglycan of cartilage.

ADSC **A**dipose **D**erived **S**tem **C**ell, also known as, Adipose Stem Cells and AD-MSC.

ALDH **Al**dehyde **deh**ydrogenase. A family of enzymes involved in metabolizing acetaldehyde to acetic acid for which there is a gene superfamily of 19 known genes and many pseudogenes in the human genome, reflecting a vital role through evolutionary history.

ALP **Al**kaline **P**hosphatase. An enzyme found throughout the body, yet predominantly in the liver, bile ducts, spleen, heart, brain, and bone. Its physiological role is the dephosphorylation of compounds. Four genes encode four iso-

J. S. Burns (ed.), *Potency Assays for Advanced Stem Cell Therapy Medicinal Products*, Advances in Experimental Medicine and Biology 1420, https://doi.org/10.1007/978-3-031-30040-0

zymes, the gene for tissue nonspecific alkaline phosphatase is located on chromosome 1, the genes for the other three isoforms are located on chromosome 2.

ALPL **Al**kaline **P**hosphatase, **l**iver/bone/kidney (tissue nonspecific), the latter encoded by the *ALPL/TNSALP* gene on chromosome band 1p36.12, a membrane bound glycosylated enzyme associated with biomineralization.

APC **A**ntigen **P**resenting **C**ell. Typically, dendritic cells, macrophages, Langerhans cells, and B cells, a heterogeneous group of immune cells that mediate the cellular immune response by processing and presenting antigens for recognition by certain lymphocytes such as T cells.

API **A**ctive **P**harmaceutical **I**ngredients. The biologically active component of a drug that is responsible for its intended therapeutic effect.

ASSURED **A**ffordable, **S**ensitive, **S**pecific, **U**ser-friendly, **R**apid and robust, **E**quipment free, and **D**eliverable to end-users. A principle for designing diagnostic tests that are economical, precise, user-friendly, and can be made available to the end user without any specialized equipment.

ATMP **A**dvanced **T**herapy **M**edicinal **P**roducts. A term used to refer to a broad range of innovative therapeutics, including gene therapy, cell therapy, and tissue engineering products.

ATP **A**denosine **trip**hosphate. A molecule carrying energy for all cellular activities. Usually, the outer phosphate of ATP is hydrolyzed to yield adenosine diphosphate releasing 30.6 kilojoules per mole under standard conditions. Released phosphate group transfer to another molecule (phosphorylation), catalyzed by specific enzymes, helps couple the release of energy from ATP to specific cellular activities.

B cell A type of white blood cell that makes antibodies. B cell maturation was first elucidated in the **B**ursa of Fabricius (a gastrointestinal tract organ in birds). In humans, B lymphocytes differentiate prenatally in the liver and subsequently in the bone marrow. B cells are primarily involved in the humoral immunity component of the adaptive immune system.

BCMA **B**-cell **m**aturation **a**ntigen. A protein found on the surface of certain cells, including cancerous B cells, that can be targeted by certain types of immunotherapies.

BK Virus The BK virus was first isolated in 1971 from the urine of a renal transplant patient, initials **B.K.** Member of the polyomavirus family, most infections are asymptomatic, but in immunocompromised individuals may cause renal dysfunction.

BLA **B**iological **L**icence **A**pplication. Submitted by any legal person or entity who is engaged in manufacture or an applicant for a license who bears responsibility for compliance with product and establishment standards.

BMP-6 **B**one **M**orphogenic **P**rotein **6** plays a critical role in bone formation and remodeling. Member of a large family of multifunctional growth factors belonging to the TGF-β superfamily, BMP proteins are encoded by 18 genes in humans. BMP-6 regulates osteoblasts by interacting with several molecules to mediate its functions, e.g., binding cell surface type I and type II serine/threonine kinase receptors can initiate osteoblast proliferation and differentiation, while interaction with members of the SMAD protein family (named after *Drosophila* protein "Mothers Against Decapentaplegic") can modulate downstream signaling events.

BMSSC **B**one **M**arrow **S**tromal **S**tem **C**ells. A type of multipotent adult stem cell found in the bone marrow that can differentiate into a wide range of cell types, such as bone, cartilage, fat. Believed to play a critical role in tissue repair, regeneration, and immunomodulation.

BREYANZI The brand name for Lisocabtagene maraleucel, a cell-based gene therapy used to treat large B-cell lymphoma.

c-Myc The human **c**ellular homologue of the Avian virus **My**elo**c**ytomatosis (v-Myc) oncogene, a nuclear phosphoprotein encoded by the *MYC* gene on human chromosome band 8q24.21. The Myc family of transcription factors activate expression of many pro-proliferative genes through binding enhancer box sequences and recruiting histone acetyltransferase. Myc controls the balance between self-renewal and differentiation and c-Myc was one of the original factors discovered to encourage generation of induced pluripotent stem cells (iPSCs). Nonetheless, it has since been demonstrated that it is possible to generate iPSCs without c-Myc.

C57BL6 "**C57** **bl**ack **6**," "C57," or "black 6" represent an inbred strain of laboratory mouse, with inherited uniformity minimizing the impact of genetic differences on results. Named by the founder of The Jackson Laboratory working at the Bussey Institute, using mice received from Miss Abby Lathrop; a female mouse (coded **57**) was crossed with a brother (coded 52) and inbred, resulting in a dark brown, nearly **bl**ack fur mouse designated **C57BL**, the capital **C** likely indicated the normal full **c**olor rather than the albino (c) allele. The number "**6**" strain survived inbreeding pressures to become the most widely used, first genome-sequenced laboratory mouse strain. Numerous sublines, e.g., C57BL/6J and C57BL/6N have been derived.

CAD **C**omputer **A**ssisted **D**esign. A time-saving approach that decreases errors, design effort and improves accuracy, but CAD often requires extensive training and can add to new system production costs.

CAR **C**himeric **A**ntigen **R**eceptor. A type of engineered receptor designed to recognize specific antigens on the surface of cells. Composed of an extracellular domain that recognizes and binds the specific antigen and an intracellular domain for cell activation.

CAR T-cells **C**himeric **A**ntigen **R**eceptor **T**-cells. A type of cell-based gene therapy where T-cells are genetically modified to recognize and proliferate in response to tumor antigens. Also found abbreviated in the literature as CAR-T.

Cas9 CRISPR **as**sociated protein **9**. An RNA-guided DNA-editing enzyme used to modify the genetic material that functions by binding to a target DNA sequence, cutting the DNA, and inducing a DNA repair process that results in a desired mutation. Applicable for a variety of research and therapeutic applications, including gene editing, gene regulation, and gene expression.

CAT **C**ommittee for **A**dvanced **T**herapies. A specialized advisory body to EMA with expertise in the evaluation of innovative therapies, such as gene and cell therapies, to assess safety, efficacy, and quality standards before they are marketed in the European Union.

CD105 **C**luster of **D**ifferentiation **105**, also known as Endoglin, a type I membrane glycoprotein on the cell surface, part of the TGF-beta receptor complex. CD antigens were originally identified by an ability to bind different monoclonal antibodies, and are used to distinguish cell types, including stem cells, T cells, B cells, macrophages, and help classify cancer cells.

CD107a **C**luster of **D**ifferentiation **107a**, also known as lysosomal-associated membrane protein 1 encoded by the *LAMP1* gene on chromosome band 13q34. A type I transmembrane glycoprotein found primarily spanning lysosomal membranes, functioning to provide selectins and carbohydrate ligands.

CD11b **C**luster of **D**ifferentiation **11b**, also known as Integrin subunit alpha M encoded by the *ITGAM* gene on chromosome band 16p11.2 encoding a protein subunit of the heterodimeric integrin alpha-M beta-2 molecule, known as macrophage-1 antigen or complement receptor 3 expressed on the surface of many leukocytes, It is involved in the innate immune system that mediates inflammation and cell migration, as well as cellular activation, phagocytosis, and chemotaxis.

CD137 **C**luster of **D**ifferentiation **137** is a member of the tumor necrosis factor receptor family. Also known as induced by lymphoma activation (ILA) and tumor necrosis factor receptor superfamily member 9, encoded by the *TNFRSF9* gene on chromosome band 1p36.23. It functions as a co-stimulatory immune checkpoint molecule with co-stimulatory activity for activated T cells, enhancing T cell proliferation.

CD14 **C**luster of **D**ifferentiation **14**. A human protein encoded by the *CD14* gene on chromosome band 5q31.3, made mostly by macrophages existing as two forms, anchored to the membrane by a glycophosphatidylinositol tail, or as a soluble form directly secreted from intracellular vesicles. Functions to help recognize pathogen-associated molecules, e.g., bacterial lipopolysaccharide as part of the innate immune system.

CD146/MCAM **C**luster of **D**ifferentiation **146**, also known as **m**elanoma **c**ell **a**dhesion **m**olecule or cell surface glycoprotein 18 encoded by the *MCAM/MUC18* gene on chromosome band 11q23.3. A receptor for human laminin alpha 4 and Galectin-1 among other miscellaneous ligands involved in cell–cell and cell–matrix interactions. Expressed in cells comprising the vascular wall, including vascular endothelial cells, smooth muscle cells, and pericytes. Actively involved in numerous physiological and pathological processes of cells, it can regulate angiogenesis and tissue organization and is believed to also influence immune system responses and inflammation.

CD19 Cluster of **D**ifferentiation **19**, encoded by the *CD19* gene located on human chromosome band 16p11.2, also known as B-Lymphocyte Surface Antigen B4, T-Cell Surface Antigen Leu-12. A transmembrane protein expressed in all B lineage cells, acting as an adaptor protein recruiting signaling molecules to the membrane and a component of the CD19/CD21 complex that decreases the threshold for B cell receptor signaling pathways.

CD20 Cluster of **D**ifferentiation **20** is encoded by the *MS4A1* geneon human chromosome band 11q12.2, a member of the **m**embrane-**s**panning **4A** gene family. Expressed on the surface of B-cells, it plays a role in the development and differentiation of B-cells into plasma cells.

CD206 **Cl**uster of **D**ifferentiation **206**, also known as **m**annose **r**eceptor **C** type 1, encoded by the *MRC1* gene on chromosome band 10p12.33, is a pattern recognition receptor primarily found on the surface of macrophages that can recognize microbial carbohydrates and mediate phagocytosis.

CD22 Cluster of **D**ifferentiation **22** is a sugar-binding transmembrane receptor, encoded by the *CD22* gene on human chromosome band 19q13.12, that binds to glycoprotein ligands present on B cells. Involved in B-cell adhesion and migration, it is considered important for maintaining B-cell tolerance and preventing autoimmunity.

CD271 Cluster of **D**ifferentiation **271**, also known as Nerve Growth Factor Receptor or the p75 neurotrophin receptor encoded by the *NGFR/p75NTR* gene on human chromosome band 17q21.33. A transmembrane glycosylated receptor with an array of biological functions through interactions with cognate ligands and co-receptors. It can mediate cell death, survival, and regulate the actin cytoskeleton to influence cell migration.

CD28 Cluster of **D**ifferentiation **28**, a protein expressed on T-cells encoded by the *CD28* gene on human chromosome band 2q33.2. It binds to the B7 family of molecules on antigen-presenting cells, providing co-stimulatory signals for T-cell activation, proliferation, cytokine production, and survival.

CD3 Cluster of **D**ifferentiation **3**, a protein complex comprised of CD3-delta, -epsilon, -gamma and -zeta polypeptides, encoded by distinct genes *CD3D*, *CD3E*, *CD3G* on human chromosome band 11q23.3 and *CD247* on human chromosome band 1q24.2. This T-cell co-receptor forms the T-cell receptor-CD3 complex activating both CD8+ naïve cytotoxic T cells and CD4+ naïve helper T cells.

CD30 Cluster of **D**ifferentiation **30**, also known as TNF receptor superfamily member 8, encoded by the *TNFRSF8* geneon human chromosome band 1p36.22, is expressed in activated but not resting T and B cells. An activator of NF-κB signaling and positive regulator apoptosis, CD30 can limit the proliferation of autoreactive CD8 effector T cells and protect against autoimmunity.

CD34 Cluster of **D**ifferentiation **34**, encoded by the *CD34* gene on chromosome band 1q32.2. A single-pass transmembrane cell surface phosphoglycoprotein that functions as a cell–cell adhesion factor.

CD4 Cluster of **D**ifferentiation **4**, encoded by the *CD4* gene on human chromosome band 12p13.31. A glycoprotein serving as a co-receptor to the T-cell receptor found on the surface of T helper cells, monocytes, macrophages, and dendritic cells.

CD40 Cluster of **D**ifferentiation **40**, encoded by the *CD40* gene on human chromosome band 20q13.12. A type I transmembrane protein found on antigen-presenting cells required for their activation.

CD45 **C**luster of **D**ifferentiation **45** antigen, originally called **l**eukocyte **c**ommon **anti**gen (LCA), also known as protein tyrosine phosphatase, receptor type C encoded by the *PTPRC* gene located on human chromosome band 1q31.3-q32.1. It is involved in regulating T- and B-cell antigen receptor signaling.

CD45RA Cluster of **D**ifferentiation **45** isoform that includes only the protein **R**egion **A**, typically found on naïve T lymphocytes.

CD62L Cluster of **D**ifferentiation **62L**, also known as L-selectin, encoded by the *SELL* gene on human chromosome band 1q24.2. A cell adhesion molecule found on the cell surface of leukocytes that acts as a "homing" receptor for lymphocyte migration to lymphoid tissues. High expression of CD62L on human bone marrow progenitor cells indicates commitment to lymphoid differentiation.

CD73 Cluster of **D**ifferentiation **73**, also known as 5′-nucleotidase. An enzyme encoded by the *NT5E* gene on chromosome band 6q14.3 that converts AMP to adenosine.

CD79a Cluster of **D**ifferentiation **79A**, encoded by the *CD79A* gene on chromosome band 19q13.2, is also known as B-cell antigen receptor complex-associated protein alpha chain or MB-1 membrane glycoprotein. Together with the related 79b protein, it forms a dimer associated with the membrane-bound immunoglobulin in B-cells to form the B-cell antigen receptor.

CD8 Cluster of **D**ifferentiation **8**. A transmembrane glycoprotein serving as co-receptor for the T-cell receptor. The co-receptor functions as either a homodimer composed of two alpha chains or as a heterodimer composed of one alpha and one beta chain, encoded by the *CD8A* or *CD8B* genes respectively, both on chromosome band 2p11.2. The cell surface CD8 antigen assists with cytotoxic T cell–antigen interactions.

CD86 Cluster of **D**ifferentiation **86**, also known as B7-2. A protein constitutively expressed on dendritic cells, Langerhans cells, macrophages, B-cells (including memory B-cells), and on other antigen-presenting cells, endoded by the *CD86* gene on chromosome band 3q13.33.

CD90 Cluster of **D**ifferentiation **90**, also known as Thy-1, a heavily glycosylated glycophosphatidylinositol-anchored conserved cell surface protein, encoded by the *THY1* gene on chromosome band 11q23.3. It is primarily involved in cell–cell and cell–matrix interactions.

CDR Circular **d**iaphragm **r**esonator. An acoustic device used to generate a resonating sound from a vibrating diaphragm.

Ce6 Chlorin **e6**. A photosensitizer that can absorb light to generate singlet oxygen that can be used to treat cancer or other diseases.

CEA **C**arcino**e**mbryonic **a**ntigen. A founding member of a highly-related family of cell surface cell adhesion glycoproteins, encoded by the *CEACAM5* gene on chromosome band 19q13.2, found in various tissues of the body. Immunologically characterised as members of the CD66 cluster of differentiation, the proteins can serve as tumour biomarkers to detect and monitor certain types of cancer, primarily in the gastrointestinal tract.

CFR **C**ode of **F**ederal **R**egulation. A compilation of the rules and regulations issued by federal agencies in the United States, in particular, procedures governing the administration of living cells or cell-derived products to repair or replace damage or diseased tissue.

CFSE **C**arboxy**f**luorescein **s**uccinimidyl **e**ster. A cell permeable fluorescent dye that covalently couples, via its succinimidyl group, to intracellular molecules. Due to stable linkage, stained cells do not transfer the dye to adjacent cells.

CFSE-diminished **C**arboxy**f**luorescein **s**uccinimidyl **e**ster fluorescent dye can trace multiple generations of proliferating labeled cells that present diminished staining, with measurement of dye dilution by flow cytometry.

CFU **C**olony-**F**orming **U**nit. Applying terminology used in microbiology to mammalian cells, this refers to cultured cells attached to a monolayer surface that produce a visible colony under controlled conditions. The visual appearance of a colony in a cell culture requires significant growth.

CFU-F **C**olony-**F**orming **U**nit-**F**ibroblasts. Used as a unit of stem cell dose for bone marrow cell cultures, whereby adherent cells form a visible colony. Harvesting methods that yield higher CFU-F are likely to result in improved patient outcomes. The quantitative measure is more applicable to comparative studies within a single laboratory, rather than comparison across independent studies, because of numerous variables related to sample preparation that can influence the absolute values obtained.

aCGH **A**rray **C**omparative **G**enomic **H**ybridization. A method for analyzing genomic DNA for unbalanced genetic alterations. Genomic DNA from the test sample is labeled and mixed with reference control DNA given an alternative color label. The hybridized mixture reveals mixture of imbalance (increased or decreased copy number). Conventional CGH analyzes metaphase chromosomes, whereas array CGH uses cloned chromosomal DNA fragments about 200 kb in size, offering greater sensitivity and resolution in detecting copy number changes.

cGMP **c**urrent **G**ood **M**anufacturing **P**ractice. Regulations promulgated and enforced by authorities such as the FDA and EMA to ensure medical products are safe and effective for consumers and patients; "current" reminds manufacturers to comply with up-to-date technologies and systems.

CGT **C**ellular and **G**ene **T**herapy. Cell therapy comprises an approach that uses a biological product derived from living cells with therapeutic effect, whereas gene therapy uses genes to treat or prevent disease, although the newest approaches forego gene delivery and instead aim to precisely repair the gene within the cell. Both technologies are evolving rapidly, requiring reappraisal of definitions. Boosting immune cell function via viral vector gene delivery constituted a powerful strategy for chimeric antigen receptor T-cell therapies to target certain types of lymphoma, leukemia, and multiple myeloma.

ChIP-Seq **Ch**romatin **I**mmuno**p**recipitation **Seq**uencing. A technique to study chromatins associated with binding proteins to explore gene silencing and DNA binding sites. Protein associated with the chromatins is precipitated and identified by using antigen–antibody interactions. The precipitated protein DNA complex is also studied by real-time PCR or sequencing to investigate the DNA linked with the histones. This combined approach detects the modifications of histones associated with DNA that regulates the transcriptional activities governing gene regulation.

CHMP **C**ommittee for **M**edicinal **P**roducts for **H**uman Use. The EMA committee conducts initial assessment of EU-wide marketing authority applications, assesses modifications to an existing marketing authorization and is responsible for authorizing medicines in the EU.

CIEMAT **C**entro de **I**nvestigaciones **E**nergéticas, **M**edio**a**mbientales y **T**ecnológicas. A Spanish public research institution in energy and the environment attached to the General Secretariat for Research of the Ministry of Science and Innovation.

CLI Critical **l**imb **i**schemia. A serious condition requiring immediate treatment to re-establish blood flow to the affected area with the priority of preserving the limb.

CMC **C**hemistry, **m**anufacturing, and **c**ontrol. The body of information that defines not only the manufacturing process but also quality control, specifications, and stability of the product together with the manufacturing facility and support utilities, including design, qualification, operation, maintenance, and release testing.

COC **C**yclic **o**lefin **c**opolymer. Materials that contain or are made from at least one cyclic monomer. Exceptionally transparent to UV and visible light and resistant to chemicals and polar materials, it is suitable for replacing glass in many applications, especially point of care products and can be extruded into films or 3D printed for development of microfluidic devices.

COGEM The Netherlands **C**ommission **o**n **Ge**netic **M**odification represents an independent scientific advisory board, advising the government on risks to human health and the environment from the production and use of genetically modified organisms, informing on ethical and societal issues associated with genetic modification.

COGS **C**ost **o**f **g**oods **s**old. How much it costs to produce or acquire the product, including the direct material and labor expenses. Note that indirect expenses such as overheads, marketing, or shipping costs are not included.

COMP **C**ommittee for **O**rphan **M**edicinal **P**roducts. The EMA committee responsible for the scientific evaluation of applications for medicines for rare diseases, termed "orphan medicines."

ConA **Con**canavalin **A**. A carbohydrate-binding protein belonging to the legume lectin family that binds specifically to structures bearing internal and nonreducing terminal alpha-D-mannose and alpha-D-glucosyl groups found in sugars, glycoproteins, and glycolipids. Widely used to characterize glycoproteins and to purify glycosylated macromolecules by lectin affinity chromatography.

CQA **C**ritical **Q**uality **A**ttributes. The predefined objectives, product, and process understanding, based on science of quality and risk management that includes all product quality characteristics and specifically the critical attributes that ensure safety and efficacy defined in the product label.

CRISPR **C**lustered **R**egularly **I**nterspaced **S**hort **P**alindromic **R**epeats. A technology used to selectively modify the DNA of living organisms by adapting naturally occurring genome editing systems found in bacteria for laboratory use.

CRO **C**ontract **R**esearch **O**rganizations. An organization contracted by another company to manage complex medical testing responsibilities, aiming to reduce the cost of research and meet the needs of the evolving medical device and pharma industry.

CTA **C**linical **T**rial **A**pplication. Prior authorization from health authorities is a requirement for clinical trials. Comprehensive information is provided to assess the benefit/risk aspects and acceptability of conducting the study.

CTIS **C**linical **T**rials **I**nformation **S**ystem. This serves as the single-entry point for submitting clinical trial information in the EU and European Economic Area. It includes a sponsor workspace for clinical trial sponsors to assist in compiling the application. The authority secure workspace will support activities and help oversee the clinical trials while a public website will allow members of the public to access detailed information on all clinical trials submitted and approved in CTIS.

CTMP **C**ell **T**herapy **M**edicinal **P**roduct. A biological product derived from or consisting of human or animal cells and their components used as a biopharmaceutical to provide therapeutic benefit.

CU **C**ompassionate **U**se. A patient centric managed access approach according to Regulation 726/2004, Article 83, whereby unapproved medicinal products are made available to patients suffering from untreatable life-threatening, chronic, or seriously debilitating disease. The novel therapeutic product needs to be subject to a Marketing Authorization Application or in a clinical trial.

Cx43 **C**onne**x**in **43**. A protein of molecular weight 43 kDa, also known as Gap Junction protein Alpha 1 encoded by the *GJA1* gene on chromosome band 6q22.31. It represents the most common and ubiquitously expressed member of the connexin family of gap junction proteins, a type of cellular junction allowing passage between cells of molecular components with weight less than 1 kDa, such as IP_3, cAMP, Ca^{2+}, and K^+ ions.

CXCL5 **C-X-C** motif chemokine **l**igand **5**, also known as epithelial-derived neutrophil-activating peptide 78 encoded by the *CXCL5* gene on chromosome band 4q13.3. Two cysteines are separated by a single amino acid in C-X-C chemokines, distinguishing them from CC chemokines with adjacent conserved cysteine residues. An inflammatory cytokine produced concomitantly with interleukin-8 that stimulates the chemotaxis of neutrophils and has angiogenic properties.

DABA **Dia**crylate **b**isphenol **A**. A resin of high hardness mainly used as an adhesive in sprays and coatings with fast curing speed.

DCN **Dec**ori**n**. Member of the small leucine-rich proteoglycan family of proteins encoded by the *DCN* gene on chromosome band 12q21.33 that interacts with fibrillar collagens to modify the extracellular matrix structure of connective tissue. It can also modulate cellular responses to growth factors. Named from its ability to "decorate" collagen type I and interact with the "d" and "e" bands of its fibrils.

ddPCR **d**roplet **d**igital **P**olymerase **C**hain **R**eaction. A relatively new form of PCR based on massive partitioning of the target sample via a water–oil emulsion system. Using a Poisson distribution, the fraction of target sequence positive droplets helps determine the concentration of template in the original sample, bypassing the need for references or extrapolation of standard curves.

DFU **D**iabetic **F**oot **U**lcer. Affects many people with diabetes, and abnormal sugar levels can prevent skin from healing itself properly because of reduced nerve function.

DLBCL **D**iffuse **L**arge **B**-**c**ell **L**ymphoma. The most common form of non-Hodgkin lymphoma, a fast-growing blood cancer.

Dlk1/FA1 **D**elta-**L**ike **1**/**F**oetal **A**ntigen **1**. A protein encoded by the *DLK1* gene on human chromosome band 14q32, a region containing an imprinted domain governing its paternal expression. Broadly expressed in human tissues during embryogenic development, its expression in adults is more confined to neuroendocrine tissues and immature stem/progenitor cells. A transmembrane protein that can be cleaved to a soluble form and may have a role in maintaining an undifferentiated cell phenotype.

DMSO **Dim**ethyl **s**ulf**o**xide. A colorless solvent of formula $(CH_3)_2OS$ commonly used as an anti-freeze product for freezing cells due to an ability to affect the thermodynamics of the freezing process. Concerns that its use in clinical practice may cause undesirable side-effects has led to the development of "DMSO-free" products for regenerative medicine cell therapy trials, and comparative trials will resolve the suitability of these new cryoprotective agents and their impact on cell behavior.

DNA **D**eoxyribo**n**ucleic **A**cid. An organic polymer of two polynucleotide chains carrying genetic instructions, coiled around each other to form a double helix.

DP **D**rug **P**roduct. A finished dosage form prepared from bulk drug substance that is ready for administration to the ultimate consumer as a pharmaceutical.

DRC **D**esign **R**ule **C**heck. A design rule is a geometric constraint imposed on circuit board and semiconductor device designers to ensure proper function.

DS **D**rug **S**ubstance. An active ingredient intended to provide pharmaceutical activity or other direct effect in the diagnosis, mitigation, treatment, prevention, or cure of disease or to influence the function of the human body, without including intermediates.

EBMT **E**uropean Society for **B**lood and **M**arrow **T**ransplantation. Founded in 1974, the organization aims to save the lives of patients with blood cancers and other life-threatening diseases by advancing the fields of blood and marrow transplantation and cell therapy worldwide.

EBV **E**pstein-**B**arr **V**irus. A member of the herpes family of DNA viruses, also known as human herpes virus 4, humans being the sole pathogen reservoir. Can cause development of infectious mononucleosis and in cases of severe immunosuppression can trigger Burkitt's lymphoma.

EC **E**uropean **C**ommission. Formed in 1967, the executive body of the European Union, initiating action in the EU, mediating between member governments.

ECFA **E**ctopic **C**artilage **F**ormation **A**ssay. An assay that enables one to assess the capacity of bioactive molecules to support cartilage formation in vivo using cartilage organoids.

ECL **E**lectro**c**hemi**l**uminescence. A kind of luminescence produced during electrochemical reactions in solutions. Electrochemically intermediates undergo an exergonic reaction, releasing free energy to produce an electronically excited state that then emits light upon relaxation to a lower level state.

ECM **E**xtra**c**ellular **M**atrix. Also known as intercellular matrix, the network of extracellular macromolecules including collagen, enzymes, and glycoproteins, plus minerals such as hydroxyapatite, that provide structural and chemical support to the associated cells.

EGFR **E**pidermal **G**rowth **F**actor **R**eceptor, also known as HER1 or ErbB-1, is a transmembrane protein that functions as a receptor to members of the epidermal growth factor family of extracellular protein ligands.

EIS **E**lectrochemical **I**mpedance **S**pectroscopy. A highly sensitive characterization technique that can assay the electrical response of chemical systems in a nondestructive manner. Quantitative measurements enable the evaluation of small-scale chemical mechanisms within the electrolyte solution at the electrode interface.

ELISA **E**nzyme-**l**inked **i**mmuno**s**orbent **a**ssay. A biological assay using an enzyme-labeled immunoreactant (antigen or antibody) and an immunosorbent (antibody or antigen bound to a solid support). Often used to determine the potency of a drug or other substance by comparison to a reference standard.

ELISpot **E**nzyme-**L**inked **I**mmuno**spot** assay. In contrast to ELISA assays, ELISpot retrieves the frequency of cells that are secreting the protein of interest, rather than only the total analyte concentration in the culture medium. After the assay, ELISpot plates can be stored or shipped for off-site counting.

EMA **E**uropean **M**edicines **A**gency. A decentralized scientific agency established in 1995, the aim of the organization is to harmonize (but not replace) the work of existing national medicine regulatory bodies.

EpCAM **Ep**ithelial **C**ell **A**dhesion **M**olecule, also known as CD326, encoded by the *EPCAM* gene on chromosome band 2p21, is a transmembrane protein mediating Ca^{2+}-independent homotypic cell–cell adhesion in epithelia. EpCAM is involved in cell signaling, migration, proliferation, and differentiation.

EU **E**uropean **U**nion. An international organization comprising 27 European countries and governing common economic, social, and security policies.

EUA **E**mergency **u**se **a**uthorization. Risk-based procedure developed by a Stringent Regulatory Authority to approve the use of a therapy under development.

EV **E**xtracellular **v**esicles. A generic term for lipid bilayer enveloped particles released by cells to the extracellular environment, including particles termed ectosomes, microparticles, micro vesicles, apoptotic bodies, and exosomes among others. Known to facilitate intercellular communication processes between cells mediated by microRNA and proteins, they represent a potential source of biomarker discovery as well as potential therapeutic agents.

EWOD **E**lectro**w**etting-**o**n-**d**ielectric. A technique for manipulating individual droplets on a single platform with high precision, configurable for micro- or nanoliter droplet actuation, controlled by an electrical stimulus, applicable to lab-on-chip fluidic operations.

FA **F**anconi **A**nemia. A very rare bone marrow failure syndrome leading to an impaired response to DNA damage, caused by a homologous recombination genetic defect in a cluster of proteins responsible for DNA repair.

FACS **F**luorescence-**a**ctivated **C**ell **S**orting. A rapid cell sorting method whereby diversely fluorescently stained live cells are analyzed together and then separated. A highly sensitive, high-throughput technique employing flow cytometry with the advantage that cell populations can be sorted to high purity and selected for several parameters concurrently.

FANCA **F**anconi **An**emia **C**omplementation group **A**. Fanconi Anemia is a clinically and genetically heterogeneous disorder causing genetic instability. Caused by homozygous or compound heterozygous mutation in the *FANCA* gene on chromosome band 16q24.

FATAL **F**luorometric **a**ssessment of **T** lymphocytes **a**ntigen specific **l**ysis. A technique employing dual staining (PKH-26 and CFSE) to identify and evaluate target cell cytolysis.

FBS **F**oetal **B**ovine **S**erum. Also misnamed as Foetal Calf Serum (the serum is not obtained from full-term calves but from a fetus at late-stage development). A popular supplement to the basal medium used in cell culture, yet prone to inter-batch variation and risk of pathogen contamination, so alternatives are being researched.

FDA **F**ood and **D**rug **A**dministration. The United States federal agency of the Department of Health and Human Services.

FET **F**ield-**E**ffect **T**ransistor. Also known as a unipolar transistor, an electric field controls the flow of current in a semiconductor. The flow of current is controlled by application of a voltage to the gate terminal, thus altering the conductivity between the drain and source terminals.

FIH **F**irst **i**n **H**uman. Such clinical trials play a critical role in bringing new interventions to clinical practice, representing studies during which a drug is administered to humans for the first time after a series of preclinical tests for safety. Mostly conducted in volunteers; however, in certain critical circumstances, patients who cannot easily benefit from available therapies are assessed. Generally, a multidisciplinary team of clinical operation specialists conduct the trials to manage inherent risks and challenges.

FR4 Flame **R**etardant **4**, the number 4 differentiates this type of woven glass-reinforced epoxy resin from other similar materials.

G6PD **G**lucose **6** **P**hosphate **D**ehydrogenase, also abbreviated G6PDH, an enzyme encoded by the *G6PD* gene on chromosome band Xq28. This cytosolic protein participates in the pentose phosphate pathway, a metabolic pathway that supplies reducing energy to cells by maintaining the level of the coenzyme nicotinamide adenine dinucleotide phosphate, which in turn maintains the level of glutathione, helping to protect red blood cells against oxidative damage.

GO **G**raphene **O**xide. A material of unique physiochemical properties, which may be derived from graphite as a single layer honeycomb lattice of graphene-analogous carbon atoms with various oxygen-containing functionalities such as epoxide, carbonyl, carboxyl, and hydroxyl groups. These are semi-randomly distributed, introducing on-plane functionalization defects (vacancies and holes) making GO more reactive than the naturally inert graphene hexagonal structure.

GQD **G**raphene **Q**uantum **D**ots. Zero-dimensional graphene derivatives with one to few layers of graphene sheets of less than 20 nm. Properties such as extremely small size, quantum confinement, biocompatibility, photostability, and water solubility make them excellent candidates for understanding molecular systems and cellular processes at the molecular scale.

GSTT1 **G**lutathione **S**-**t**ransferase **t**heta **1**. The protein encoded by the *GSTT1* gene on chromosome band 22q11.23 is haplotype-specific and absent from 38% of the population. It catalyzes the conjugation of reduced glutathione to a variety of electrophilic and hydrophobic compounds enhancing detoxification.

GTMP **G**ene **T**herapy **M**edicinal **P**roduct. An emerging class of biopharmaceutical that aims to modify or manipulate the expression of a gene or cure disease by replacing a disease-causing gene with a healthy copy, inactivating the mal-functional disease-causing gene or introducing a new or modified gene to help treat a disease.

GTP **G**ood **T**issue **P**ractice. Requirements include standards for facilities, environmental control, equipment, supplies, reagents, recovery, labeling controls, storage, receipt and distribution, and donor screening and donor testing. The aim is to ensure cells and tissues maintain their integrity and function.

GvHD **G**raft **v**ersus **H**ost **D**isease. A clinical complication that can arise after allogeneic bone marrow and stem cell transplantation whereby associated T cells recognize the recipient patient's body as foreign and react immunologically against the recipient's antigens, attacking cells and tissues.

hBM-MSC **h**uman **B**one **M**arrow **M**ultipotent **S**tromal **C**ell, also described as human Bone Marrow-derived Mesenchymal Stem Cells.

hBMSC **h**uman **B**one **M**arrow **S**tromal **C**ells also described as human Bone Marrow-derived Mesenchymal Stem Cells.

HE **H**ospital **E**xemption. Principles that allow for the use of an ATMP without a marketing authorization under certain specific circumstances. This only applies to a hospital setting on a nonroutine basis, when no centrally authorized treatment or clinical trial is available.

Hedgehog The Hedgehog signaling pathway transmits information to embryonic cells required for proper cell differentiation. Its name derives from its polypeptide ligand, an intracellular signaling molecule called **Hedgehog** (Hh) found in fruit flies of the genus *Drosophila*.

HEK 293T **H**uman **e**mbryonic **K**idney 293T cells express a mutant version of the SV40 large **T** antigen. They are a derivative of a human immortalized cell line HEK **293**, grown in tissue culture from a spontaneously miscarried female fetus.

Hemgenix Brand product name for etranacogene dezaparvovec-drlb, an adeno-associated virus vector-based one-time gene therapy for treatment of adults with Hemophilia B, the first FDA-approved gene therapy.

HER2 **H**uman **E**pidermal growth factor **R**eceptor **2**, also known as CD 340, is a human protein encoded by the *ERBB2* gene on chromosome band 17q12. HER2 was so named because of similar structure to the human epidermal growth factor receptor HER1, but contrary to other members of the ERBB family, HER2 does not directly bind ligand.

hESC **h**uman **E**mbryonic **S**tem **C**ells. Derived from the inner cell mass of a mammalian embryo at a very early blastocyst stage of development, composed of a hollow sphere of dividing cells. The first hESC line was created in 1998, demonstrably capable of forming a wide variety of different tissue phenotypes in vitro, forming teratomas when grafted into immunosuppressed mice.

HGF **H**epatocyte **G**rowth **F**actor, also known as Scatter Factor, is a paracrine cellular growth, motility, and morphogenic secreted factor, encoded by the *HGF* gene on human chromosome band 7q21.11, a protein with a major role in wound healing.

HHV-6 **H**uman **h**erpes **v**irus **6**, the collective name for the double-stranded DNA viruses HHV-6A and HHV-6B, the latter a ubiquitous virus, etiologic agent of the childhood illness exanthema subitem (roseola infantum).

hiPSC **h**uman-**i**nduced **P**luripotent **S**tem **C**ell. These may be generated from patients of any genetic background and possess the capacity to differentiate into almost any desired terminal cell type.

HLA-DR **H**uman **L**eukocyte **A**ntigen-**DR** isotype, an MHC class II cell surface receptor encoded by the human leukocyte antigen complex on chromosome band 6p21.31.

HLA-G5 **H**uman **l**eukocyte **a**ntigen, class **G5**, also known as histocompatibility antigen. The fifth of at least seven alternatively spliced isoforms, it can present a narrower variety of peptides than its classical HLA I counterpart, a major immune checkpoint, it downregulates the immune system's response.

hMSC **h**uman **M**ultipotent **S**tromal **C**ells also frequently described as human Mesenchymal Stem Cells, the moniker Medicinal Signaling Cells, has also been proposed.

Holoclar The brand name of the first stem cell therapy approved in Europe, consisting of ex vivo expanded autologous human cornea epithelial cells containing stem cells. The name Holoclar derives to the Greek words "**holo**s" and "**clar**us" meaning "whole" and "clear or bright". The cell therapy aims to restore long-term corneal integrity, providing clarity to patients experiencing vision loss.

HPC **H**ematopoietic **p**rogenitor **c**ell. The cell types derived from hematopoietic stem cells that give rise to one or more types of mature blood cells. Unlike true stem cells, which can develop any blood cell line, progenitor cells are more restricted; multipotent HPC produce most subsets of blood cells, oligopotent HPC include lymphoid and myeloid lineage restricted progenitor cells, and unipotent HPC produce one set of blood cells, e.g., unipotent megakaryocytic progenitor cells.

HPV **h**uman **p**apilloma**v**irus. A large number of double-stranded DNA papillomaviruses of numerous genotypes cause various human warts and some HPV have been associated with induction of cervical cancer.

HSC **H**ematopoietic **S**tem **C**ell. Present in blood and bone marrow, they are capable of forming mature blood cells, such as red blood cells, platelets, and white blood cells and can be transplanted to replace or rebuild a patient's hematopoietic system.

hTERT **h**uman **Te**lomerase **R**everse **T**ranscriptase. A catalytic subunit of telomerase encoded by the *TERT* gene on chromosome band 5p15.33. It is essential for telomere maintenance and overexpression of the gene can suffice to induce the unlimited proliferation of cells.

hUC-MSC **h**uman **u**mbilical **c**ord tissue derived **m**ultipotent **s**tromal **c**ells. Wharton's Jelly derived MSC cells to be sourced from otherwise discarded tissue, demonstrate high proliferative capacity and improved immunopotency when compared to similar MSC from other tissue sources.

HUVEC **H**uman **U**mbilical **V**ein **E**ndothelial **C**ells. Cells derived from the vein of the umbilical cord and often used for physiological and pharmacological investigations related to macromolecular transport, blood coagulation, and angiogenesis.

ICH The **I**nternational **C**ouncil of **H**armonization of Technical Requirements for Pharmaceuticals for Human Use, uniquely gathers regulatory authorities and pharmaceutical industry to discuss scientific and technical aspects of drug registration.

IDE **I**nter-**d**igitated **e**lectrode. Fabricated by combining two separately addressable electrode arrays, with the resulting electrode structure having a comb-shaped or zipper-like arrangement. IDE are used for diverse electrochemical sensor operations, favored for high sensitivity specialized sensors that have relatively straightforward, low-cost fabrication.

IDO/IDO1 **I**ndoleamine 2,3-**dio**xygenase. A haeme-containing enzyme encoded by the *IDO1* gene found on human chromosome band 8p11.21. Physiologically expressed in a number of tissues and cells, it catalyzes the first rate-limiting step in tryptophan catabolism to *N*-formyl-kynurenine.

IFN-γ **Interferon** gamma, encoded by the *IFNG* gene on human chromosome band 12q15, is a dimerized soluble cytokine, member of the type II class of interferons with a role in regulating the immune response of the target cell.

IGF-1 **I**nsulin-like **g**rowth **f**actor 1, also known as somatomedin C, is encoded by the *IGF1* gene found on human chromosome band 12q23.2. A hormone with similar molecular structure to insulin, playing an important role in childhood growth, with anabolic effects in adults.

IL-10 **I**nter**l**eukin-**10**, also known as human cytokine synthesis inhibitory factor, an anti-inflammatory cytokine encoded by the *IL10* gene on human chromosome band 1q31–1q32.

IL-15 **I**nter**l**eukin-**15**. An inflammatory cytokine encoded by the *IL15* gene on human chromosome band 4q31.21, that regulates T and natural killer cell activation and proliferation.

IL-1β **I**nter**l**eukin-**1b**eta, also termed leukocytic pyrogen, leukocytic endogenous mediator, mononuclear cell factor or lymphocyte activating factor. A cytokine protein encoded by the *IL1B* gene on chromosome band 2q14.1. An important mediator of the inflammatory response with involvement in cell processes including proliferation, differentiation, and apoptosis.

IL-2Rα **I**nter**l**eukin-2 **r**eceptor **al**pha chain also called CD25, the protein encoded by the *IL2RA* gene on human chromosome band 10p15.1. A type I transmembrane protein present on activated T cells, activated B cells, some thymocytes, myeloid precursors, and oligodendrocytes.

IL8 **I**nter**l**eukin **8**. A small soluble-activating peptide encoded by the *CXCL8* gene on human chromosome band 4q13.3. This member of the CXC chemokine family attracts neutrophils, basophils, and T cells to an inflammatory site.

Imlygic The brand name for talimogene laherparepvec, the first and only FDA-approved viral therapy injected directly into melanoma tumors, where it multiplies inside the cancer cells and destroys them.

IMP **I**nvestigational **M**edicinal **P**roduct. A pharmaceutical substance or placebo being tested or used as a reference in a clinical trial. This includes products that already have marketing authorization in place.

iMSC **i**nduced pluripotent stem cell-derived **MSC**. Human iMSC lines are being investigated as potentially a clinically relevant source of MSC that circumvent ageing-associated aspects pertaining to adult bone marrow derived MSC.

IND **I**nvestigational **N**ew **D**rug. Refers to a drug or biological product that is or will be used in a clinical investigation, having already been approved by the FDA or EMA for use in humans in a research setting, after successful review of a submitted application.

IOVANCE Biotherapeutics A biopharmaceutical start-up based in San Carlos, California, focusing on tumor-infiltrating lymphocyte-based therapies against cancer.

IPC **I**n-**p**rocess **C**ontrols. Checks performed during a production process to monitor and if necessary, adjust the process to ensure the product conforms to its specifications.

iPSC **i**nduced **P**luripotent **S**tem **C**ells. Generated from cells easily obtained from living organisms and humans, i.e., blood or skin cells, that have been genetically reprogrammed to become stem cells with the ability to differentiate into any cell type in the body.

ISCT **I**nternational **S**ociety for **C**ell and Gene **T**herapy. Established in 1992, a global society for researchers, regulators, technologists, clinicians, and industry partners aiming to translate cell and gene research into safe and effective therapies. A collaborative forum for academia, regulatory authorities and industrial commercialization.

ISEV **I**nternational **S**ociety for **E**xtracellular **V**esicles, the largest professional society for researchers and scientists involved in the study of extracellularly secreted vesicles.

ISFET **I**on **s**ensitive **f**ield-**e**ffect **t**ransistor. A type of field effect transistor that measures ion concentrations in solution. When the ion concentration changes, the current through the transistor changes accordingly.

ITP **I**so**t**acho**p**horesis, analytical chemistry technique for selective separation and concentration of ionic analytes.

JACIE **J**oint **A**ccreditation **C**ommittee **I**SCT-Europe & **E**BMT is Europe's only official accreditation body in the field of hematopoietic stem cell transplantation and cellular therapy. A profession-led voluntary accreditation scheme promotes high-quality patient care and medicinal as well as laboratory practice.

KIAA1199/CEMIP **Ce**ll **m**igration-**i**nducing and hyaluronan-binding **p**rotein. Encoded by the *CEMIP* gene found on human chromosome band 15q25.1. Genes identified in the Kazusa DNA Research Institute cDNA sequencing project were systematically designated **KIAA** plus a **four-digit number**, as a temporary symbol for genes of unknown function, to be changed when a function is identified.

Klf4 **K**ruppel-**l**ike **f**actor **4**, a member of the KLF family of zinc finger transcription factors, belonging to the relatively large family of Specificity Protein 1-like transcription factors. Encoded by the *KLF4* gene on chromosome band 9q31.2, it is important in regulating chromosome number, genetic stability, response to DNA damage, and cell survival. It can influence Wnt signaling pathway genes, regulating differentiation.

KYMRIAH The brand name for Tisagenlecleucel, a CAR T-cell medication for the treatment of B-cell acute lymphoblastic leukemia that uses the body's own T cells to fight cancer by adoptive cell transfer.

Lab-on-PCB **Lab on** **P**rinted **C**ircuit **B**oard. Adopting the lab-on-a-chip concept, a device that integrates one or several laboratory functions on a single printed circuit board, with the advantage of established PCB production facilities providing an inherent upscale potential.

LAL **L**imulus **a**moebocyte **l**ysate. An aqueous extract of blood cells from the Atlantic horseshoe crab (*Limulus polyphemus*) that reacts with bacterial endotoxin lipopolysaccharide, a membrane component of gram-negative bacteria.

LAV **L**ive **a**ttenuated **v**irus. This form of virus has been used to generate vaccines containing infectious agents of virulence weakened by a series of treatments.

LoC **L**ab-**o**n-a-**C**hip. A device that integrates laboratory functions on a single integrated circuit that can achieve high-throughput screening.

LOD **L**imit **o**f **d**etection. The lowest concentration or quantity of a component or substance that can be reliably distinguished and measured by an analytical method.

M1 Phenotype **M**acrophages of **M1**-type represent a classically activated "killer" form that have pro-inflammatory, bactericidal, and phagocytic functions.

M2 Phenotype **M**acrophages of **M2**-type represent an alternatively activated "repair" form that function in constructive processes such as wound healing and tissue repair turning off damaging immune system activation by producing anti-inflammatory cytokines.

MAA **M**arketing **A**uthorization **A**pplications. An application submitted to EMA to market a medicinal product in the EU Member States.

MACI **M**atrix-induced **A**utologous **C**hondrocyte **I**mplantation. An autologous cell therapy procedure whereby a patient's own cells are harvested from the knee and used to regenerate new cartilage for the knee joint.

MB **M**olecular **B**eacon, also known as molecular beacon probes, are hairpin-shaped oligonucleotide hybridization probes designed to report the presence of specific nucleic acids in homogeneous solutions. An internally quenched fluorophore has its fluorescence restored when the beacon probe binds to a target nucleic acid sequence.

MCB **M**ixed **C**ircuit **B**oard. Also known as mixed signal-integrated circuits, contain both digital and analogue circuitry in the same chip.

MCP1 **M**onocyte **C**hemoattractant **P**rotein **1**, also known as Chemokine (CC-motif) ligand 2, has a vital role in the process of inflammation by attracting and enhancing the expression of inflammatory factors and cells.

MEA **M**icro**e**lectrode **a**rrays. A promising device to detect electrical activities of tissues or organs in vivo or in vitro and measure extracellular electrical activity, for example monitoring cardiac and neural cellular electrophysiology.

MEMS **M**icro**e**lectro**m**echanical **s**ystem. A miniature machine, micro mechatronic device with both mechanical and electronic components. Some lack mechanical parts yet miniaturize structures of conventional machinery, e.g., cavities, holes, and membranes. Some MEMS act as transducers, since they convert mechanical signals into electrical or optical signals.

MERTK **MER** **t**yrosine **k**inase proto-oncogene. An enzyme that in humans is encoded by the *MERTK* gene located on chromosome band 2q13. A type I receptor tyrosine kinase and member of the TAM family of homologous tyrosine kinases, it has important roles in homeostasis of normal cells and is often overexpressed in a wide range of cancers.

MHC **M**ajor **h**istocompatibility **c**omplex, also known as human leukocyte antigens. A large genetic locus on chromosome 6 contains a set of polymorphic genes encoding cell surface proteins essential for the adaptive immune system.

MHLW **M**inistry of **H**ealth, **L**abour and **W**elfare. A cabinet level ministry of the Japanese government that provides services on health, labor, and welfare.

miRNA **Mi**cro **R**ibose **N**ucleic **A**cid. Small single-stranded noncoding RNA molecules containing 21–23 nucleotides that base-pair to complementary sequences in mRNA molecules, to then activate processes that result in silencing and post-transcriptional regulation of gene expression.

MISEV **M**inimal **i**nformation for **s**tudies of **e**xtracellular **v**esicles. Guidelines first released in 2014 by the International Society for Extracellular Vesicles to provide standardization of protocols and reporting in the extracellular vesicle field. MISEV 2018 guideline were subsequently published to update the topics of nomenclature, separation, characterization, and functional analysis.

MLR **M**ixed **l**ymphocyte **r**eaction. A test to demonstrate the safety of a drug or implantable material, by assessing how T-cells react to external stimuli. In the ex vivo cellular immune assay, allogeneic populations of T-lymphocytes are mixed together with measurement of the reaction that occurs. In a one-way MLR, only one T-lymphocyte population can respond or proliferate, in two-way MLR both can.

MMP **M**atrix **m**etallo**p**roteinase, also known as matrix metallopeptidases or matrixins, a family of calcium dependent, zinc-containing endopeptidases, that collectively are capable of degrading all kinds of extracellular matrix proteins, yet can also process several bioactive proteins.

MMP-13 **M**atrix **M**etallo**p**roteinase **13**, also known as Collagenase-3, a protein encoded by the *MMP13* gene on chromosome band 11q22.2 involved in the breakdown of extracellular matrix in normal physiological processes such as tissue remodeling as well as pathological processes such as arthritis and metastasis.

MoA **M**echanism **o**f **A**ction. A description of the detailed understanding at the biochemical and molecular level, of changes within the host that bring about the specific action of the administered substance. The mechanism of action involves altered specific biochemical reactions that consequently influence the mode of action, i.e., how the action is brought about by physiological, chemical and functional changes that occur in the cell. Needed to fulfil knowledge of biological pathways that underlie a given disease and an estimate of the degree or amount of disruption each pathway can tolerate without evoking pathway-specific toxicity.

MRL/MpJ **M**urphy **R**oths **L**arge (MRL/MpJ) is a strain of laboratory mouse developed in 1999 at The Wistar Institute in Philadelphia, Pennsylvania. It demonstrates a remarkable capacity for cartilaginous wound closure, plus an ability to regenerate cardiac tissue.

mRNA **M**essenger **r**ibose **n**ucleic **a**cid. A single-stranded molecule generated by the process of transcription in correspondence to the genetic sequence of a gene, read by a ribosome in the process of synthesizing a protein.

MSC **M**esenchymal **s**tem **c**ell. One of the most widely studied cell types for advanced therapy medicinal products used to explore tissue engineering and immune therapy strategies. Mesenchymal stem cell represents the most widely used moniker among **me**senchymal **s**tromal **c**ell, **m**ultipotent **s**tromal **c**ell, and **m**edicinal **s**ignaling **c**ells for non-hematopoietic multipotent, self-renewable cells capable of trilineage differentiation.

MT1-MMP **M**embrane-**t**ype-**I** **m**atrix **m**etallo**p**roteinase, a transmembrane protein encoded by the *MMP14* gene on chromosome band 14q11.2. A tethered collagenase and important modifier of the pericellular microenvironment, it has an important role in extracellular matrix degradation for both normal physiological and disease processes such as metastasis.

MUC1 **Muc**in short variant S**1**, also known as polymorphic epithelial mucin (PEM), epithelial membrane antigen (EMA), or CD227, a high molecular weight type I membrane tethered glycoprotein encoded by the *MUC1* gene on chromosome band 1q21–24. It is multitasked with immunosuppressive properties, a role in protection against infections and involvement in oncogenic processes as well as cell-signaling, including interactions with cancer antigens such as HER2.

MV **M**icro**v**esicles. Also known as ectosomes or microparticles, MV represent a type of extracellular vesicle delimited by a phospholipid bilayer released from the cell membrane into the interstitial space between cells. Generally considered to be larger than exosomes, they also play a role in intercellular communication by transporting molecules that include mRNA, miRNA, and proteins.

MVB **M**ulti**v**esicular **b**odies. Membranous 0.5–1.0 μM wide organelles found in the cytoplasm of cells containing small vesicles and hydrolases. MVB transport material from early endosomes to late endosomes via intracellular sorting organelles and accumulated intraluminal vesicles, resulting in a multivesicular appearance.

NALM6 cells A B-cell precursor leukemia cell line initiated from a 19-year-old male patient with **a**cute lymphoblastic **l**eukemia. Characterization indicated the leukemia was derived from **n**on-T and non-B cells and positive in a one-way **m**ixed lymphocyte reaction. A CD19+ cell line ideal for testing novel CAR T-cell or other immunotherapies for efficacy in vivo.

NBM **N**ano-**Bio**m**aterials. Nanotechnology-derived biomaterials, converge material science and biology with a regulatory size threshold of 1–100 nm applied to at least one external dimension of the material containing particles or agglomerates larger than 100 nm formed by such particles. Diverse types include metallic, ceramic, semiconductor-based, organic/carbon-based, silica-based, polymeric, and biological.

NCA **N**ational **C**ompetent **A**uthority. The authority of the member state of the European Union primarily responsible for the authorization of medicines available in the EU that do not pass through the centralized procedure.

NELL **N**eural **E**pidermal Growth Factor-**l**ike **l**ike proteins 1 and 2 are protein kinase C-binding enzymes that have six EGF-like repeats and constitute a family of multimodal extracellular glycoproteins. The heterotrimeric protein may be involved in cell growth regulation and differentiation.

NF-κB **N**uclear **f**actor **k**appa-light-chain-enhancer of activated **B** cells, represents a protein complex controlling DNA transcription, cytokine production, and cell survival. Residing in an inactive state, it can serve as a rapid acting transcription factor that does not require new protein synthesis. It has a key role in regulating the acute immune response to infection and inflammation.

NHL **N**on-**H**odgkin's **l**ymphoma. A collective term for a number of subtypes of cancer that usually originate in lymph nodes or other lymph tissue whereby Lymphocytes (white blood cells) grow abnormally and can form tumors of indolent or aggressive nature. NHL are distinguished from Hodgkin lymphoma that usually starts in B lymphocytes, yet spread and respond to treatment differently.

NK **N**atural **k**iller cells or large granular lymphocytes (LGL) represent a type of cytotoxic lymphocyte critical to the innate immune system.

Notch The human proteins constitute a family of four single-pass transmembrane receptors that are homologues of the *Drosophila* development gene **Notch** (the name is derived from a characteristic mutation-induced *Drosophila* wing phenotype, a nick or **notch** in the wingtip). Members of the Notch gene family mediate cell-fate decisions of multipotent precursors in a number of different species.

NSCLC **N**on-**s**mall **c**ell **l**ung **c**ancer. The most common type of lung cancer that grows slowly in comparison to small cell lung cancer; nonetheless, its metastatic potential augments need for early detection and treatment.

NY **N**ew **Y**ork. A state in the north-east US, capital Albany. The seaport in south-eastern NY, New York City at the mouth of the Hudson River, comprises the boroughs of Manhattan, Queens, Brooklyn the Bronx, and Staten Island.

OA **O**steo**a**rthritis. The most common form of arthritis. Associated with trauma or ageing, the cartilage within a joint begins to break down leading to chronic pain with risk of stiffness reduced function and disability.

OCT 3/4 **Oct**amer-binding transcription factor **3/4** represents a key transcriptional factor encoded by the *POU5F1* (POU Class 5 Homeobox 1) gene on chromosome band 6p21.33. The transcription factor contains a POU homeodomain (derived from the names of three types of transcription factors: **P**ituitary-specific Pit-1, **O**ctamer transcription factors (octamer sequence is ATGCAAAT), and the neural **U**nc-86 transcription factor). Oct 3/4 has a key role in embryonic development and stem cell pluripotency. Expression levels govern the fate of primitive inner mass and embryonic stem cells.

OTAT **O**ffice of **T**issues and **A**dvanced **T**herapies. One of three product offices within the Centre for Biologics Evaluation and Research in consultation with FDA leadership, overseeing biological products through a data-driven process to provide regulatory oversight to ensure medical products are safe and effective.

PAX6 **Pa**ired bo**x** **6** protein, also known as aniridia type II protein or oculorhombin, is a protein encoded by the *PAX6* gene on human chromosome 11p13. Transcription factor activity of this protein is key in the development of neural tissues, particularly the eye.

PBMC **P**eripheral **B**lood **M**ononuclear **C**ells. Any blood cell with a single nucleus, including lymphocytes, monocytes, and dendritic cells.

PC **P**oly**c**arbonate. A group of thermoplastic polymers containing carbonate esters with planar cores that confer rigidity in their chemical structures.

PCR **P**olymerase **C**hain **R**eaction. A fast and inexpensive technique that can amplify specific DNA sequences in vitro by incubating separated DNA strands with oligonucleotide primers and DNA polymerase. Thirty to 40 cycles of PCR reaction can yield more than one billion precise copies of the original DNA segment.

PD-1 **P**rogrammed Cell **D**eath Protein **1**, also known as CD279, an immune checkpoint protein encoded by the *PDCD1* gene on chromosome band 2q37.3. Found on the surface of T and B cells with a role in regulating immune system responses by suppressing T cell inflammatory activity. It promotes apoptosis of antigen-specific T cells in lymph nodes and reduces apoptosis in anti-inflammatory regulatory T cells.

PDGF **P**latelet-**d**erived **g**rowth **f**actor. A protein family of growth factors that bind and activate PDGF receptor tyrosine kinases to regulate growth and division. For example, the product of the *PDGFA* gene on chromosome band 7p22.3 represents a potent mitogen for cells of mesenchymal origin.

PDL1 **P**rogrammed **d**eath-**l**igand **1**, also known as CD274 or B7 homolog, is a 40 kDa protein encoded by the *CD274* gene on chromosome band 9p24.1. It functions to suppress the adaptive arm of the immune system.

PDMS **P**oly**dim**ethyl**s**iloxane, also known as dimethylpolysiloxane or dimethicone, a polymeric organosilicon compound widely used in silicon-based organic polymers.

PEG **P**oly**e**thylene **g**lycol. Any of a family of polymers $H(OCH_2CH_2)_nOH$ where n is greater than 3, derived as condensation polymers of ethylene and glycol of that have high molecular weight and are soluble in water as well as many organic solvents. Ranging from viscous liquids to white solids, they are used as emulsifiers, lubricants, and plasticizers.

PET **P**oly**e**thylene **t**erephthalate or poly(ethylene terephthalate), consisting of repeating ($C_{10}H_8O_4$) units, is the most common plastic polymer resin of the polyester family, used in fibers for clothing, food, or liquid containers and composite resins.

PFSB **P**harmaceutical and **F**ood **S**afety **B**ureau of the Japanese Ministry of Health, Labour and Welfare and its review agency, the Pharmaceutical and Medical Devices Agency, or any successor agency thereof.

PGE2 **P**rosta**g**landin **E2**, also known as dinoprostone, a potent inflammatory mediator generated by conversion of arachidonic acid by the enzyme cyclooxygenase 2.

PHA **P**hyto**ha**em**a**gglutinin. A lectin found in plants consisting of two closely related proteins, leucoagglutinin and PHA-E, that can cause blood cells to clump together and can represent a toxin in improperly prepared uncooked legumes.

PKH-67 A green-fluorescent lipophilic membrane dye (excitation 490 nm, emission 504 nm) invented by **P**aul **K**arl **H**oran, widely used for general cell membrane staining. It has a longer aliphatic carbon tail than alternative PKH green dyes previously described for in vitro and in vivo cell tracking. Thus, PKH-67 has reduced cell–cell transfer with an extended half-life and stability useful for long-term in vivo studies.

PLGA **P**oly(**l**actic-*co*-**g**lycolic **a**cid), a biodegradable and biocompatible FDA-approved copolymer, used in many therapeutic drug delivery devices and tissue engineering applications.

PMDA **P**harmaceuticals and **M**edical **D**evices **A**gency, the government organization in Japan overseeing the unique set of processes and agencies for the regulation of drugs. Established in 2004, it consolidated the services of the Pharmaceuticals and Medical Devices Evaluation Centre of the National Institute of Health Sciences, the Organization for Pharmaceutical Safety and Research and part of the Japan Association for the Advancement of Medical Equipment.

PMMA **P**oly(**m**ethyl **m**eth**a**crylate). A transparent thermoplastic, also known as acrylic glass, often used in sheet form as a lightweight shatter-resistant alternative to glass.

PNA **P**eptide **n**ucleic **a**cid. An artificially synthesized polymer similar to DNA or RNA but with repeating *N*-(2-aminoethyl)-glycine units linked by peptide bonds. Used in molecular biology procedures and for diagnostic applications as well as antisense therapies. High binding strength properties, enhanced binding specificity characteristics, resistance to nucleases or proteases and stability over a wide pH range, convey advantages and obviate need for long PNA oligomers.

PONT **Po**int **o**f **n**eed **t**esting, also known as Point-Of-Care Testing (POCT), involves screening and tests at or near the point of care, aiming to provide accurate and rapid diagnostics that allow prompt actionable care.

PRIME **Pri**ority **Me**dicines. A strategic initiative launched by EMA to enhance support for the development of voluntary schemes for medicines that target an unmet medical need, promoting development with accelerated assessment of medicine product applications, building on existing resources of scientific advice and accelerated assessment at the time of application for marketing authorization.

PROCHYMAL Also known as rememstemcel-L, a stem cell therapy made by Osiris Therapeutics, the first of its kind approved by Canada. Renamed Ryoncil.

QbD **Q**uality **b**y **D**esign. A concept outlined by quality expert Joseph M. Juran, indicating that quality could be planned and adopted by the FDA for the discovery, development, and manufacture of drugs.

QC **Q**uality **C**ontrol. Process procedures that product quality is maintained or improved through well-defined controls and safety measures to minimize any risk that a person might be adversely affected by a product.

QMS **Q**uality **m**anagement **s**ystem. A formalized system that documents processes, procedures, and responsibilities for achieving quality objectives.

qPCR **q**uantitative **P**olymerase **C**hain **R**eaction. A laboratory method for determining the amount of a specific DNA sequence in a sample, involving amplification of the target DNA sequence and measure the amount of product generated.

QTPP **Q**uality **T**arget **P**roduct **P**rofile. Described in the 2009 ICH Q8 guide (R2) as a prospective summary of the quality characteristics of a drug product. Considerations include dosage and drug product quality criteria, e.g., sterility, purity, stability, and drug release.

RCL **R**eplication-**c**ompetent **l**entivirus. Used in biomedical research, it advantageously offers the possibility of infecting both dividing and nondividing cells, an effective tool in gene therapy. Guidance recommendations seek to minimize the biosafety risk of self-replication through a process of recombination.

RCR **R**eplication-**c**ompetent **r**etrovirus. Retroviral vectors engineered to deliver genes are usually replication defective, yet it is possible for recombination to generate replication competent viruses during the manufacturing process and this represents a testable safety concern for individuals treated with retroviral vector gene therapy.

Remestemcel-L A stem cell therapy developed by Osiris Therapeutics. Also known by specific brand names Prochymal or Ryoncil. An allogeneic stem cell therapy based on mesenchymal stem cells from the bone marrow of adult donors, used as a treatment for acute graft-vs-host disease.

RGO **R**educed **G**raphene **O**xide. The form of graphene oxide processed by chemical, thermal, and other methods to reduce the oxygen content. This change in chemical composition influences electrical conductivity, hydrophobic behavior, mechanical strength, and dispersibility, extending utility in engineering and biomedical applications.

RhoA **R**as **ho**molog family member **A**, also known as transforming protein RhoA, is a small GTPase protein encoded by the *RHOA* gene on chromosome band 3p21.31. It is primarily involved in actin organization, myosin contractility, cellular morphological polarization, and transcriptional control – important processes governing stem cell commitment and cytoskeletal maintenance.

RNA **R**ibo**n**ucleic **a**cid, a polymeric molecule assembled as a chain of nucleotides, found as a single strand folded onto itself. It can be found in diverse forms, a prefix is used to discriminate the different types, e.g., mRNA (messenger), rRNA (ribosomal), tRNA (transfer), aRNA (antisense), ncRNA (noncoding), etc.

ROCK **Rho**-associated **c**oiled-coil **k**inase, also known as Rho-associated coiled-coil containing protein kinase 1, encoded by the *ROCK1* gene on chromosome band 18q11.1, belongs to a family of serine–threonine specific protein kinases involved in regulating the shape and movement of cells by acting on the cytoskeleton. It is a downstream effector of the small GTPase RhoA.

RYONCIL Brand name for the first stem cell therapy approved in Canada, rememstemcel-L for acute graft versus host disease. Originally branded Prochymal, the name was changed to Ryoncil after transfer of ownership to the Australia-based company Mesoblast Limited in 2013.

S100A **S100** calcium-binding protein **A**, encoded by a family of genes whose symbols use the *S100A* prefix – the *S100A1* gene is on human chromosome band 1q21.3. Member of a family of at least 21 low molecular-weight proteins characterized by two calcium-binding sites that have helix-loop-helix ("EF-hand-type") motifs. Intracellular functions involve interaction with intracellular receptors, membrane protein recruitment and transportation, transcriptional regulation and integration with enzymes or nucleic acids, as well as DNA repair.

S100B **S100** calcium-binding protein **B**, encoded by the *S100B* gene on human chromosome band 21q22.3. A protein of the S-100 protein family, localized in the cytoplasm and nucleus of most but not all astrocytes, that acts as a neurotrophic factor and neuronal survival protein involved in a number of cellular processes including axonal proliferation and differentiation.

SAW **S**urface **A**coustic **W**ave. An acoustic wave travelling along the surface of a material that has elasticity, with an amplitude that typically decays exponentially with depth into the material. Sensors take advantage of the fact that sonic acoustic waves excited by an electrical signal at the resonance frequency can propagate mechanical vibrations under piezoelectric solid surfaces. Changes in SAW velocity can correlate to changes in surface temperature, with high sensitivity to changes in mass loading.

sCTMP **S**omatic **c**ell **t**herapy **m**edicinal **p**roducts contain cells or tissues that have been manipulated to change their biological characteristics, or represent cells not intended for the same essential function in the body, but repurposed for prevention, diagnosis, and/or treatment of diseases due to their pharmacological, immunological, or metabolic actions.

sFRP-1 secreted **F**rizzled-**R**elated **P**rotein **1**. A protein encoded by the *SFRP1* gene on chromosome band 8p11.21. A member of the SFRP family containing a cysteine-rich domain homologous to the putative Wnt-binding site of Frizzled proteins. SFRP proteins can bind Wnt proteins and Fz receptors in the extracellular compartment, thereby acting as a soluble modulator of Wnt signaling.

SLRP **S**mall **l**eucine-**r**ich **p**roteoglycans. Non-collagenous proteins that constitute the major bone glycoproteins. SLRPs include decorin, the major SLRP produced by osteoblasts, biglycan, osteoadherin, lumican, fibromodulin, and mimecan.

sox2 **S**ex determining region Y-b**ox** **2**, a member of the Sox family of transcription factors encoded by the *SOX2* gene on human chromosome band 3q26.33. It is essential for maintaining self-renewal of undifferentiated embryonic stem cells. Sox2 binds DNA cooperatively with Oct4 at non-palindromic sequences to activate transcription of key pluripotency factors.

SSC **S**keletal **s**tem **c**ells. Tissue-resident self-renewing and multipotent cells that continuously provide chondrocytes, bone cells, marrow adipocytes, and stromal cells for the development and continuous homeostasis of the skeletal system.

STARS **St**rengthening **T**raining of **A**cademia in **R**egulatory **S**cience. An EU initiative to reach out to medical innovators in academia to bridge the regulatory knowledge gap, enhancing knowledge that facilitates the implementation of academic research findings in clinical practice.

STC2 **St**annio**c**alcin **2**. A protein encoded by the *STC2* gene located on human chromosome band 5q35.1. Also known as hypocalcin, teleocalcin, or parathyrin, it belongs to a family of proteins that regulate calcium and phosphate balance in the body. A homodimer glycoprotein expressed in a broad spectrum of tumor cells and tumor tissues that is significantly stimulated under various stress conditions, including hypoxia and nutrient deprivation and correlates with tumor growth, invasion, and metastasis.

STR **S**hort **t**andem **r**epeat. A pattern in DNA sequence where two or more nucleotides are repeated and the repeated sequences are directly adjacent to each other in a head-to-tail manner. They are generally present in non-coding DNA yet can serve as markers to trace inheritance in families.

SU-8 **S**tructured by **U**V-**8**. A commonly used epoxy-based negative photoresist whereby the parts exposed to UV become cross-linked while the remainder of the film remains soluble and can be washed away during manufacturer.

T cell A type of white blood cell originating from hematopoietic stem cells in the bone marrow that then migrates to the **t**hymus gland to mature to several distinct types of T cells that have important roles in orchestrating the immune response.

T7E1 **T7** **E**ndonuclease **1**. A structure-selective T7 phage-derived enzyme that catalyzes the cleavage of most but not all types of DNA mismatch and non-β DNA structures, leading to deformities in heteroduplex DNA. Useful for assessment of genome editing and mutation detection.

TAGLN **Transgelin**. A protein encoded by the *TAGLN* gene found on human chromosome band 11q23.3. An actin cross-linking/gelling protein found in fibroblast and smooth muscle cells sensitive to cell shape changes.

TALEN **T**ranscription **a**ctivator-**l**ike **e**ffector **n**ucleases represent restriction enzymes that can be engineered to cut specific sequences of DNA. Made by fusing TAL effector DNA binding domain, that can be engineered to bind a DNA sequence of choice, with a DNA nuclease that cuts DNA strands. TALEN restriction enzymes can thus be introduced into cells for use in gene editing or genome editing in situ, making it a versatile gene editing tool in addition to zinc finger nucleases and CRISPR/Cas9.

TAM **T**yro-3, **A**xl, and **M**er family of receptor tyrosine kinases. Three homologous type I receptor tyrosine kinases that are activated by endogenous ligands protein S and growth arrest specific gene 6. They have key roles in the resolution of inflammation and restoration of homeostasis.

TAZ **T**ranscriptional co**a**ctivator with PD**Z** binding motif. A protein known to bind a variety of transcription factors as a transcriptional coactivator (no effect on transcription alone) to control cell differentiation and organ development. Encoded by the *WWTR1* gene on chromosome band 3q25.1.N.B.: Disambiguation, the gene symbol *TAZ* is used for the gene encoding the unrelated protein Tafazzin.

TECARTUS The brand name for Brexucabtagene autoleucel, a cell-based therapy medication for the treatment of mantle cell lymphoma and acute lymphoblastic leukemia.

TEMCELL HS Inj. Brand product name for mesenchymal stem cell derived from human bone marrow by JCR Pharmaceuticals Co., Ltd. The first allogeneic cell therapy to be fully approved in Japan.

TEP **T**issue **e**ngineered **p**roduct. A medicine containing engineered cells or tissues, aiming to regenerate, repair, or replace a human tissue.

Terc **Te**lome**r**ase **RNA** **c**omponent. A long noncoding RNA found in eukaryotes, which is a component of the ribonucleoprotein polymerase telomerase. The core domain of Terc contains the RNA template from which telomerase synthesizes the nucleotide sequence TTAGGG constituting the telomeric repeats.

TGF-β1 **T**ransforming **g**rowth **f**actor **b**eta **1**, a secreted protein, member of a family of potent cytokines influencing many functions including cell growth, proliferation, differentiation, and apoptosis. First identified as a protein of 25 kDa in human platelets with a potential role in wound healing, it also plays an important role in controlling the immune system, and most immune cells secrete TGF-β1.

Th1 **T** **h**elper cell type **1**. Can be triggered to release cytokines that increase cell-mediated response, primarily by macrophages and cytotoxic T cells.

Th17 **T** **h**elper cell type **17**. A subset of pro-inflammatory T helper cells defined by production of interleukin 17.

TIL **T**umor **i**nfiltrating **l**ymphocytes. White blood cells, both T and B cells, that have left the bloodstream and migrated toward a tumor to become a component of the tumor-infiltrating immune cells, which also includes mononuclear and polymorphonuclear immune cells.

TNFR1 **T**umor **N**ecrosis **F**actor **R**eceptor **1**, also known as CD120a, encoded by the *TNFR1* gene on chromosome band 12p13. A member of the tumor necrosis factor receptor superfamily, one of the major receptors for tumor necrosis factor alpha that can subsequently activate the transcription factor NF-κB, mediate apoptosis, and regulate inflammation.

TNFα **T**umor **n**ecrosis **f**actor **al**pha, also known as cachexin or cachetin, encoded by the *TNF* gene on human chromosome band 6p21.33. An adipokine and cytokine associated with insulin resistance and immune system cell signaling.

TNRF1 **T**umor **n**ecrosis **f**actor **r**eceptor **1**, member of a TNF receptor **s**uper**f**amily of proteins, specifically member **1A** encoded by the *TNRSF1A* gene on human chromosome band 12p13.31. Ubiquitously found in many cell types, one of the major receptors for tumor necrosis factor-alpha, functioning as a regulator of inflammation.

TPP **T**arget **p**roduct **p**rofile. A planning tool for therapeutic candidates based on FDA guidelines with considerations regarding primary product indication, patient population, treatment duration, delivery mode, dosage form, regimen, efficacy, risk/side effect, and therapeutic modality.

TSG-6 **T**umor necrosis factor-**s**timulated **g**ene **6** protein, also known as TNF-alpha-induced protein 6, a 30 kDa secreted protein encoded by the *TNFAIP6* gene on chromosome band 2q23.3. It contains a hyaluronan-binding domain involved in extracellular matrix stability and cell migration with tissue protective and anti-inflammatory properties.

TYRO3 **TYRO3** protein tyrosine kinase. An enzyme that in humans is encoded by the *TYRO3* gene located on chromosome band 15q15.1. Part of a 3-member transmembrane receptor tyrosine kinase that can bind several ligands, including GAS6 (growth arrest specific 6) and PIK3R1 (phosphatidylinositol 3-kinase regulatory subunit alpha), regulating many physiological processes including cell survival, migration, and differentiation.

UK NEQAS **UK** **N**ational **E**xternal **Q**uality **A**ssessment **S**cheme. Aims to improve patient care by independently monitoring the quality and reporting of tests on a not-for-profit basis, promoting comparable, safe, and clinically useful tests.

US **U**nited **S**tates of America

VCN **V**ector **c**opy **n**umber. A critical parameter that measures the genetic dose of a transgene in gene-modified cells. A widely adopted assay in the development and testing of gene therapy products that combined with the number of gene-modified cells helps determine the dose of the medicinal product.

VEGF **V**ascular **e**ndothelial **g**rowth **f**actor. A family of signal proteins, a principal form encoded by the *VEGFA* gene on human chromosome band 6p21.1. A glycosylated mitogen that is synthesized by many cell types acting prominently on vascular endothelial cells to stimulate blood vessel formation.

VIP **V**iability, **i**dentity, and **p**otency. Together these constitute very important parameters for ATMP development.

VITAL ASSAY Assay of specific cytotoxicity. A fluorescent-based assay that can be used to assess the cytotoxic activity of cytotoxic T lymphocytes and Natural Killer T cells with an invariant T-cell receptor in vitro and in vivo. Direct assessment of cytotoxicity in vivo may be achieved by monitoring survival of injected fluorescent targets relative to a differently labeled internal control population without specific antigen.

VLP **V**irus-**l**ike **p**article. Molecules that mimic viruses but are not infectious.

VLU **V**enous **l**eg **u**lcer. Wounds thought to occur through improper functioning of venous valves. They develop mostly along the medial distal leg and can be painful chronic wounds that negatively affect the quality of life.

VST **V**irus-**s**pecific **T** lymphocytes. The key component of an immunotherapeutic approach to viral disorders involving adoptive transfer of VST that rapidly reconstitute antiviral immunity post-transplantation, with low rates of adverse events, without causing graft-versus-host disease.

WHO **W**orld **H**ealth **O**rganization. Created in 1948 by Member States of the United Nations. The directing and coordinating authority for health within the United Nations.

Wnt **W**ingless-related **int**egration site. A portmanteau acronym from the words **W**ingless and **Int**-1 (integration 1), a proto-oncogene that is highly conserved in humans and *Drosophila*. The int-1 protein encoded by the *INTS1* (integrator complex subunit 1) gene on human chromosome band 7p22.3, corresponds to a *Drosophila* gene homologue known as Wingless, a segment polarity gene involved in the formation of the body axis during embryonic development. Wnt signaling represents one of most important developmental pathways controlling cell fate choices and tissue patterning during early embryonic phases and in later development.

WNT/A **Win**gless-**t**ype MMTV integration site family, member **A**, also known as Wnt1, encoded by the *WNT1* gene on human chromosome band 11q13.2. Member of a large family of structurally related and highly conserved Wnt growth factors secreted as palmitoylated glycoproteins that can bind membrane receptor complexes composed of a frizzled G-protein coupled receptor and a low-density lipoprotein receptor-related protein. They activate distinct intracellular cascades, often referred to as canonical (β-catenin dependent) or noncanonical (β-catenin independent) pathways.

YAP **Y**es **A**ssociated **P**rotein, discovered from an ability to interact with the SH3 domain of Yes and Src protein tyrosine kinases, encoded by the *YAP1* gene on human chromosome band 11q22.1. It functions as a transcriptional regulator that can act as both a coactivator and a corepressor, the critical downstream regulatory target in the Hippo signaling pathway playing a key role in organ size control by restricting proliferation and promoting apoptosis.

YESCARTA A prescription medicine used in the treatment of two types of non-Hodgkin lymphoma, large B-cell lymphoma, or follicular lymphoma when other forms of treatment have failed to control the cancer.

Zolgensma A prescription gene therapy in a one-time injectable form used to treat children less than 2 years old with spinal muscular atrophy.

γδ T-cells **G**amma **d**elta **T-cells** bearing this form of T-cell receptor on their surface are less common than T cells with alpha beta T cell receptors, but are found predominantly in the gut mucosa, within intraepithelial lymphocytes and are likely to have a prominent role in recognition of lipid antigens.

μPCR **M**icro **P**olymerase **C**hain **R**eaction. A technique for amplifying small amounts of DNA using the polymerase chain reaction. Advantages include the possibility of amplifying small limited-size samples, less time to complete the reaction, reduced reagent costs and integration with lab-on-chip devices so that assays may be performed at the point of care.

μTAS **M**icro **T**otal **A**nalysis **S**ystem. Devices that automate and include all the necessary steps for a chemical analysis of a sample, suitable for fluid processing in a microchannel structure with microliter volumes of test sample. Unique microstructure properties allow miniaturized fluidic lab-on-a-chip systems to consume negligible amounts of sample, reduce process costs, and provide fast analysis times. Nonetheless, detection of analytes at trace levels is a constrain requiring preconcentration strategies.

Index

J. S. Burns (ed.), *Potency Assays for Advanced Stem Cell Therapy Medicinal Products*, Advances in Experimental Medicine and Biology 1420, https://doi.org/10.1007/978-3-031-30040-0

MIX
Papier aus verantwortungsvollen Quellen
Paper from responsible sources
FSC® C105338

If you have any concerns about our products,
you can contact us on
ProductSafety@springernature.com

In case Publisher is established outside the EU,
the EU authorized representative is:
Springer Nature Customer Service Center GmbH
Europaplatz 3, 69115 Heidelberg, Germany

Printed by Libri Plureos GmbH
in Hamburg, Germany